Zoophysiology Volume 33

Springer
Berlin
Heidelberg
New York
Barcelona
Budapest
Hong Kong
London
Milan
Paris
Santa Clara
Singapore
Tokyo

Zoophysiology

Volumes already published in the series:

Volume 1: *P.J. Bentley*
Endocrines and Osmoregulation (1971)

Volume 2: *L. Irving*
Arctic Life of Birds and Mammals Including Man (1972)

Volume 3: *A.E. Needham*
The Significance of Zoochromes (1974)

Volume 4/5: *A.C. Neville*
Biology of the Arthropod Cuticle (1975)

Volume 6: K. *Schmidt-Koenig*
Migration and Homing in Animals (1975)

Volume 7: *E. Curio*
The Ethology of Predation (1976)

Volume 8: *W. Leuthold*
African Ungulates (1977)

Volume 9: *E.B. Edney*
Water Balance in Land Arthropods (1977)

Volume 10: *H.-U. Thiele*
Carabid Beetles in Their Environments (1977)

Volume 11: *M.H.A. Keenleyside*
Diversity and Adaptation in Fish Behaviour (1979)

Volume 12: *E. Skadhauge*
Osmoregulation in Birds (1981)

Volume 13: *S. Nilsson*
Autonomic Nerve Function in the Vertebrates (1983)

Volume 14: *A.D. Hasler*
Olfactory Imprinting and Homing in Salmon (1983)

Volume 15: *T. Mann*
Spermatophores (1984)

Volume 16: *P. Bouverot*
Adaption of Altitude-Hypoxia in Vertebrates (1985)

Volume 17: *R.J.F. Smith*
The Control of Fish Migration (1985)

Volume 18: *E. Gwinner*
Circannual Rhythms (1986)

Volume 19: *J. C. Rüegg*
Calcium and Muscle Activation (1986)

Volume 20: *J.-R. Truchot*
Comparative Aspects of Extracellular Acid-Base Balance (1987)

Volume 21: *A. Epple* and *J.E. Brinn*
The Comparative Physiology of the Pancreatic Islets (1987)

Volume 22: *W.H. Dantzler*
Comparative Physiology of the Vertebrate Kidney (1988)

Volume 23: *G.L. Kooyman*
Diverse Divers (1989)

Volume 24: *S.S. Guraya*
Ovariant Follicles in Reptiles and Birds (1989)

Volume 25: *G.D. Pollak* and *J.H. Casseday*
The Neural Basis of Echolocation in Bats (1989)

Volume 26: *G.A. Manley*
Peripheral Hearing Mechanisms in Reptiles and Birds (1989)

Volume 27: *U.M. Norberg*
Vertebrate Flight (1990)

Volume 28: *M. Nikinmaa*
Vertebrate Red Blood Cells (1990)

Volume 29: *B. Kramer*
Electrocommunication in Teleost Fishes (1990)

Volume 30: *W. Peters*
Peritrophic Membranes (1991)

Volume 31: *M.S. Kaulenas*
Insect Accessory Reproductive Structures (1992)

Volume 32: *A.L. Val* and *V.M.F. de Almeida-Val*
Fishes of the Amazon and their Environment (1995)

Volume 33: *R. Wiltschko* and *W. Wiltschko*
Magnetic orientation in Animals (1995)

R. Wiltschko W. Wiltschko

Magnetic Orientation in Animals

With 93 Figures
and 17 Tables

Springer

Dr. Roswitha Wiltschko
Prof. Dr. Wolfgang Wiltschko
J.W. Goethe-Universität Frankfurt
FB Biologie, Zoologisches Institut
Siesmayerstraße 70, D-60054 Frankfurt/Main

Library of Congress Cataloging-in-Publication Data

Wiltschko, R. (Roswitha), 1947–
Magnetic orientation in animals / R. Wiltschko, W. Wiltschko.
p. cm. – (Zoophysiology ; v. 33)
Includes bibliographical references and index.

1. Animal orientation. 2. Geomagnetism. I. Wiltschko, W. (Wolfgang), 1938– ,
II. Title. III. Series.
QL782.5.W54 1995
591.19'17–dc20 95-24610-CIP

ISBN-13: 978-3-642-79751-4 e-ISBN-13: 978-3-642-79749-1
DOI: 10.1007/978-3-642-79749-1

Typesetting: Mitterweger Werksatz GmbH, 68723 Plankstadt, Germany
Production: PROEDIT GmbH, 69126 Heidelberg, Germany
Cover design: Springer-Verlag, Design & Production

SPIN: 10497819 31/3132 – 5 4 3 2 1 0 – Printed on acid-free paper

Preface

Biological effects of magnetic fields have been studied in many animals and plants. The magnetic fields were of a wide intensity range and, as alternating fields, of a wide frequency range and of a variety of impulse shapes. Effects on the cellular level, on biochemical processes, growth and development, interactions with physiology, sensory input, reflexes and rhythm control, to name just a few, have been reported. Numerous magnetically induced changes in behavior have also been described. Recently, the amount of literature covering biological effects of magnetic fields has been rapidly increasing. By now it has grown to such an extent that it can no longer be covered in one volume. Most reviews specialize and focus on particular aspects and/or types of fields or effects. For example, the book edited by MARET et al. (1986) gives an overview on biological effects of steady magnetic fields, MISAKIAN et al. (1993) reviewed those of extremely low frequency magnetic fields, focusing on in vitro effects. BERNHARD (1992) reported on 'electromagnetic smog' in view of possible effects on human health and well-being, and a series of papers edited by AMEMIYA (1994) summarizes Japanese research on effects of electromagnetic fields ranging from extemely low to ultra-high frequencies. TENFORDE (1979) and ADEY (1981) summarized and discussed tissue interactions, REITER (1993a) neuroendocrine and neurochemical changes associated with various kinds of electromagnetic fields. The book edited by KIRSCHVINK et al. (1985a) concentrates on magnetite, its biomineralization and potential role in magnetoreception; MANN et al. (1990) reviewed the findings on magnetotactic bacteria with all their side aspects. Comprehensive reviews on a wide variety of effects of electric and magnetic fields were published by WADAS (1991) and CARPENTER and AYRAPETYAN (1994).

The present book is devoted to *magnetic orientation*, i.e to the problem of how animals use magnetic information to control their behavior in space. Hence, we restrict our considerations to findings and hypotheses which are relevant to this question, which means (1) we deal with responses of animals as a whole only, not covering changes observed in specific organ systems or at the cellular level, and (2) we concentrate on spatial behavior and (3) on static magnetic fields in the range of the natural geomagnetic field, except when behavior that is known to be magnetically controlled is directly effected.

The use of the magnetic cues in orientation has traditionally met with great interest by scientists and laymen alike, as man has always been fascinated by the mysterious magnetic forces. In Europe, man himself has made practical use of a technical magnetic compass for direction finding from the early Middle Ages onward. Hence, it is not surprising that already in the last century the magnetic field was discussed as a potential orientation cue when feats like bird migration or homing were to be explained.

However, although the magnetic compass as a technical device has become matter of course by sailors and surveyors, it cannot be denied that magnetic fields remain an alien sensory quality to us, as they escape our conscious perception. This has not been without consequences for the research on magnetic orientation. Several points proved to be a considerable handicap: (1) A certain technical effort is necessary to define and control magnetic conditions. (2) A limited understanding of the natural magnetic phenomena and the way they might be used has often led to ill-designed experimental approaches, which, in turn, led to negative results. (3) The inability of man to consciously perceive the magnetic field caused a certain reluctance towards magnetic research. Many scientists hesitated to accept the idea that animals might make widespread use of a factor which man is so obviously unable to detect. Findings on magnetically controlled behavior were always met with great scepticism and faced the demand for painstaking elimination of other potential cues and careful guard against possible artifacts. This turned out to be an advantage: magnetic effects are often far better controlled than many effects in the more 'conventional' fields. – Additionally, magnetic effects are seldom eye-catching and spectacular, and thus might look disappointing when compared with the effects of other cues.

Despite the increasing literature, our general knowledge on magnetic orientation is still rather limited, mainly because there has been a certain tendency to neglect behavioral aspects. When an effect was established, many authors rather focused their attention on physiological problems, namely the detection of magnetic fields. The major question was not "how is magnetic information *used*?", but "how is magnetic information *perceived*?" The literature provides some impressive examples: When, in the 1970s, a mechanism of magnetoreception based on electric induction in the *ampullae of* LORENZINI was described for elasmobranch fish, most authors working with fishes turned to the question whether or not magnetic parameters were perceived via electric receptors. Earlier attempts to analyze magnetic orientation in its behavioral context were discontinued. Likewise, when permanent magnetic material was discovered in animal tissue, numerous authors began at once to speculate that it might be involved in magnetoreception. This led to a search for magnetic material in various kinds of animals. This search was highly successful, and scientific interest centered on the existence or nonexistence of magnetite in the various species. When magnetite was found, it was often discussed as suggesting the ability to orient with the help of the magnetic field – even if there were no indications that the animals in question might make use of magnetic information at all. Today, many current studies on magnetic orientation are also carried out with the intention to find a means by which the physiology of magnetoreception may be approached, and the behavioral relevance of the findings is frequently not given due attention.

However, orientation by the magnetic field (or by any other factor) is not a phenomenon per se. It helps the animals to cope with their environment, find favorable sites, save time and energy, avoid predators and adverse conditions, in short, it increases fitness. Like any behavior, it is the result of an evolutionary process, and hence it can only be understood in its behavioral context. In view of this, we will emphasize the ecological background of magnetic orientation. Attention will be devoted to questions of motivation and the design of the test apparatus, as meaningful results can only be expected when animals are tested in situations which reflect, in an adequate way, natural situations where magnetic information plays a role in the control of behavior.

The central questions will concern circumstances that determine the utilization of magnetic parameters, the parameters that are used and how they are used, their advantages and limitations, and how magnetic information is integrated with other available orientation cues. The main emphasis will be on compass orientation, since this use of magnetic information has been extensively studied and is fairly well understood. Orientation behavior based on spatial variations of the earth's magnetic field is less well documented, and many of the observed effects are still rather unclear. This is even more true for other magnetic phenomena. These findings will be briefly summarized, without going into too much detail about speculative interpretations.

The reader will soon realize that the various animal groups are not equally represented. Birds are by far the best-studied group; the available data suggest that they make use of magnetic information in more than one way. The orientation of birds has been thoroughly analyzed, not only with respect to the magnetic field, but also with respect to various other orientation cues. As a result, the role of magnetic information and its interaction with other factors in birds are better understood than in most other animals. Furthermore, findings on birds frequently inspired the search for particular phenomena in other species. For these reasons, a considerable portion of the text is devoted to the orientation of birds. Amphipods, teleost fish, sea turtles, and honeybees have also been fairly well studied. Evidence for magnetic orientation in other groups is far more scarce. We tried to collect the widely scattered literature in the field and give an overview in which species magnetic orientation has been demonstrated. We hope that this will help to encourage our colleagues to look for magnetic effects in the groups so far neglected, and to analyze some of the cases that are still poorly understood.

Altogether, the book has five parts. Part 1 provides general information and considerations which form the framework for an understanding of the findings presented in the later parts of the book. Chapter 1 gives a brief overview on the various kinds of magnetic information available in nature, together with some comments on experimental magnetic fields, their generation and description. Chapter 2 defines 'orientation' and discusses the functional background of various types of mechanisms in view of a potential role of magnetic parameters. Part 2, the central part of the book, is devoted to orientation controlled by the *direction* of the magnetic field: Chapter 3 deals with alignment responses; Chapter 4 reviews findings on magnetic compass orientation in the various animal groups, including the functional properties as far as they are known, the behavioral situation, and the origin of the set direction. Chapter 5 presents selected examples of how magnetic compass information is integrated with other cues in multifactorial orientation systems. Part 3 covers non-compass effects on orientation: Chapter 6 discusses hypotheses on orientation by the spatial distribution of magnetic parameters and the findings related to them, whereas Chapter 7 gives an overview on behavioral changes associated with temporal variations of the magnetic field, together with responses to changing magnetic conditions. Part 4 is concerned with magnetoreception: Chapter 8 describes numerous attempts to condition animals to magnetic stimuli, discussing possible reasons why they were so often unsuccessful, while Chapter 9 briefly summarizes the current hypotheses on magnetoreception, together with the corresponding electrophysiological and behavioral findings. In view of the research presently in progress in this field, this chapter will soon be outdated by new results. In

Part 5, the closing Chapter 10 gives a brief overview on past and future research on magnetic phenomena.

The idea to write a book on magnetic orientation was originally suggested by the late Donald S. Farner, who encouraged us to gather the findings in this field and make them easier accessible to a wider scientific public. In the process of compiling the book, we received help and support from many of our colleagues. They pointed out literature, sent us yet unpublished manuscripts, informed us about their current research, discussed findings and hypotheses, read early versions of chapters of this book, pointed out ambiguities and made many helpful suggestions. We are grateful to Kenneth P. Able, Robin R. Baker, Robert C. Beason, Wilfried Haas, Peter Jacklyn, Joseph L. Kirschvink, A. Peter Klimley, Varis Liepa, Kenneth J. Lohmann, Stephan Marhold, Ursula Munro, Nikolai Petersen, John B. Phillips, Michael Salmon, Felicitas Scapini, Roland Sandberg, David Schmitt, Peter Semm, Ulrich Sinsch, Alberto Ugolini, Charles Walcott and many others not listed here personally, especially our colleagues of the Frankfurt Zoology Department and the members of the Frankfurt orientation group. We also thank Nikolai Petersen and Dirk Schüler for providing photos of magnetotactic bacteria, Martin Becker for providing recordings of the geomagnetic field, Wincenty Harmata, Varis Liepa, and Stephanie Thamm for their help with Polish, Russian and French literature, respectively, and Gisela Schmiedeskamp for her valuable help in preparing the manuscript.

Sincere thanks are also due to Friedrich W. Merkel, who was major professor, *Doktorvater*, to both of us. Being an enthusiastic ornithologist working in the field of bird migration, he directed our interest to questions of how birds cope with the problems arising from having to find their way over thousands of kilometers. In 1965, he initiated the experiments showing that the European Robin, a small passerine, orients its migration with the help of the magnetic field. This was the first demonstration that animals can use magnetic information for direction finding and marks the beginning of modern research on magnetic orientation.

This book is dedicated to him.

Frankfurt a.M., Spring 1995

Roswitha Wiltschko
Wolfgang Wiltschko

Contents

Part 4 Perception of Magnetic Fields

Part 5 Outlook

Part 1: Background

Chapter 1

Magnetic Fields

In order to assess what kinds of magnetic information are available in nature, we will first outline the properties of the geomagnetic field. As we ourselves are unable to perceive magnetic fields consciously, all our knowledge is acquired indirectly through technical means. Today, the advancement of technology has supplied us with extensive knowledge on the magnetic field of the earth and its variations in time and space. At the same time, modern techniques enable us to change magnetic conditions in a controlled way and to perform tests in experimental fields of defined properties.

1.1 The Magnetic Field of the Earth

Detailed descriptions of the geomagnetic field have been published in various books and articles on geophysics (e.g. Chapman and Bartels 1940; Chernosky et al. 1966; Merrill and McElhinny 1983; Skiles 1985). Here, we will briefly summarize the most important features relevant for orientation.

1.1.1 The Vector Field

In essence, the earth's magnetic field is a dipole field. Its origin is now generally attributed to dynamo effects in the earth's fluid outer iron core, although several details of this theory are still unclear. The two poles do not coincide with the geographic (rotational) poles, yet lie in their vicinity. Since the north-seeking pole of a compass needle is traditionally defined as a magnetic *north pole*, the geomagnetic pole in the arctic, north of North America, is a *south pole* by physics nomenclature. The physical north pole lies in the antarctic region south of Australia; here, the field lines leave the ground going straight up. They curve around the earth and reenter its surface at the pole in the arctic, going straight down. The zone where the field lines are parallel to the earth's surface is defined as the *magnetic equator*; it runs north of the geographic equator in Africa and south of it in South America. Thus, the magnetic vector points northward-upward in the (magnetically) southern and northward-downward in the northern hemisphere (see Fig. 1.1).

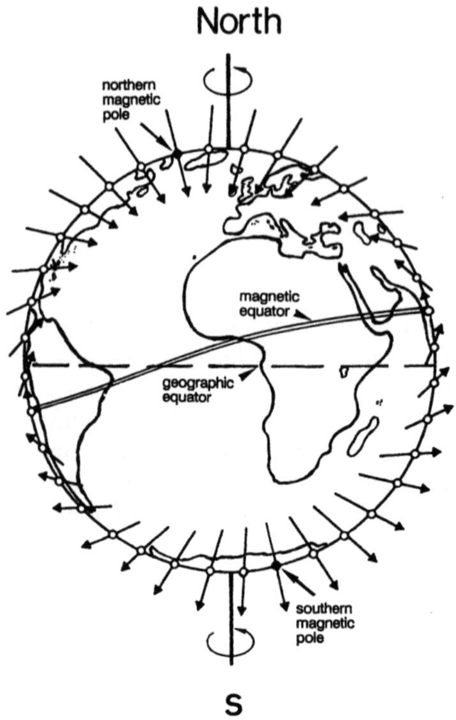

Fig. 1.1. Schematic view of the earth and the geomagnetic field. The *arrows* at the circumference surface are proportional to the local total intensity, their *deviation from the vertical* represents local inclination. The two magnetic poles are indicated, together with the magnetic equator

The magnetic field at a given location is described by its *total intensity* and its direction, which is given as *declination* and *inclination*. Declination[1] indicates the direction in the horizontal by stating the deviation of magnetic North from true geographic North. Since the geographic and magnetic poles do not coincide (and because of irregularities of the geomagnetic field), the respective north directions normally deviate by a certain angle. This deviation is less than 30 ° in most parts of the world (Fig. 1.2), but it reaches extreme values near the magnetic poles (and at local anomalies, see Sect. 1.1.2.2). Inclination or 'dip' indicates the direction in the vertical by giving the angle between the magnetic vector and the horizontal plane, with the downward dip of the northern hemisphere traditionally defined as positive. Lines of equal inclination form a system of roughly parallel lines comparable to the lines of equal latitude; their intervals decrease toward the magnetic equator (Fig. 1.3). The total intensity of the field is highest near the magnetic poles, with more than 60 000 nT. It decreases to values of about 30 000 nT at the magnetic equator, reaching a minimum with values below 26 000 nT at the east coast of South America (Fig. 1.4)[2]. – In Frankfurt a.M., Germany (50 ° 08'N, 8 °40'E), for example, the geo-

[1] In aeronautical and similar charts, magnetic declination is designated as 'variation'.

[2] nT = nanoTesla, a unit of the international standardized units, tesla (T) replacing the old unit gauss (G). 1 G = 100 000 γ = 100 000 nT. 1 T = 1 V·s/m²; 1 nT = 10^{-9} T.

magnetic field had the following values in 1990: ca. 47 800 nT, declination 1 °W, +66 ° inclination (see also Fig. 1.5).

The above-mentioned values are given in polar coordinates. Another way of describing the magnetic field uses Carthesian coordinates, giving the strength of components in the three spatial directions, X (north), Y (east) and Z (downward). Most observatories that monitor the geomagnetic field measure these components by probes fixed in the respective directions. Both types of data are equivalent and can be easily transformed into each other. For orientational consideration, it has to be remembered, however, that the three components are based on human conventions and require the definition of reference directions. It is unclear to what extent they are meaningful for animals.

1.1.2 Spatial Distribution of the Geomagnetic Field

Figures 1.2 to 1.4 give data on the spatial distribution of declination, inclination and total intensity. The distributions of these parameters show some regularity, but they are not as symmetrical and regular as one might expect if the earth's magnetic field were exclusively a dipole field.

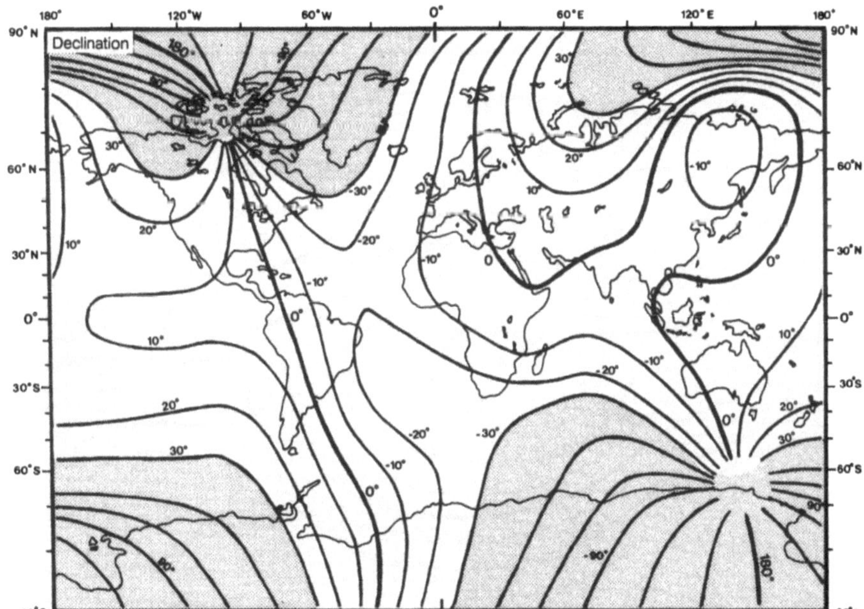

Fig. 1.2. Declination or variation of the magnetic field of the earth (epoch 1965); *negative signs* indicate deviation from geographic North to the west. Regions with declination exceeding 30 ° are *shaded*. (After SKILES 1985)

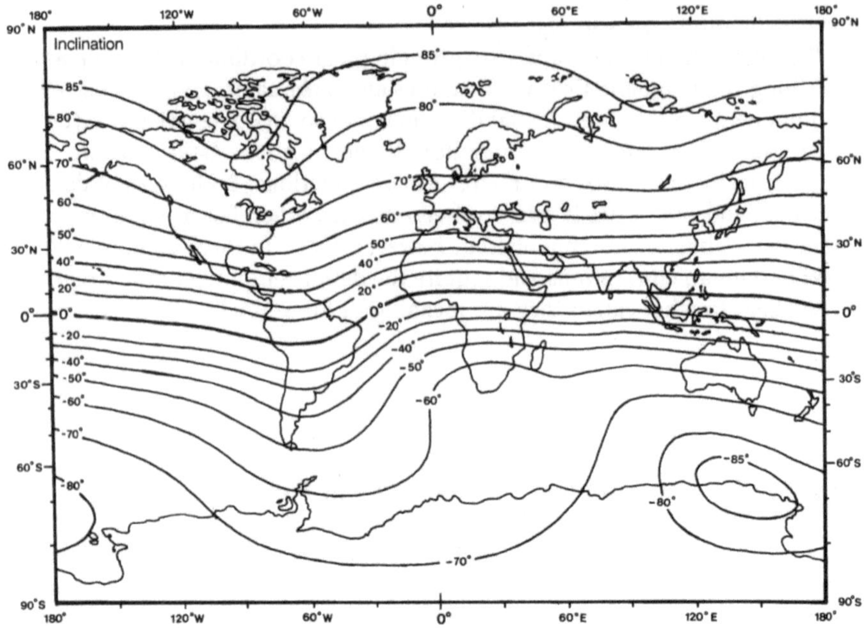

Fig. 1.3. Inclination or dip of the magnetic field of the earth (epoch 1965); *negative signs* designate upward inclination. (After SKILES 1985)

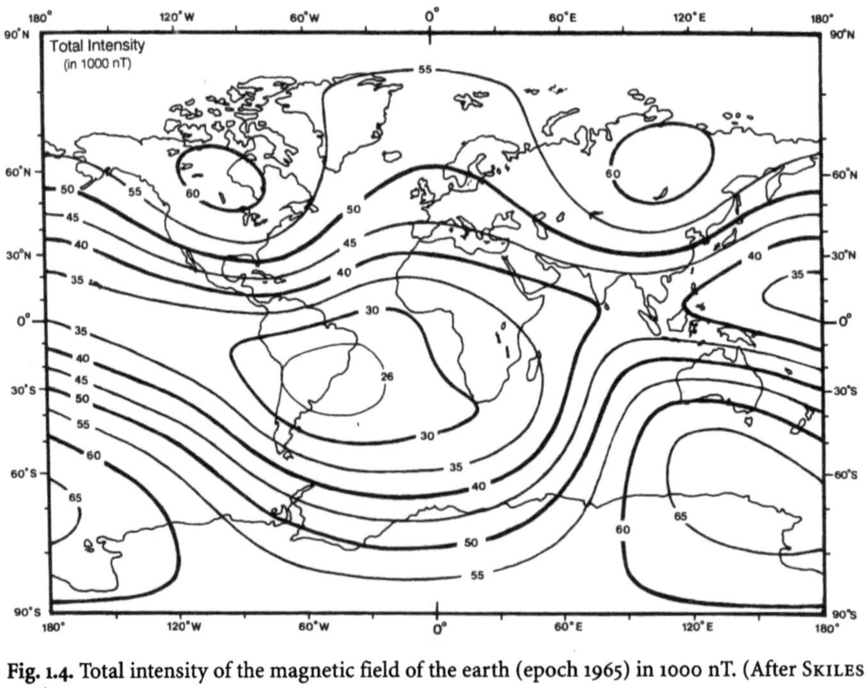

Fig. 1.4. Total intensity of the magnetic field of the earth (epoch 1965) in 1000 nT. (After SKILES 1985)

1.1.2.1 *Worldwide Gradients and the Non-Dipole Field*

Total intensity and inclination are characterized by large-scale gradients running from the poles towards the equator. Extended deviations from the ideal values, mostly not very large, are referred to as *non-dipole field* and are summarized in worldwide maps (see e.g. SKILES 1985 for examples). These fields originate also from within the earth itself, but their causes (and temporal changes, see below) are still largely unknown.

The steepness and the direction of the magnetic gradients vary between the various regions of the world. For example, in northeastern North America, total intensity has an average change of about 3.4 nT/km along an axis running NNW-SSE (see LEDNOR 1982), whereas the regional gradient in central Europe is 2.5 nT/km along an axis slightly east of north – slightly west of south.

1.1.2.2 *Magnetic Topography and Anomalies*

Aside from these worldwide variations, different magnetizations of rock units cause magnetic deviations on a more local scale, which are found all over the earth. An exact analysis reveals that the geomagnetic field is practically nowhere completely regular. Magnetic parameters show spatial variations that can be described as magnetic *topography* consisting of 'hills' with locally higher total intensities and corresponding 'valleys'. The size and the pattern of these deviations vary greatly. In most regions they represent slight fluctuations around regional means, but at some locations they exceed more than 10% of the total intensity of the regional field. A most prominent magnetic *anomaly* is the Kursk anomaly, 450 km SSW of Moscow, Russia, with local intensities above 190 000 nT (i.e. more than three times the normal value) and changes in declination between 60° E and 110° W within less than 1 km. Such anomalies are exceptional, however. Deviations of more than 1000 nT are seldom, even if steep and variable gradients can occasionally be observed within limited areas. For example, in the northeastern USA, where the regional gradient is 3.4 nT/km, an extended area around Ithaca, New York, shows gradients up to 16 nT/km (LEDNOR 1982).

At sea, the magnetic field tends to be less irregular than on land. The process of seafloor spreading leads to series of extended, roughly linear magnetic hills and valleys running parallel to the mid-oceanic ridges on both sides, thus forming fairly regular, linear patterns of magnetic topography in large parts of the oceans. In the Atlantic Ocean, these linear contours are mostly aligned along a north-south axis (VACQUIER 1972; KIRSCHVINK et al. 1986).

1.1.3 Temporal Variations of the Magnetic Field

Temporal changes in the geomagnetic field occurring at three levels of time are important for biological considerations: (1) daily variations and magnetic storms, (2) secular variations, and (3) variations in geological time.

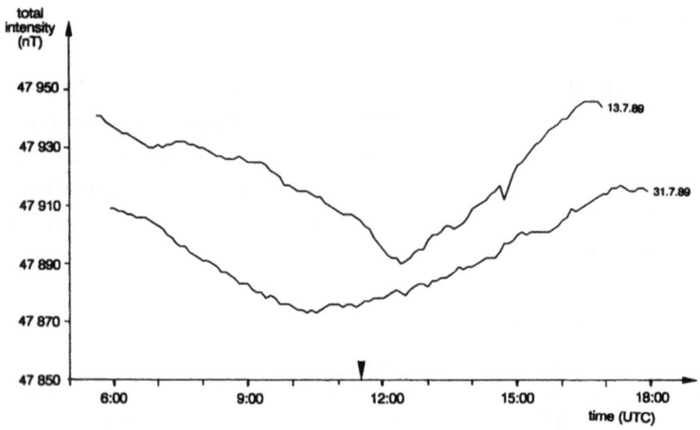

Fig. 1.5. Variation of magnetic total intensity in the course of morning and afternoon on two magnetically quiet days at a rural site 20 km east of Frankfurt a.M. (50° N, 9° E), based on readings every 10 min. Time is given in UTC (time of the 0 ° meridian); an *arrow* marks local noon as defined by the culmination of the sun. The *K*-values (see text for details) recorded by the Geophysikalisches Observatorium at Fürstenfeldbruck, approximately 300 km south-southeast from the recording site, were 2,1,2,2 on 13.7.1989, and 1,1,1,1 on 31.7.1989 for the four 3-h periods beginning at 6:00 UTC. (Courtesy of M. Becker)

1.1.3.1 Short-Term Variation

Regular daily variations are caused by the solar electromagnetic radiation acting on the sun-facing side of the earth and creating currents in the upper atmosphere and ionosphere. They form a complex pattern over the various continents which changes with latitude, season and, to a smaller extent, with lunar phases (Matsushita and Maeda 1965; Matsushita 1967). At northern temperate latitudes where most orientation experiments have been performed, the intensity decreases until local noon by 30 to 100 nT, then increases again (Fig. 1.5), the differences being slightly larger in summer. At lower latitudes, the size and the course of intensity changes are different; increases toward noon and subsequent decreases may be observed. Declination and inclination are also involved, with changes in the order of up to 20' (0.33 °).

Magnetic storms are irregular fluctuations associated with sun spot activity, solar flares, etc. superimposed on the regular patterns. Normally, these fluctuations are rather small, but they may reach considerable values in severe magnetic storms. These storms, too, differ with latitude; they are smallest at lower latitudes and largest between 60 ° and 80 °.

Several *geomagnetic indices* have been established to characterize temporal variations of the earth's magnetic field. For example, the *K-index* expresses the size of magnetic fluctuations based on the maximum differences observed in defined 3-h intervals (cf. Fig. 1.5). *K* has a quasi-logarithmic scale, i.e. an increase in *K* by 1 means approximately doubling the range of fluctuations. The K-indices are normalized to take geographic differences into account: e.g. *K 6* indicates fluctuations in the range of 240 to 399 nT at the Sitka observatory,

Alaska (57° N, 135° W), and fluctuations in the range between 120 and 190 nT at Niemegk, Germany (52° N, 13° E). Other indices characterize magnetic fluctuation on different temporal and local scales (see LINCOLN 1967 for summary).

K-indices are readily available in published form for many observatories. This might be the reason for their frequent use in correlations, etc. (see Sect. 6.1.1.2). Their crucial disadvantage, however, is that they characterize fluctuation only, without specifying how the individual components are affected and how much they deviated from normal (CHAPMAN and BARTELS 1940; SKILES 1985). Hence, it is doubtful whether they best reflect biologically relevant temporal variability in magnetic parameters. Other recordings may prove to be more suitable in the future.

1.1.3.2 Long-Term Variation

Secular variations are slow changes in intensity, inclination, and declination of the earth's magnetic field, involving the dipole and the non-dipole field. The dipole field varies in intensity by about 0.05% per year with periods of several thousand years. The non-dipole fraction of the field shows a westward drift of about 0.2° longitude per year. The rate of change in any given parameter varies in time and space. For example, the declination of London changed from ca. 10° E in 1575 to 23° W in 1800 and back to 3° W in 1965. Within the same time, the inclination assumed values between 75° and 67° (GAIBAR-PUERTAS 1953). These changes are so slow that they are negligible within the life span of an individual animal, but they become important for evolutionary considerations.

The same is true for *variations in geological times*. These include changes in total intensity with values possibly up to twice as high as today, and, more important, *reversals of polarity*. At some times during the geological past, the magnetic vector pointed toward geographic South instead of geographic North as today. These reversals, which are associated with a decrease in intensity of the dipole field, have occurred frequently in Cenozoic and Mesozoic periods, as the data from the magnetization of oceanic sediments and volcanic rocks indicate (VACQUIER 1972; SKILES 1985). The last one occurred about 730 000 years ago.

1.1.4 Man-Made Magnetic Noise

Today, most of the technical equipment is metallic and/or is powered by electricity. This adds a certain amount of electromagnetic noise to the natural magnetic field, which may lead to considerable small-scale changes of the local magnetic field in heavily populated areas. Sources like high voltage transmission lines and generating stations mainly produce spatial variations in the magnetic field, other sources such as technical equipment operating during various times of the day; moving objects, etc. also cause temporal fluctuations.

Even if the direction of the magnetic field used for compass orientation is little affected, this 'electromagnetic smog' might not be negligible any longer in industrialized areas (see BERNHARD 1992 for review).

Any change in the local properties of the magnetic field by technical noise would necessarily also affect the characteristics of magnetic information. There has been some concern about anthropogenic interference with electromagnetic fields, and some studies were initiated to examine possible effects, but, with few exceptions, they focus on questions of whether human health might be endangered by technical, medical and household equipment (for literature, see BERNHARD 1992; AMEMIYA 1994; CARPENTER and AYRAPETYAN 1994). Effects on orientation behavior have been studied so far only in birds. The number of such studies is still limited (see Sect. 7.2) so that it remains largely unknown what man-made electromagnetic noise might mean to animals living in such an environment and how they might cope with it.

1.2 Magnetic Manipulation

For the experimental analysis of magnetic orientation, it is necessary to subject animals to various magnetic conditions. Establishing experimental magnetic fields, even measuring magnetic fields, requires a certain technical effort, however. For a long time, this seems to have handicapped research on magnetic effects. It was a common technique to disrupt the available magnetic information rather than provide the animals with meaningful different information in order to change their behavior in a predicted way.

1.2.1 Altering Magnetic Conditions

In most cases, it is necessary to change the magnetic conditions in a more or less extended test cage or arena. This can be achieved in various ways.

1.2.1.1 Magnets and Shielding

In early studies, experimentalists often used one or several bar magnets to alter the magnetic field. This technique appears simple, yet it has the crucial disadvantage that the field changes rapidly with increasing distance from the magnets, and the animal is subjected to magnetic conditions that can hardly be defined. Only rarely, the course of field lines was documented by using iron dust so that at least the direction of the test field at various points of the test arrangement was known (e.g. ALTMANN 1981). In general, magnets seem to be inferior to other methods when the test animals are confined within the limited space of a cage. When animals are released, however, applying magnets to their body still seems to be a most practicable method to disrupt magnetic information.

Other attempts to test animals in the absence of meaningful magnetic information involved experiments under better controlled conditions in shielded rooms. Various types of *shielding*, like Mu-metal, purified iron, Permalloy, etc., and their advantages and disadvantages are discussed by SCOTT and FROHLICH (1985). Coil systems (see below) producing static or alternating magnetic fields have also been used to block access to natural magnetic information.

1.2.1.2 Coil Systems

Studies aiming at the analysis of the specific role of magnetic information require that animals be tested under various defined magnetic conditions. For example, one parameter is to be changed while the others are held constant. Such fields are usually produced electromagnetically by various types of coil systems. Any problem arising from their need of direct current can be solved in the laboratory by using suitable rectifiers and stabilizers. In outdoor tests, the use of a strong car battery solves the problem.

Types of Coils. One of the most frequently used coil systems are HELMHOLTZ coils, consisting of a pair of round coils with an equal number of wire turns on each coil. The gap between the coils is equal to their radius. In the center of the coils, the generated field[3] H_i in nT is given by $H_i = 0.899 \cdot 10^3\ n \cdot I/r$, with I = electric current in ampere (A), n = number of wire turns, and r = radius of the coils in m. The induced field increases with increasing flow of electric current and increasing number of wire turns, while it is inversely proportional to the radius of the coils. For example, in order to deflect magnetic North by 90 ° to 120 °, sets of coils with a radius of 1 m and 30 or 40 windings were used; depending on the local magnetic field and the size of the desired deflections, the current was in the range of 0.7 to 1.5 A (e.g. W. WILTSCHKO 1968; W. WILTSCHKO and WILTSCHKO 1975a,b). The field produced by HELMHOLTZ coils is fairly homogeneous in the center of the coil system, but inhomogeneity increases rapidly towards the edges (see BELL and MARINO 1989). The RUBENS coil system (RUBENS 1945) works in a similar way. It is a cubic structure of five square coils (side length d in m, spaced $d/4$); the ratio of turns is 19:4:10:4:19, producing a field of $H_i = 3.569 \cdot 10^4\ I/d$, i.e. the generated field increases with increasing electric current and is inversely proportional to the side length of the cube.

RUBENS coils and other systems involving more than two coils are discussed by KIRSCHVINK (1989). They produce a homogeneous field in a larger volume of space. This is highly advantageous when small changes in magnetic parameters are involved and homogeneity is crucial. In the two-coil HELMHOLTZ system, on the other hand, the space directly above the gap of the arrangement is free of equipment. This might be important when interactions between magnetic and

[3] The unit tesla (and nanoTesla) actually designates magnetic induction, which is proportional to the field strength.

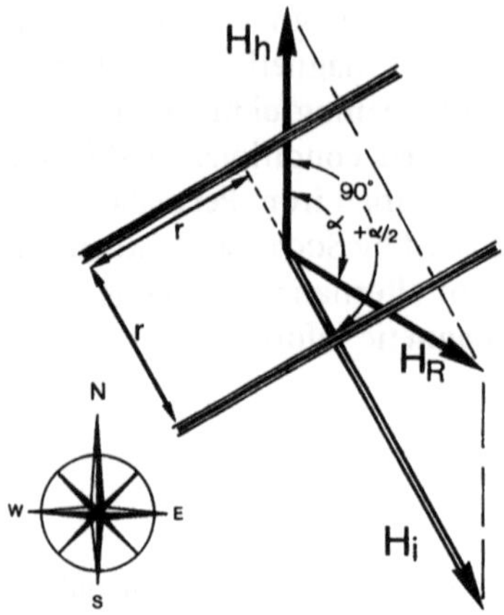

Fig. 1.6. Coil arrangement designed to deflect magnetic North by the angle α, leaving total intensity and inclination unchanged; view from above. The axis of the coils is horizontal, inducing a magnetic field of $H_i = H_h \sqrt{(1 - \cos\alpha)^2 + (\sin\alpha)^2}$ in the direction (90°+ α/2). The induced field H_i and the horizontal component H_h of the ambient field add to form the resultant horizontal component H_R which has the same strength as H_h, pointing towards (mN + α)

celestial cues are to be studied. Miniature battery-operated HELMHOLTZ-type coils have also been used to expose pigeons to various magnetic conditions during free flight (e.g. C. WALCOTT and GREEN 1974; see Fig. 4.7).

Coil Alignment. The magnetic field produced by coil systems adds to the ambient magnetic field. The resultant field depends on the alignment of the coils, i.e. the ambient field can be compensated, augmented or altered in specific ways. Figure 1.6 gives as an example one of the most frequently used test situations, namely, the magnetic North is deflected by a certain angle α, while total intensity and inclination remain unchanged. In this case, the axis of the coils has to be horizontal and directed towards 90 ° + α/2. With this coil position, when the desired shift in magnetic North is reached, total intensity and inclination have the same values as in the original field.

1.2.2 Describing Magnetic Fields

Early authors often found it difficult to describe their test conditions adequately, giving only vague and incomplete descriptions (e.g. "*at a distance of 30 cm, a compass needle was deflected by 50°*"). This makes it impossible to estimate the conditions that the test animals experienced. Normally, static magnetic fields ought to be described giving their total intensity, declination and inclination, and possibly also their inhomogeneity, if it is no longer negligible. For alternating fields, frequency and impulse shape should be given in addition.

1.2.2.1 Measuring Direction and Intensity

For biological purposes, the direction of magnetic fields in the horizontal is still measured with the traditional magnetic compass. Information on local values of magnetic declination can be found in area charts and aeronautical charts; the latter are updated every year. In the vertical, a free swinging dip needle that aligns itself with the axis of the field lines indicates inclination or dip.

For measuring intensity, a number of modern instruments are available. They make use of various magnetic effects like induction, HALL effect and optical pumping, etc. Among the most commonly used ones are *fluxgate magnetometers* based on changes in permeability, and *proton magnetometers* utilizing the precession of spinning nuclei of the hydrogen atom (e.g. BREINER 1973). For an overview on the various types of magnetometers, see THOMPSON and OLDFIELD (1985).

1.2.2.2 Carthesian and Polar Coordinates

When altering the magnetic field for experimental purposes, one additional problem remains. For technical reasons, it is often convenient to add to the horizontal or vertical component only. The two different systems of describing the magnetic field mean that a change in *one* variable of Carthesian coordinates results in a change in *two* variables in polar coordinates and vice versa. For example, when the horizontal component is increased, this means an increase in total intensity *and*, at the same time, a smaller angle of inclination. Likewise, when total intensity is increased, while keeping inclination constant, both the horizontal *and* the vertical component are increased. Consequently, any response should be carefully tested in view of whether it truly results from a change in the component in question, or whether associated changes in other components are crucial.

1.2.3 Treatments Designed to Affect Magnetic Receptors

Another approach has been to interfere with magnetic orientation by subjecting test animals to treatments designed to interfere with the magnetic receptor system(s). These experiments generally suffered from lack of knowledge about the mechanisms of magnetic detection; it was not possible to predict how the treatments worked and what they specifically effected. So the applied fields were usually rather strong, 0.05 T (i.e. 1000 times the intensity of the geomagnetic field) and more, following the idea: "*The stronger the field, the greater a possible effect.*"

When magnetite was found in bees and pigeons in the late 1970s (GOULD et al. 1978; C.WALCOTT et al. 1979), a potential role of these small permanent magnets in magnetoreception was widely discussed (see Sect. 9.3). After C. WALCOTT et al. (1979) and KIRSCHVINK and GOULD (1981) had determined the magnetic properties of magnetite particles, specific treatments were designed that would alter their magnetization. This involved fields of intensities above 0.2 T, oscillating fields or fields with steep gradients. However, the possibility remained that magnetite particles would simply align in a field and return to their original position when the field was turned off and/or the animal was moved out of the apparatus at the end of treatment. To avoid such problems, a very short, strong magnetic pulse was applied. With 0.5 T, it was strong enough to change the magnetic moment of magnetite particles, while the duration of only 0.5 ms seemed too short to allow movements of the particles because of their inertia (see Sect. 9.3.3).

Another type of treatment involved irregular alternating fields of varying intensity and direction (e.g. Papi et al. 1983), with maximum intensities of only 104 000 nT (see Sect. 6.1.1.3). The basis of their effectiveness remains unclear.

1.3 Summary

In essence, the geomagnetic field is a dipole field. The magnetic poles lie in the vicinity of the geographic poles. The deviation of local magnetic North from geographic North, magnetic declination or variation, is fairly small in most parts of the world, with the exception of the regions near the poles. Inclination or dip is vertical at the poles and horizontal at the equator. Total intensity also shows a gradient, with the highest values at and near the poles and the lowest values around the equator.

The large-scale regular picture of the magnetic field is disrupted by continentwide, regional and local irregularities. These deviations from the ideal dipole field form a 'magnetic topography', with areas of higher intensity and those of lower intensity forming magnetic 'mountains' and 'valleys'. Occasionally, local deviations from the regional values, caused by strongly magnetic rocks, exceed 10% of the normal field. Such anomalies are rare, however.

Temporal variations of the earth's magnetic field occur as regular daily variations, as irregular fluctuations (magnetic storms), and, on a long-term level, as secular variations and changes in geological time. Regular daily variations are caused by electromagnetic radiation of the sun acting on the ionosphere; their shape changes with geographic latitude and with season. Magnetic storms are associated with sun spot activity and solar flares; the amount of these irregular fluctuations is characterized by K-indices on a quasi-logarithmic scale. Secular variations mean slow changes in local magnetic parameters over the course of years. In geological times, the magnetic field underwent dramatic changes, the most significant ones being reversals of polarity.

In industrialized areas, man-made magnetic fields from transmission lines, various kinds of technical equipment, etc. may add local irregularities and fluctuations to the magnetic field.

Several methods were used to change the ambient magnetic field for experiments. Bar magnets have the disadvantage that fields so produced are difficult to define. Various kinds of shielding are also used to deprive animals of magnetic information. Controlled magnetic fields with optional directions and intensity are usually produced electromagnetically with the help of coil systems, the most frequently used ones being Helmholtz coils, a system of two round coils, and Rubens coils, a system of five square coils. The generated magnetic field adds to the local field.

Any description of magnetic fields ought to include total intensity, inclination and declination, and in alternating fields, frequency and impulse shape. The direction of magnetic fields is usually measured with the traditional compass and a dip needle. For measuring intensity, various types of magnetometers based on different magnetic effects are available.

One type of magnetic treatment aimed at interfering with the magnetic receptors. It involved very strong fields and strong magnetic pulses of short duration. Irregular alternating fields also proved to be effective in changing behavior.

Chapter 2

Orientation

For further consideration, it is important to circumscribe the theme of the book. What does 'magnetic orientation' mean, and how does it differ from other magnetically controlled responses? – Orientation is based on the use of *information*; in particular, magnetic orientation means that animals use information from the magnetic field to direct and control their behavior.

2.1 Passive Responses Versus Active Behavior

From the above definition, it follows that orientation requires *active information processing*, and that the organism must have faculties adapted to perceive and treat this information in an adequate way. The energy for responses is provided by the organism. This is self-evident when cues like light, sound, etc. are considered. However, in the case of the magnetic field, the field itself exerts a torque on any particle with magnetic moment. If an organism is small enough and contains a sufficient amount of magnetic material, these forces may become sufficiently strong to rotate it and align it along the magnetic field. Thus, in the case of magnetic orientation, it is important to distinguish between active orientation processes based on information processing and mere passive effects.

2.1.1 Magnetotactic Bacteria

In view of this, a now classic example of 'magnetic orientation', the famous *magnetotactic bacteria* first described by BLAKEMORE (1975) become a borderline case. The bacteria move 'north-seekingly', following the magnetic field lines northward (Fig. 2.1). The magnetic response is caused by small particles of magnetite, Fe_3O_4, arranged in chains (see Fig. 2.2). Their magnetic moment rotates these bacteria, *alive as well as dead*, into a position parallel to the field lines with the front end toward the north. Because of this alignment, their own movement will propel living cells on a path along the field lines in the direction of the magnetic vector.

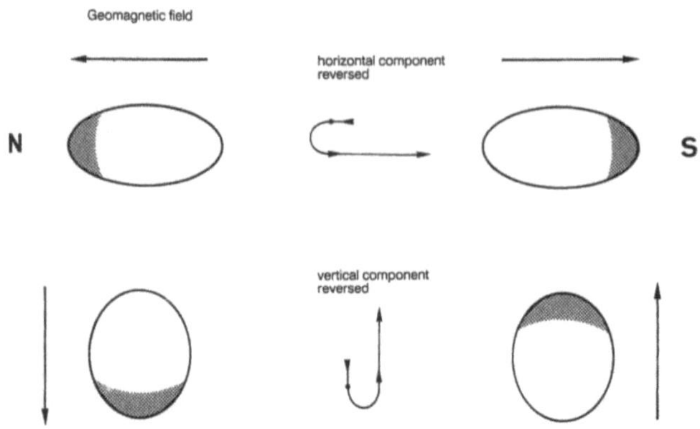

Fig. 2.1. Orientation of magnetic bacteria. The bacteria collect at the side of the water droplet which is pointing towards magnetic North. On reversal of one component of the geomagnetic field, they migrate to the opposite end. *Upper diagram* horizontal droplet; *lower diagram* hanging droplet. (After KALMIJN and BLAKEMORE 1978)

The magnetotactic bacteria first decribed were from marine sediments, but soon magnetotactic bacteria were also found in freshwater sediments (e.g. FRANKEL et al. 1979; KIRSCHVINK 1980). As they are normally anaerobic or microaerobic, the biological significance of these movements is easily understood. The first studies (BLAKEMORE 1975) were performed with bacteria from marine sediments at the east coast of North America, where the inclination is positive, pointing to the ground. This means that when the bacteria are stirred up from the bottom in their natural environment, their 'north-seeking' movement will guide them from the open, oxygen-rich water down to the sediment and thus into more favorable conditions. In the southern hemisphere, where the field lines run upward, bacteria with reverse polarization, responding 'south-seekingly', have been found (KIRSCHVINK 1980). Both types are present at the magnetic equator, but when the inclination angles exceed ± 6 °, one type of bacteria clearly predominates, 'north-seeking' ones at positive and 'south-seeking' ones at negative inclinations (FRANKEL et al. 1981; TORRES DE ARAUJO et al. 1990). For anaerobic and microaerobic organisms too small to make use of gravity, the directing forces of the magnetic field provide a useful alternative for moving downward. Under these circumstances, magnetotaxis is of considerable survival value. An extended review on magnetic bacteria and their biomineralization is given by MANN et al. (1990); FRANKEL (1986) summarizes corresponding findings in single-celled euglenoid algae.

In these bacteria and algae, the movement is active, consuming energy supplied by the organism. The 'orientation', however, is not an active process, as the alignment along the field lines occurs passively by the magnetic force acting on magnetite particles. Yet the prerequisites of this response, the synthesis and the correct arrangement of the magnetite particles within the cell (Fig. 2.2), involve biological processes and have thus been subject to natural selection in the course of evolution. Here, actions of the living cell produce the means for

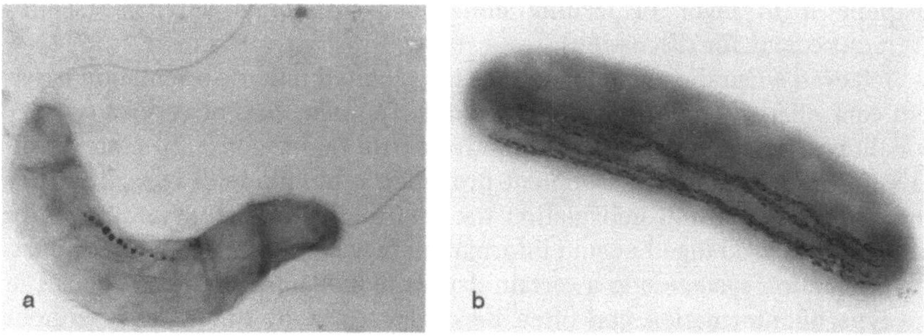

Fig. 2.2. Electron micrograph of two micro-aerobic magnetic bacteria from freshwater sediments. *a Magnetospirillum gryphiswaldense*, young cell about 3 μm long, from river sediments near Greifswald, northeastern Germany; one chain of magnetosomes (diameter 42 nm) is clearly visible. (Courtesy of D. SCHÜLER). b *"Magnetobacterium bavaricum"*, 8 to 10 μm long, from Lake Chiemsee sediments, southern Germany; it is characterized by multiple chains and a high number of magnetosomes. (Courtesy of N. PETERSEN)

passive alignment which then leads to movements in a desired direction – a case that is difficult to classify. Functionally, the reaction is equivalent to other orientation responses, but the way in which it is achieved is fundamentally different from the orientation processes of animals (see below). Because of this, it leads to an unfortunate confusion of terms when this alignment is called 'navigation' or 'navigational compass', as some authors enthusiastically did (e.g. BLAKEMORE and FRANKEL 1981; FRANKEL 1986).

2.1.2 Orientation in Animals

The orientation behavior of animals is of very different nature. Any relationship mediated by information is much more complex, because in principle, information is always subject to interpretation by the recipient, and the recipient may have various degrees of freedom in his response, including the option not to respond at all. In the case of animal orientation, three aspects add to complicate the general picture, namely, (1) variability of motivation, (2) the use of multiple cues, and (3) variability in information processing, in particular in the rating and ranking of various cues.

Orientation does not occur per se, but is always a component of a specific behavior, and thus it can only be observed if the respective behavior is performed. The motivational state – whether the animal intends to behave in a specific way – is of crucial importance. A reliable estimate is not always possible, however. It should be kept in mind that when a certain oriented behavior is expected, but not found, the reason may lie in lack of motivation just as much as in lack of orientation. For example, when displaced birds upon release do not take off towards home, this does not necessarily indicate disorientation. It might simply mean that they do not start the return flight immediately, but postpone it in favor of feeding and other sustaining activities (see R. WILTSCHKO 1992 for discussion).

Yet, even when the animal performs the required task, its orientation is seldom controlled by just a single factor alone. Different cues of various sensory modalities may be integrated and jointly control the response. In nature, animals live in a complex environment providing a multitude of cues, and they tend to make use of all information the environment offers. Several types of cues might be used together, and information may be transferred between cues. The orientation system is to a certain degree redundant so that deficiencies in one type of information can often be compensated by other kinds without apparent loss. This is not only true for highly complex behaviors like migration and navigation of birds, marine turtles or migrating fishes (ABLE 1991a; SALMON and WYNEKEN 1994; R. WILTSCHKO and WILTSCHKO 1994a), but also for seemingly simpler tasks like that of littoral amphipods returning to the border between land and sea (e.g. PARDI and ERCOLINI 1986). Chapter 5 will present examples of such interactions of magnetic information with other cues.

Additional problems arise from the fact that the behavior in a given situation is not solely determined by the present combination of cues. Animals do not respond *quasi* standardized like automatons, but rate and evaluate the available information in a variable manner which may depend on their *genetic disposition* as well as on their *previous experiences*. Examples for both in homing pigeons have been discussed by W. WILTSCHKO et al. (1991). Modifications of orientation behavior based on experience have also been reported in many other animal groups so that they may be assumed to be a general phenomenon.

Consequently, animal orientation cannot be understood on the basis of sensory input alone. Motivation, genetic disposition and individual experience must always be considered. In practice, this renders the analysis of the role of any single orientation factor more difficult. In particular, when the magnetic field is concerned, the situation is complicated by the fact that its use may be restricted to certain behavioral situations, and that it is normally used together with other cues (W. WILTSCHKO and WILTSCHKO 1991a).

The KIRSCHVINKS describe the situation of magnetic orientation as presented in the literature by stating:
"*None of these reports have yet approached the level of clarity and simplicity displayed in experiments with magnetotactic bacteria, which is the best example of geomagnetic sensitivity in any living organism.*" (KIRSCHVINK and KOBAYASHI-KIRSCHVINK 1991, p. 169 verbatim).

This is an evident misunderstanding. What they really comment is the basic difference between passive alignment and active orientation based on information processing. The bacteria are merely rotated by their magnetic moment – they are no more 'sensitive' to the magnetic field than a weather vane to the wind.

Table 2.1. Magnetic effects on orientation behavior in the animal kingdom

Phylum	Classis	Ordo	Type of response
Plathelminthes	Turbellaria	Seriata	mbeh
"	Trematoda, Digenea	Echinostomatida	(comp)
Nemathelminthes	Rotatoria, Eurotatoria	Ploima	mbeh
Annellida	Clitellata, Oligochaeta	Lumbricida	(tvar)
Mollusca	Polyplacophora	Ischnochitonida	(comp), mbeh
"	Gastropoda, Prosobranchia	Neogastropoda	mbeh
"	" , Opistobranchia	Nudibranchia	comp
Arthropoda	Crustracea, Malacostraca	Decapoda	comp
"	" , "	Amphipoda	comp
"	" , "	Isopoda	comp
"	Insecta, Paurometabola	Isoptera	comp
"	" , Holometabola	Hymenoptera	comp, mbeh
"	" , "	Coleoptera	comp, grad, tvar, mbeh
"	" , "	Diptera	comp, (grad)
"	" , "	Lepidoptera	comp, (tvar)
Vertebrata	Chondrichthyes	Carcharhiniformes	(comp), grad, mbeh
"	"	Myliobatiformes	comp
"	Osteichthyes	Osteoglossiformes	mbeh
"	"	Anguilliformes	comp, mbeh
"	"	Cypriniformes	[tvar]
"	"	Siluriformes	mbeh
"	"	Salmoniformes	comp, [grad]
"	"	Gasterosteiformes	mbeh
"	"	Perciformes	mbeh
"	Amphibia	Anura	(comp), [grad], (tvar)
"	"	Urodela	comp, [tvar], mbeh
"	Reptilia	Testudines	comp, [grad], mbeh
"	"	Crocodylia	[comp], (tvar)
"	Aves	Procellariiformes	(comp), (grad)
"	"	Charadriiformes	(comp), (tvar)
"	"	Columbiformes	comp, (grad), (tvar), mbeh
"	"	Passeriformes	comp, (grad), (tvar), mbeh
"	Mammalia	Rodentia	comp, [grad], (tvar)
"	"	Perissodactyla	(comp)
"	"	Primates	comp
"	"	Cetacea	(grad), (tvar)

comp = compass orientation, grad = response to magnetic gradients, tvar = response to temporal variations, mbeh = other responses and modifications of behavioral responses. - Without parentheses: experimentally demonstrated; ( ): response demonstrated, discussed as belonging to respective category, or correlative evidence; [ ]: possibility discussed.

2.2 Various Uses of Magnetic Information

The earth's magnetic field represents a very reliable, omnipresent source of information for all living organisms. Various parameters provide different types of information, and animals that are sensitive to the respective parameters can, at least in theory, obtain information on direction, position and time. A considerable number of publications, most of them published within the last 20 years, suggest indeed that many animals make use of magnetic information to control their behavior in space and time. The species range from protozoans, various invertebrates to the vertebrate classes of fishes, amphibians, reptiles, mammals and birds. Their orientation can be roughly classified in compass and non-compass responses, whereby the latter have to be further divided into several subcategories. Table 2.1 lists the groups, together with the various types of responses that have been described or are discussed.

2.2.1 Orientation Via an External Reference

The majority of orientation processes discussed in this book concern spatial orientation, i.e. animals moving towards a goal (or away from adverse conditions, which is the reciprocal task). This goal may be spatially closely defined, like a bird's nest which is situated at a specific site. In other cases, the animal may seek certain conditions which are found at more than one site or in a more or less extended area. In the latter cases, factors indicating these conditions directly may control the spatial behavior. Finding the desired conditions may turn into a navigational task, however, when the animal knows from previous experience where such conditions can be found.

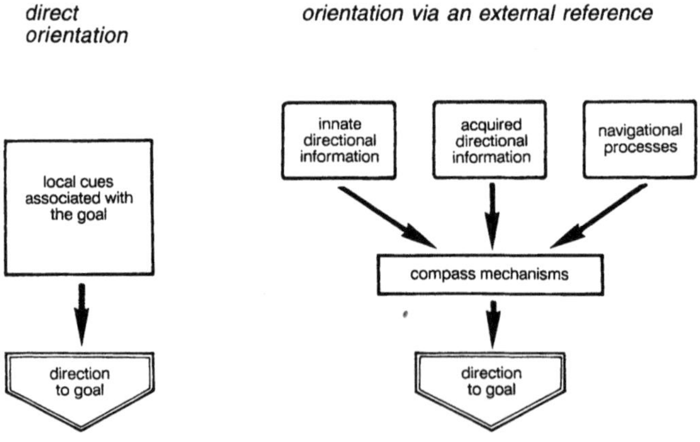

Fig. 2.3. Schema illustrating two different types of orientation. *Left* Direct orientation by local cues associated with the goal, which indicate the direction to the goal directly; *right* indirect orientation, a two-step process which establishes the direction to the goal via an external reference. The direction is first represented as a compass course; in the second step, this course is located with the help of a compass and thus transformed into a direction of movement towards the goal (for further explanations, see text)

The direction of movement towards a goal can be determined in two different ways: either it is indicated *directly* by local factors regularly associated with the goal, or it can be established *indirectly* via an *external reference* (see Fig. 2.3).

Direct orientation means that the animals can perceive local cues emitted by the goal itself or regularly associated with the goal. These cues are pointing towards the goal, specifying something equivalent to "*go there*". Particular stimuli indicating the position of a goal over a certain distance are required; such cues may also help animals in the first place to discover sites where the desired conditions occur. Very often, however, local cues indicating the goal directly are not available. When this is the case, the direction towards a goal may be established via an *external reference*, a compass. This is a common procedure in spatial orientation, at least when distant destinations outside the animals' sensory range are involved. It means that the direction toward the goal is first established as a compass course[1] in terms equivalent to e.g. "*go 225 ° SW*", and then a compass is used to locate this course and transform it into a direction of movement, leading to a specification like "*go there*".

This latter type of orientation represents a complex process with two components, namely, (1) mechanisms supplying the direction toward the goal as a compass course, and (2) compass mechanisms transforming it into an actual direction of movement (cf. Fig. 2.3, right). The distinction between the two components is of theoretical importance. Unfortunately, there is no commonly used English term for the concept which, in German, is called *Soll-Richtung*, a direction (*Richtung*) which is the one to be pursued (*Soll*). Several different terms are in use. PARDI and ERCOLINI (1986) called it 'guide value', but more commonly, authors speak of 'desired directions' or 'intended directions', which seem to imply some conscious action and thus are maybe not entirely adequate. We will use the term *set direction*, as an analogue to the term 'set point' in kybernetics and control engineering, or simply talk about a *compass course* or a *course* that is to be located with a compass.

The two components – the set direction and the compass – are basically independent of each other, i.e. a compass can locate any course regardless of its origin, at least in theory. As indicated by Fig. 2.3, set directions may be innate directional tendencies, like the migratory directions of birds, or acquired preferences of various natures. The latter range from long-term, rather constant directions like in the Y-axis orientation of amphibians and littoral amphipods via various induced directional tendencies acquired by experiencing certain directional relationships in the environment to training directions in conditioning experiments (see Chap. 4). They also include compass courses towards a specific goal determined by navigational processes, i.e. courses whose relevance

[1] The process of establishing a relation to a distant goal which is outside the range of the sensory organs is frequently referred to as 'navigation'. Traditional definitions of navigation (e.g. GRIFFIN 1952, 1955; SCHMIDT-KOENIG 1965; KEETON 1974; WALLRAFF 1974) explicitly exclude the use of direct cues, without defining any specific procedure. Perhaps one might define as 'navigation' those orientational processes which establish the relation to a specific goal indirectly via an external reference and then make use of a compass.

is limited to the momentary situation, being terminated when the animal has reached the goal. Studying *homing* after displacement, KRAMER (1953, 1961) recognized the two-step nature of the orientation process involved and described it by his classic 'map-and-compass' model.

2.2.2 The Role of Magnetic Cues

Considering potential uses of magnetic information, it becomes obvious that the geomagnetic field can only rarely serve as a direct factor. The sole occasions are when organisms make use of the fact that the inclination of the field is reliably pointing downward or, in the southern hemisphere, upward. The magnetotactic bacteria have already been mentioned; these anaerobic or microaerophil organisms must reach specific conditions found in the mud at the bottom of water bodies, and they rely on the magnetic force to rotate them into the right position. The only case of magnetotactic orientation in animals known so far involves the larvae of a fluke which also must move downward as they infect bottom-dwelling aquatic snails (STABROWSKI and NOLLEN 1985; see Sect. 3.1). For animals orienting in the horizontal, however, the magnetic field cannot be a direct factor, because magnetic cues do not indicate any specific conditions.

However, when orientation is established via an external reference, magnetic information may be used in several ways. The most important function of the magnetic field is to provide the reference system itself. Perceiving the field per se results in an anisotropy of space: for animals that are able to detect the direction of the field lines, the spatial directions are no longer equal. In this sense, the magnetic field structures space in the horizontal just like gravity structures space in the vertical by allowing the distinction of up and down. Hence, based on its vector quality, animals can utilize the magnetic field as a reference for various types of courses, innate, acquired or determined by navigational processes, and locate these courses with the help of a *magnetic compass* (see Chap. 4). This is equivalent to our own use of a technical magnetic compass.

Additionally, there are indications suggesting the involvement of magnetic information in the processes determining the home course after displacement. Here, the nature of information depends on the navigational strategy: for route reversal (see Sect. 4.1.4.2), animals utilize the vector quality by recording the net direction of the outward journey, which is another use of the magnetic compass. Navigation by local information, in contrast, may employ the spatial distribution of total intensity or inclination. Any location on earth is characterized by a specific set of magnetic parameters which might give animals information about their position, provided the animals are sensitive enough to measure these values with sufficient accuracy (see Sect. 6.1).

The spatial distribution of magnetic parameters may be used in yet other ways. These functions of magnetic information cannot be classified as described previously in Section 2.2.1, because they involve non-goal-oriented movements. One hypothesis suggests that animals move along magnetic contours, using them as invisible guidelines (see Sect. 6.2). Such a strategy does not lead the animals towards a specific goal; however, if suitable contours are available, it may control large-scale movements between more or less extended areas. Also, specific magnetic conditions may serve as local triggers, starting the next step in innate programs, etc. (see Sect. 6.3.1 and 6.3.2).

Altogether, it is obvious that the magnetic field may provide relevant information for various kinds of orientation processes. Aside from this, spatial and temporal variations of the magnetic field have been found to affect orientation in a number of yet unexplained ways (see Chap. 7).

2.2.3 Demonstrating Compass Orientation

Compass orientation is by far the most widespread with regard to the use of magnetic information among animals (cf. Table 2.1). To show that a species uses the magnetic field as a compass for direction finding (or as a basis for alignments), it must be shown that the directional tendencies of the animals in question truly depend on the direction of the ambient magnetic field. The observation that orientation persists in the absence of all other known cues is not sufficient; even a breakdown of orientation in the absence of meaningful magnetic information is indirect evidence only. In a strict sense, magnetic compass orientation can only be accepted when altering magnetic North results in a deflection of the animal's directional preference that *qualitatively* as well as *quantitatively* corresponds to the change in the magnetic direction. There must be a roughly 1:1 relationship between the deflection of magnetic North and the directional shift in the animal's response. Hence, testing for a magnetic compass is only possible when the test animals show a suitable orientation that is reliably found and, in control tests, provides the reference against which the response in the experimentally altered magnetic field can be compared.

This basic requirement is not always easy to meet. A number of species have thus far escaped testing for magnetic orientation, because experimentalists have not yet found a reliable directional tendency for experimental manipulation. This is a natural handicap, because in most situations, the various directions are equivalent, and there is no reason why an animal should prefer one over another. Sometimes, however, specific behaviors like the migration of birds or the movements of arthropods at the seashore, lead to strong spontaneous preferences for specific directions which provide useful starting points for analysis. In other cases, experimentalists have tried to induce directional preferences by specific spatial arrangements in the animal's housing quarters or by directional training (see Sect. 8.2). For the compass response per se, the origin of the set direction is irrelevant, as long as the animals reliably show the respective behavior. However, as a baseline for the interpretation of any results, the experimentalists must understand the motivational situation of the behavior and the mechanisms which determine the courses. Since behavior as a whole is

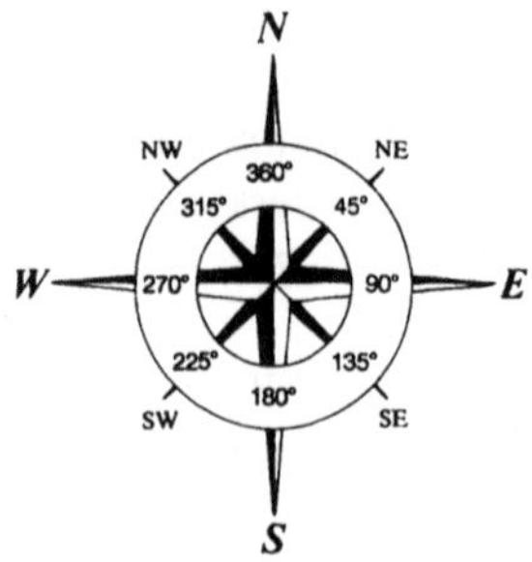

Fig. 2.4. Geographic reference system, designating directions in terms of 360 °, starting in the north and counting clockwise, via 90 ° east, 180 ° south, 270 ° west to 360 ° north. N, north; NE, northeast; E, east; SE, southeast; S, south; SW, southwest; W, west; NW, northwest

observed, one must be able to assess how any change in conditions interferes with the various steps, and whether motivation, establishing the compass course or the compass mechanism itself, is affected.

2.3 Analysis of Orientation Data

Normally, the orientation is measured by recording directional tendencies. Animals are released in round arenas, various types of circular cages, bowls, etc. or set free in the open. The direction of their movements is usually recorded after a predefined time period or after the animal has crossed a predefined border. The data obtained in this way represent a number of bearings given either as geographic (or magnetic) directions or in relation to some reference, like the home direction, the migratory direction, the direction towards the shore, etc. They represent a circular distribution, since bearings are usually recorded as angles ranging from 1 ° to 360 °, following, as a rule, geographic convention (Fig. 2.4).

The statistical treatment of such data, together with the most frequently used tests for non-uniformity and for differences between distributions, has been described in detail by BATSCHELET (1981), which became the standard work in this field. Here, we give a brief summary of the first common steps.

The individual bearings α_i of the n animals of a test group must be processed in order to obtain summary data characterizing the behavior of the group as a whole. This is done by calculating the *mean vector* with the direction α and the vector length r by vector addition. In the case of unimodal (circular normal) distributions, the mean direction $\alpha = \arctan(\Sigma \sin \alpha_i / \Sigma \cos \alpha_i)$ gives an estimate of the average direction. The vector length $r = 1/n \sqrt{(\Sigma \sin \alpha_i)^2 + (\Sigma \cos \alpha_i)^2}$ is a measure of concentration; it is 1 when all α_i coincide and it approaches 0 when they are evenly distributed.[2] In linear data, the

[2] In cases of axially bimodal (or quadrimodal) distributions, the procedure is similar: The angles have to be doubled (or multiplied by four) so that the ends of the axes fall together. The resulting distribution can be treated like before, with the calculated mean direction retransformed into the axis at the end.

mean always represents a characterizing quantity. With regard to circular data, however, it has to be decided first whether the group of animals truely shows a common directional preference, or whether their data are more or less randomly distributed. The standard test to answer this question is the RAYLEIGH test. It uses the vector length, r, or derived quantities like $z = nr^2$. Significance according to the RAYLEIGH test means that the assumption of a uniform distribution can be rejected with a given probability; a directional preference around α may be accepted.

Diagrams giving circular distributions of bearings α_i and their mean vectors, and/or specifications like n (number of data), α (direction) and r (length of mean vector), together with the significance level of the RAYLEIGH test, are found frequently throughout this book to describe orientation.

2.4 Summary

Magnetic orientation means that animals use information from the magnetic field to control their behavior. Hence, the movements of magnetotactic bacteria, which are passively aligned along the field lines by their magnetic moment, are fundamentally different from the orientation responses of animals.

Orientation behavior of animals is characterized by a complex relationship between stimulus and response, which is mainly caused by three factors, namely, variability in motivation to perform the respective behavior which cannot always be adequately controlled, the use of multiple cues of which the magnetic field is but one, and a variability in information processing, in particular differences in rating and ranking of various cues based on genetic dispositions and individual experience.

In spatial orientation, the direction towards the goal may be established directly when the animal is within sensory contact to the goal, or it may be established indirectly via an external reference. The latter case means that the direction towards the goal is first established as a compass course, and then a compass is used to locate this course and transfer it into an actual direction of movement. The origin of the course, the *set direction*, may vary according to the behavioral situation; a compass may be used to locate innate courses, courses acquired by experience on a long-term or short-term basis, or courses determined by navigational processes.

As a direct factor, the magnetic field can indicate a goal only when organisms move downward along the inclination of the field lines. However, it can be used in several ways when the relationship to the goal is established via an external reference. Because of its vector quality, it provides a reference for innate or acquired courses, and it can be used as a compass to locate directions. It may also be involved in the processes determining the home course after displacement; here, the type of magnetic information depends on what navigational strategy is used. Additionally, aspects of the spatial distribution of magnetic parameters may be used as guidelines or as triggers.

In order to demonstrate that the magnetic field is used as a compass, it must be shown that the animal responds to a deflection of magnetic North with a corresponding shift of its directional preferences. Hence, a crucial precondition of such tests is that the animals in question show a reliable preference for a specific direction.

Orientation data represent circular distributions and are analyzed using the methods introduced by BATSCHELET (1981). Normally, the orientation behavior of a group of animals is characterized by giving the direction and the length of the mean vector. The RAYLEIGH test is used to decide whether a directional preference can be assumed.

Part 2: Directional Orientation

Chapter 3

Magnetotaxis and Alignment Behaviors

Magnetotaxis designates movements along the field lines. Alignments place animals in positions that are symmetrical with respect to the stimulus, i.e. in the case of the magnetic field, parallel, antiparallel and often also perpendicular to the field lines. Both types of behavior represent the simplest directional response to the magnetic field; they roughly correspond to a *Tropotaxis* as defined by KÜHN (1919).

On a horizontal plane, the prominent positions are along the magnetic north-south axis and the east-west axis, which mostly results in quadrimodal preferences for these cardinal directions. Occasionally, the secondary directions NE, SE, SW and NW between the cardinal axes are also preferred, but to a lesser extent. On the vertical plane, the situation is complicated by the fact that the field lines are inclined. This means that on a plane described by the gravity vector and the magnetic north-south axis, the local angle of inclination marks the magnetically symmetrical position, whereas on all other vertical planes the respective position depends on the projection of the course of the field lines on that plane (see Fig. 3.1). This results in angles exceeding the angle of inclination. On an east-west plane, the angle is 90°, i.e., it corresponds to the vertical.

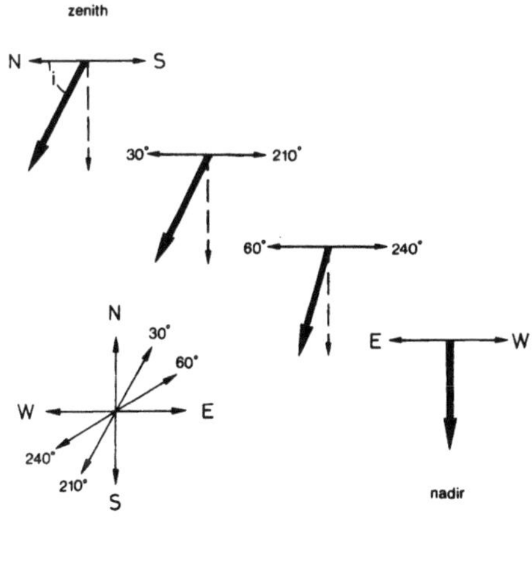

Fig. 3.1. On vertical planes, positions symmetrical to the magnetic vector depend on the orientation of the plane: section through the geomagnetic field viewed from above (*lower left diagram*) and sideways: on a N-S-oriented plane, the symmetrical position corresponds to the angle of inclination, *i*, and its reverse (*left*); on an E-W-oriented plane it corresponds to *up* or *down* (*right*). Two intermediate positions are included

3.1 Magnetotactic Orientation

Magnetotactic behavior, i.e. movement along the field lines, is biologically meaningful only when the direction of the magnetic field reliably indicates certain conditions which the animal intends to seek out. This is, for example, the case when the inclination is sufficiently steep to indicate the downward direction. 'Magnetotactic' bacteria make use of this relationship when they return to the anaerobic or oxygen-poor mud layer after being stirred up (see Sect. 2.1.1).

In metazoan animals, a possible case of true magnetotactic orientation was reported by STABROWSKI and NOLLEN (1985). When released in an elongated test tube, miracidia larvae of the eyefluke *Philophthalmus gralli* (Plathelminthes, Trematoda: Echinostomatidae) preferably moved towards the magnetic North. The ratio between the number of miracidia found in the northern and southern end was 162 to 84 or 66% to 34%. In nature, such a response would lead the larvae of *P. gralli*, which infect bottom-dwelling aquatic snails, along the field lines towards the ground of water bodies. The miracidia also show a strong positive geotaxis; the authors suggest that the observed north-seeking behavior might increase the geotactic tendency and help them to reach the ground. Another *Philophthalmus* species which does not show positive geotactic behavior did not show any preference for magnetic directions when tested in the same apparatus (STABROWSKI and NOLLEN 1985).

These findings suggest that in *Philophthalmus gralli* miracidia, magnetotaxis and geotaxis are coupled to help the larvae to reach the habitat where potential hosts are to be found. Formally, this magnetotactic response is very similar to the one described for magnetotactic bacteria and algae and thus represents an interesting parallel. However, in bacteria, a purely passive alignment is the basis for magnetotactic movements (BLAKEMORE 1975; cf. Sect. 2.1.1). The miracidia of *P. gralli*, in contrast, are not passively aligned, but must be assumed to actively process magnetic information, even if details are not yet known.

3.2 Early Reports on Alignments in Insects

First indications that some insects show a tendency to align their body axis with the cardinal axes of the magnetic field came from isolated reports on the resting positions of the *imagines* of termites in their natural environment (cf. ROONWAL 1958). Laboratory studies beginning in the 1960s confirmed a certain preference for symmetrical alignment to the magnetic vector for a variety of insect groups.

3.2.1 Alignments in Termites

In nature, the queens of *Odontotermes (redemanni ?)* and *O. obesus* var. *oculatus* (Isoptera: Termitidae) were often found in N-S or E-W positions. Another subspecies of *O. obesus*, however, did not seem to prefer any specific directions (cf. ROONWAL 1958). Alignment in the N-S axis was also reported from *Termes malabaricus* and *Odontotermes* sp. in the Bombay area (DEORAS 1962). This led to a further investigation of the phenomenon.

In the laboratory, pairs of imagines of *Odontotermes* and *Macrotermes* (Termitidae) that spend most of their time quietly resting were found to prefer N-S and E-W oriented positions. When the cage was turned, the animals resumed their former directional position after a while. The orientation was not maintained when the nest was put in an iron container which distorted the ambient magnetic field. A strong magnet over their nest caused the termite pairs to change their position; they now oriented themselves parallel or perpendicular to the new field lines. Interestingly, the exposure to the strong magnet seemed to have after-effects, as it took several days after the removal of the magnet before the termites aligned to the natural geomagnetic field again (for a summary, see G. BECKER 1964).

Another behavior of termites seems to include alignments, namely, the building of galleries leading away from the nest. These galleries are covered tubes, forming long tunnels on a surface, but also bizarrely shaped branching systems, depending on the species involved. They are built in search for food and water, and when completed, they allow movement between the nest and these goals avoiding the open. In nature, the course of the galleries is determined by the position of the food sources, by the smell of certain fungi, by the terrain to be crossed etc. G. BECKER (1976) studied the behavior of small colonies of *Coptotermes* and *Heterotermes* species (Rhinotermitidae) kept in Petri dishes in the laboratory. When food sources of equal attractiveness were distributed all around the nest, most termites (but apparently not all) built

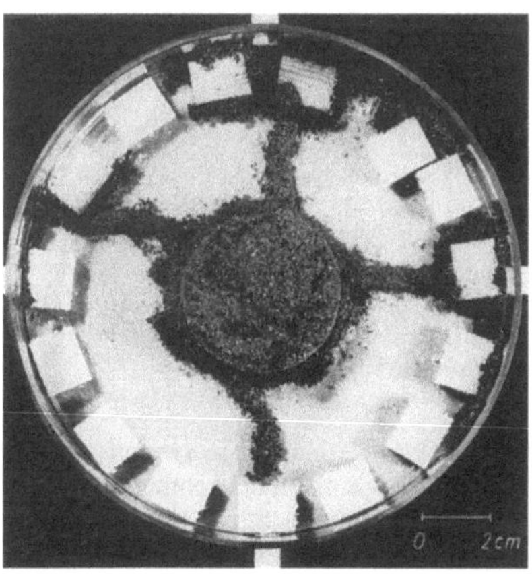

Fig. 3.2. Galleries built in a Petri dish by the termite *Coptotermes amanii* point in the cardinal directions of the magnetic field. (G. BECKER 1975)

their galleries preferably in the cardinal compass directions (Fig. 3.2). In the group that showed this behavior, compensation of the ambient magnetic field by HELMHOLTZ coils resulted in disordered directions of galleries, while turning magnetic North by HELMHOLTZ coils led to the construction of galleries along the new magnetic axes (G. BECKER 1976). These findings suggest that building as a group activity can be directed by the ambient magnetic field.

The spectacular mounts of the compass termites have also been described as being aligned with the magnetic N-S axis (e.g. von FRISCH 1968). Recent reports (DUELLI and DUELLI-KLEIN 1978; JACKLYN 1990), however, suggest that they might represent magnetic compass orientation rather than a mere alignment (see Sect. 4.3.3.1).

3.2.2 Aligned Resting Positions of Diptera

Several common species of flies seemed to prefer landing directions which corresponded to the N-S or the E-W axis (G. BECKER 1963). These field observations led to laboratory studies in which the body axes of flies resting on a horizontal plane were recorded by photography under isotropic light conditions (G. BECKER and SPECK 1964; G. BECKER 1965). The analysis showed a preference for the cardinal N-S and E-W axes for the blowfly *Calliphora erythrocephala*, and the housefly *Musca domestica*. This tendency continued even when sunlight was allowed to enter the room. The meatfly *Sarcophaga* additionally showed a certain minor preference for the secondary axes NE-SW and SE-NW (G. BECKER 1964). The extent of the preference varied between individuals and between species, with the ratio of the number of flies directed in a 45 ° sector around the cardinal axes to the number of those in the other 45 ° sectors usually in the order of 65:35% for *Calliphora* and up to 80:20% for *Musca*. When the magnetic field was compensated, this relationship changed significantly, approaching 50:50% (Fig. 3.3).

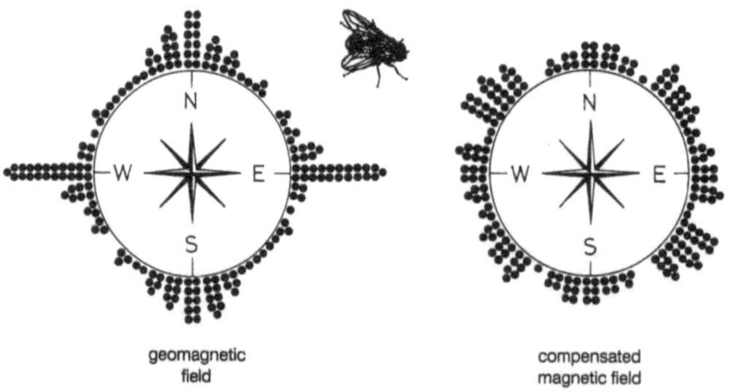

Fig. 3.3 Axial directions of the body axes of flies, *Musca domestica*, marked as *symbols on both ends of the axis* at the periphery of the circle. *Left* Data recorded in the local geomagnetic field; *right* data recorded in a partly compensated magnetic field (about 5% of the local intensity). (Data from G. BECKER 1965)

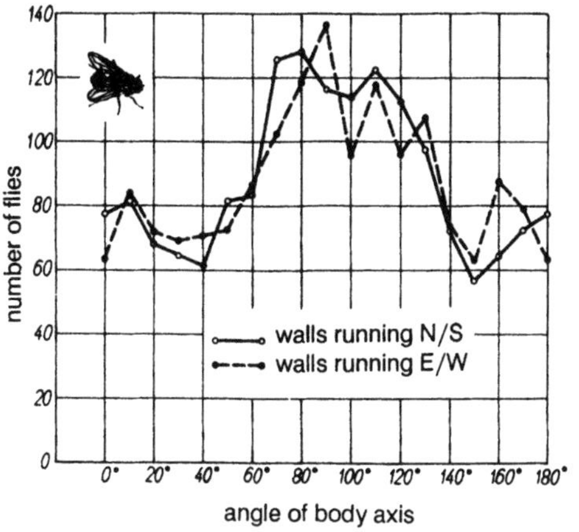

Fig. 3.4. Frequency distribution of body axes of blowflies, *Calliphora erythrocephala*, landing and resting on vertical walls. Based on data of 1526 flies; the directions are given to the nearest 10 °. (G. BECKER and SPECK 1964)

On vertical walls, G. BECKER and SPECK (1964) observed that *Calliphora* preferentially rested in vertical positions when the walls ran in E-W directions; whereas on N-S running walls, the body axes of the flies showed certain maxima 70 ° from the horizontal which corresponded to the local magnetic inclination (Fig. 3.4).

WEHNER and LABHART (1970) reported that the fruitfly *Drosophila melanogaster*, too, significantly preferred resting positions along the cardinal axes of the geomagnetic field. The authors also studied the negative geotactic response of *Drosophila* on an inclined plane in connection with its directional relationship to the magnetic vector. When the plane was inclined 30 ° to the south, i.e. the relationship was symmetrical, the flies moved more or less straight up. When the plane was inclined 30 ° to the east or west, they showed a 5 ° counterclockwise deviation in both cases. This is surprising, because the relation gravity-magnetic vector in one case was the mirror image of the other. How the observed behavior is to be interpreted and whether it involves an alignment to the magnetic field is unclear.

Altogether, the data on Diptera suggest a marked tendency to align their body axes along the N-S and E-W axes when resting. It seems to persist even when it shifts the animals out of balance with other environmental factors, like incident sunlight on the horizontal plane or gravity in the vertical. The behavior on the vertical wall and the fact that the behavior disappeared when the magnetic field was compensated suggest that it is indeed an alignment with the magnetic field lines. Strangely enough, a straightforward attempt to alter magnetic North and see how this affects the flies' behavior was made only once. The results of this short series were inconclusive, as the behavior changed significantly, but without leading to a pronounced preference for the new magnetic axes (G. BECKER 1965).

3.2.3 Magnetic Alignments in Other Insects

A preference for the N-S and E-W axes by the cockchafer *Melolontha melolontha* (Coleoptera: Scarabaeidae) that could be affected by magnets was already mentioned by SCHNEIDER (1960). G. BECKER (1964) reported that, in the absence of other cues, resting long-horned beetles *Hylotrupes bajulus* (Coleoptera: Cerambycidae) preferred the cardinal compass directions, with the most pronounced peak in the north. Similar preferences are suggested in the cockroach *Blatta americana* (Blattodea: Blattidae), the cricket *Acheta domestica* (Ensifera: Gryllidae) and yellow jackets *Paravespula* sp. (Hymenoptera: Vespidae). The respective observations were summarized by G. BECKER (1964); they were discussed by the author as alignments to magnetic field lines, as they were taken to be analogous to the behavior of Diptera.

This might suggest that alignment behavior to the magnetic field lines is a rather widespread phenomenon among insects. However, most of the last-mentioned cases are based on few data only, and hardly any of them was thoroughly analyzed, so that their magnetic nature cannot be accepted without confirmation. Further studies on the problem are badly needed.

3.3 Alignment Behaviors in Honeybees

In Honeybees, *Apis mellifera* (Hymenoptera: Apidae), alignment responses have been described in connection with preferred resting positions and in connection with the *waggle dance* by which bees inform their nestmates about the position of food sources.

3.3.1 Resting Positions

ALTMAN (1981) reported that Honeybees, too, prefer resting positions along the cardinal axes of the magnetic field. In his experiments, he changed the magnetic field in a bee cage with a large permanent magnet, documenting the course of the field lines by particles of iron dust. Photos of the body axes of bees resting at night revealed a pronounced preference for the four cardinal magnetic directions, i.e. the bees aligned preferably parallel, antiparallel or perpendicular to the course of the field lines (Fig. 3.5). With 65:35%, the proportion along the major axes and the directions in between was very similar to the ratio found in flies (G. BECKER and SPECK 1964). No such relationship was found during daytime, when the bees were active. Bees of different origin showed the same alignment, which was obviously independent of the orientation of the hive from which the bees were taken.

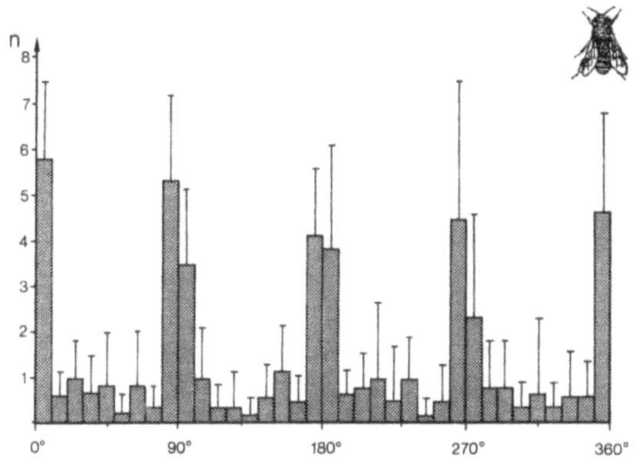

Fig. 3.5. Resting positions of Honeybees, *Apis mellifera*, with respect to the magnetic field lines. The diagram gives the mean number of bees (*n*) and the standard deviation (*plotted upward*) of two groups of 70 bees measured 5 nights each. (After ALTMANN 1981)

3.3.2 Prominent Directions of the Waggle Dance

Successful bee foragers urge their nestmates to join them in exploiting rich food sources. The position of a food source is conveyed to the helpers by a complex movement pattern, known as 'dancing', which has been described in detail by VON FRISCH (1968) and LINDAUER (1975). For food at distances of more than 100 m, this is done by the *waggle dance*: The dancing bee moves in a specific way on a path forming two small semicircles which share a common diameter (see Fig. 3.6, inset). The direction of the straight part of the dance corresponds to the direction to the food source. When walking along that line, the bees waggle their abdomen in a particular way and produce characteristic sound signals, while the periphery of each semicircle is alternately used to return to the starting point. The information on distance is indicated by the frequency of the waggling. The information on direction depends on the circumstances of the dance: outside, with a direct view of the sun, the bees indicate the true direction. However, when they dance inside the dark hive on vertical sheets of comb, which is the usual situation, the angle between sun and food source is transformed into an angle to gravity, with the upward direction representing the direction towards the sun. Other bees pick up this information by following the dancer closely and retransform it into a flight direction for their foraging trips. Odorous substances, carried home by the successful forager, help the newcomers to locate the new food source (von FRISCH 1968).

The specific relationship of these dances to the total intensity of the ambient magnetic field is described in more detail in Section 7.1. Here we focus on the alignments involved, namely, (1) the *zero crossings* of misdirection (TOWNE and GOULD 1985 for the German *Nulldurchgänge* by LINDAUER and MARTIN 1968) and (2) the dances on a horizontal comb in the absence of directing light cues.

3.3.2.1 Zero Crossings of 'Misdirection'

As the sun moves across the sky during the day, the angle between sun and food source changes. The bees alter the angle of their dances accordingly, compensating rather precisely for the sun's changing azimuth (LINDAUER 1954). Hence, based on the direction of the food source and the sun's azimuth, the varying angles danced during the day can be predicted. The observed angles, however, differ from the predicted ones in a variable fashion. The deviations are small – mostly in the range of up to 15 °, with maxima of about 25 ° (Fig. 3.6). They are not just randomly scattered, since all bees dancing simultaneously show the same kind of 'error', which varies during the course of day and also from day to day (see also Fig. 7.2). These deviations were first observed by von FRISCH (1948) who introduced the German term *(Rest)-Mißweisung* for this phenomenon, which was later translated as *(residual) misdirection*. The fact that the misdirections disappear when the ambient magnetic field is compensated to less than 5% of its intensity (LINDAUER and MARTIN 1968; see Sect. 7.1.1, Fig. 7.2) indicated the magnetic nature of the phenomenon.

In view of alignment responses, it is important that certain angles between sun azimuth and food source are indicated correctly, i.e. the 'misdirection' disappears and the curve of misdirection passes zero. On sheets of comb facing N or S, such 'zero crossings' are observed when the direction indicated is directly towards or directly away from the sun, i.e. when the bees dance along a vertical line. On sheets of comb along the N-S axis, facing E or W, the situation is different. In Germany, on a comb facing E, the bees showed no misdirection when the dancing angle was 25 ° left or 155 ° right of the upward direction, while on a

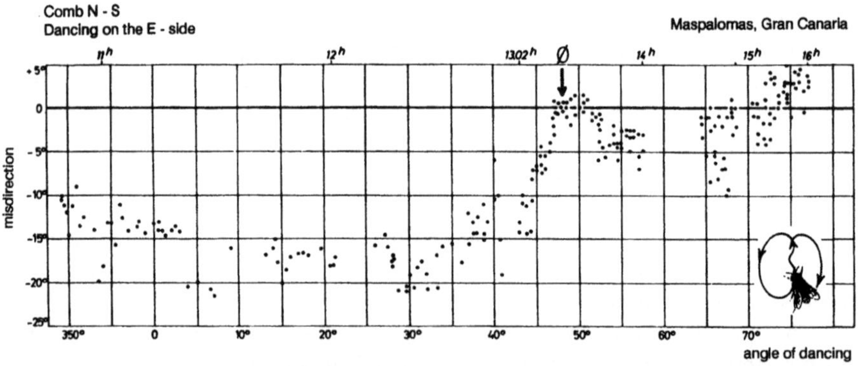

Fig. 3.6. Curve of 'misdirection' recorded from bees dancing on the east side of a comb aligned along the N-S axis in Maspalomas, Gran Canaria, indicating food ca. 510 m south of the hive. *Abscissa* Dancing angle calculated according to the changing sun azimuth; angles are counted counterclockwise, starting with upward = 0 °; *ordinate* 'misdirection', i.e. deviation from the predicted angle. Ø marks the point where the dancing angle 48 ° coincides with the inclination of the local magnetic field, which is 42 °. *Upper abscissa* Local zonal time. (MARTIN and LINDAUER 1977)

comb facing W, the misdirection disappeared at dancing angles that were the mirror images of the ones mentioned above, 25 ° right and 155 ° left (LINDAUER and MARTIN 1968). Since the inclination at the test site was 65 °, these findings mean that the bees dance without error when their dancing directions run parallel or antiparallel to the angle of inclination (or its projection, cf. Fig. 3.1), i.e. when the dancing bee is in a symmetrical position to the magnetic vector. The dances of bees on the vertical comb thus reveal an interesting parallel to the resting positions of flies on vertical walls reported by G. BECKER and SPECK (1964).

At locations with different inclinations, the misdirection curve passes the zero line at different positions. For example, during tests in Maspalomas, Gran Canaria, with a local angle of inclination of 42 °, zero crossings were observed on an E-facing comb when the dancing angle was 48 ° left of the sun (Fig. 3.6; see also Fig. 7.2). MARTIN and LINDAUER emphasize this point in their 1977 paper. The accuracy of the 'zero crossings' is reported to be extraordinary; LINDAUER and MARTIN (1972) claim that the average deviation from the expected values is 0.2 ° ± 3.7 °.

However, the situation seems to be considerably more complex. There are zero crossings at other points of the curve which remain unexplained (see Sect. 7.1.1). In addition, several of the published curves do not show 'zero crossings' at the predicted positions, e.g. none of three curves recorded on E-facing combs in Meknes, Marocco (local inclination 49 °), show a zero corssing at 41 ° (LINDAUER and MARTIN 1968, Figs. 3 and 6). MARTIN and LINDAUER (1977, p. 149) attribute this to an uneven distribution of light at the times in question. Yet the

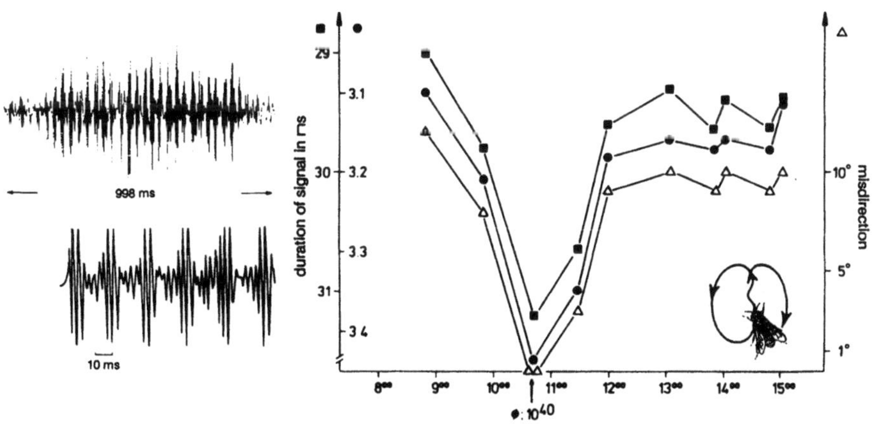

Fig. 3.7. 'Misdirection' of the waggle dance of an individual bee on the west side of the comb, and the associated sound signals, recorded in the local geomagnetic field. *Left* Analogous recording of the entire sound signal (*above*) and of six single sounds (*below*). *Right* Size of 'misdirection' and duration of sounds recorded at various times of the day; *abscissa* time of day; 10:40, when the angle of the waggle dance is parallel to the magnetic vector, is indicated by Ø and marked with an *arrow*; *ordinates on the left* duration of sound signals in ms for the entire signal (outside) and single sounds (inside); *right ordinate* angular size of 'misdirection'. (After KILBERT 1979)

curves look rather smooth and lack sudden changes as one might expect from temporary artifacts. The alignment part, like many aspects of 'misdirection', is not yet understood in detail (see Sect. 7.1.1).

One phenomenon associated with dancing shows a similar relationship to the ambient magnetic field: during the waggling part of the dance along the straight line, bees emit typical sound signals consisting of sequences of vibrational pulses, which in turn consist of several single pulses (e.g. ESCH 1961; WENNER 1962). The durations of both of these elements, 30 to 35 ms and 3.0 to 3.8 ms, respectively, show significant variations in the course of the day which are closely correlated with the variation of 'misdirection'. Maximal duration is observed when the misdirection passes zero at the predicted angles (Fig. 3.7), i.e. when the bees dance in a symmetrical position with respect to the magnetic field (KILBERT 1979). Whether a similar effect is observed at other zero crossings, is unclear.

3.3.2.2 Dancing on a Horizontal Comb

One of the best-documented alignments to the magnetic field involves dances of bees on a horizontal comb. Outside, when the sun is visible, the bees indicate the course leading to the food source directly. In diffuse light, i.e. without visual cues to direct their dances, bees normally do not show oriented dances on a horizontal comb. However, after an adaptation time of 2 weeks or more, directed dances can be observed again, with a strong preference for the cardinal compass directions and a weaker preference for the secondary axes (LINDAUER and MARTIN 1972). This behavior was found to depend on the intensity of the magnetic field. The preferences almost disappeared when the local magnetic field (intensity not given; probably about 47 000 nT) was compensated to less

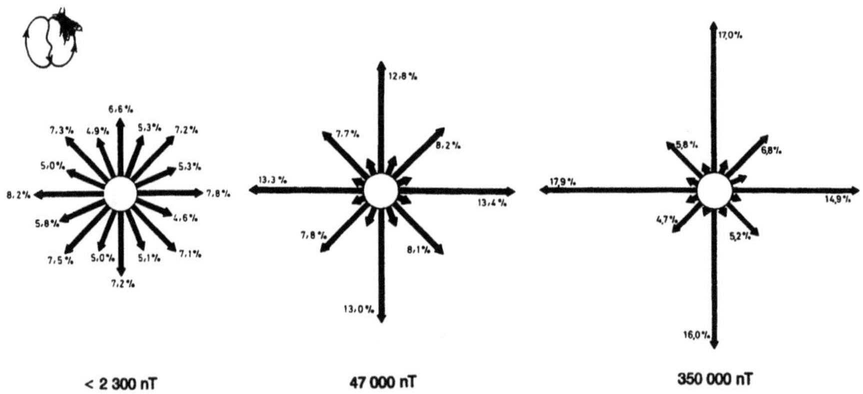

Fig. 3.8. Dances of bees on a horizontal comb in the absence of directing visual stimuli. *Left* 10 541 dances recorded in a field compensated to less than 5%; *center* 24 601 dances recorded in the local geomagnetic field; *right* 5032 dances recorded in a field of increased total intensity. The *arrows* are proportional to the percentage of dances in the various directions; the percentage is also given numerically. (After MARTIN and LINDAUER 1977)

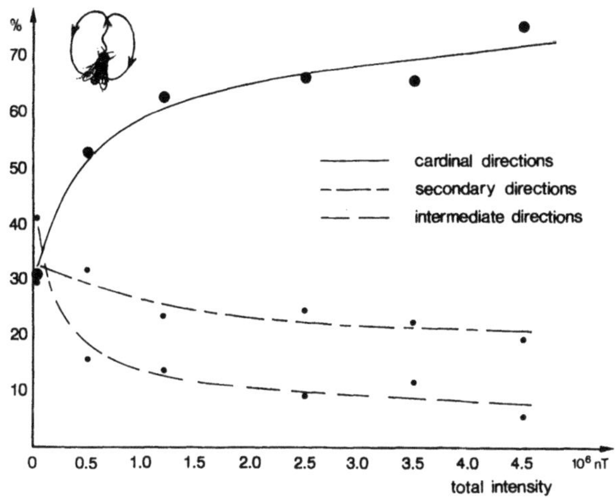

Fig. 3.9. Portion of dances in the cardinal directions N, E, S, and W, in the secondary directions NE, SE, SW, and NW, and the intermediate directions at various total intensities. Note that the circle was subdivided into 16 sectors, with the cardinal and secondary directions including four sectors each, while the intermediate directions consisted of the remaining eight sectors. (Data from MARTIN and LINDAUER 1977)

than 5%, whereas the tendencies in the cardinal directions became more pronounced when the total intensity was increased (Fig. 3.8). Staying in a strong magnetic field of 450 000 nT had some after-effects, however: back in the local geomagnetic field, the bees first produced dances similar to those recorded in a reduced field. After about 1 h, the dances approached the pattern regularly observed in the geomagnetic field, and those recorded on the next day looked normal (MARTIN and LINDAUER 1977).

The portion of dances in the cardinal directions, the secondary directions between them and the other sectors are given as a function of the ambient intensities in Fig. 3.9. An increasing preference for the cardinal directions becomes obvious. This curve differs from a similar one published by KIRSCHVINK (1981a) and by TOWNE and GOULD (1985) on the basis of the same data. While those curves seemed to suggest saturation already between 100 000 and 200 000 nT, the curve in Fig. 3.9 indicates an increase even beyond 450 000 nT. The reason for this difference lies in the fact that the graphics by KIRSCHVINK (1981a) and TOWNE and GOULD (1985) do not present the percentage of runs in the cardinal directions, but some index based on the relation of 'aligned' and 'non-aligned' dances. As 'aligned dances' the authors pooled, oddly enough, the data from the cardinal *and* from the secondary directions, although the latter do not represent symmetric positions with respect to the field lines[1]. From Fig.

[1] Additionally, the value at 450 000 nT should be 0.89 instead of the 0.82 given in the graphs by KIRSCHVINK (1981a) and TOWNE and GOULD (1985). When calculating the data on which the curves are based, KIRSCHVINK accidentally took the data of MARTIN and LINDAUER (1977), Fig. 32, which were recorded in the geomagnetic field after the bees had stayed in a 450 000 nT field, to be recorded at an intensity of 450 000 nT (see Table 1 in KIRSCHVINK 1981a).

3.8, it appears more meaningful to separate the dances in the cardinal directions from all other data. The alignment behavior recorded by ALTMANN (1981; cf. Fig. 3.5) also points out that only the cardinal directions are prominent, while all other directions are similar.

The dances of Honeybees on a horizontal comb represent an interesting phenomenon, although they have not been documented very often. The data from LINDAUER and MARTIN involve more than 64 500 dances altogether, but all published data were recorded within 18 days between 10 August and 13 September 1968, possibly all with bees from the same hive. Only the basic finding, namely, the preference for the cardinal axes and a weaker one for the secondary axes in the geomagnetic field, has been confirmed independently by GOULD et al. (1980) so far.

3.4 Alignment Behavior in Vertebrates

Among vertebrates, fishes are the only group for which alignment behaviors have been reported. In the laboratory, goldfish *Carassius auratus* (Cyprinidae) oriented their body axis preferably along the N-S or E-W axes when resting in a round Petri dish. These directions deviated from external cues such as the walls of the room or the directions of the fluorescent light. The behavior was reported to disappear when the ambient magnetic field of 45 000 nT was decreased to 11 000 nT (G. BECKER 1974), but unfortunately, no data have been published.

An orientation that looks like an alignment has also been described for the stationary phase of European Eels, *Anguilla anguilla* (TESCH and LELEK

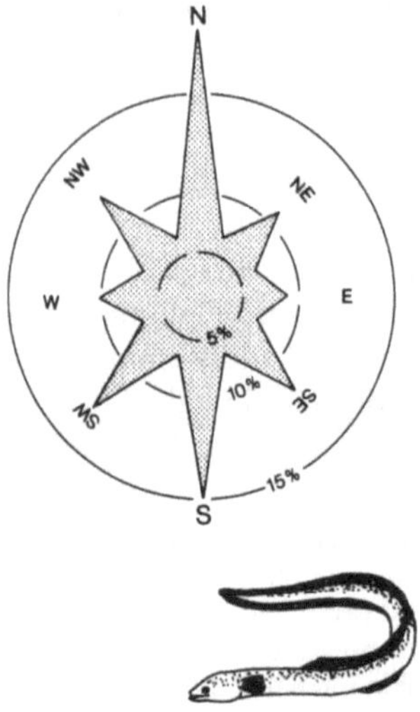

Fig. 3.10. Directional preferences of stationary yellow eels of the European Eel, *Anguilla anguilla*. The diagram is based on 3648 recorded headings, indicating the percentage in each of the 45 ° sectors. (After TESCH and LELEK 1973a)

1973a,b). When the fish were kept in a round tank and the directions of their heads photographically recorded, a strong preference for the N-S axis emerged (Fig. 3.10). As the tank had been covered and the fish were deprived of all other known cues except the geomagnetic field, a magnetic orientation is not unlikely, but whether it truly depended on the magnetic field was never tested (see also Sect. 4.2.1.1).

Another case discussed as an alignment was observed in the Rainbow Trout, *Oncorhynchus mykiss* (= *Salmo gairdneri*, Salmonidae), by Chew and Brown (1989). The fish selected positions heading towards magnetic North when they were held in circular tanks in the local geomagnetic field. When the tank was placed in a room shielded with Mu metal, the preferences could no longer be observed, suggesting that the magnetic field was the directing cue. Chew and Brown (1989, p. 642) described the behavior as tending "*to adopt stations that tracked the horizontal component of the local magnetic field*". However, in contrast to the alignment behavior of invertebrates and the Goldfish, the orientation was unimodal rather than bimodal or quadrimodal. Possibly, it is not an alignment, but represents a compass orientation by the magnetic field, although the origin of the northerly course and its biological significance are unknown.

3.5 Biological Significance

When considering the biological significance of magnetotactic orientation and alignments, one must be aware that these responses are not very strongly directed. The ratio of miracidia found in the north and in the south was 66:34% only, in contrast to the passively aligned bacteria which 'respond' 100%. Interestingly, authors report the same ratio for aligned resting positions along the cardinal axes and the intermediate axes in flies and bees. Even the dances of Honeybees on the horizontal comb recorded in the local geomagnetic field (cf. Fig. 3.8, center) show a similar ratio of dances within the 45° sectors along the cardinal axes and the intermediate sectors. This seems to suggest that a ratio of about 2:1 might generally reflect the expression of magnetotaxis and alignments in the geomagnetic field.

The biological significance of magnetotaxis in miracidia is clear: in nature, magnetotaxis and a positive geotaxis combine to enable the larvae to move downward and reach the habitat of potential hosts as fast as possible. The significance of many of the alignments is less obvious. Especially when they involve resting behavior, it is unclear why some directions should be preferred above others. With respect to a magnetic sensor, they represent maximal, minimal and symmetric stimulation. It has been speculated that maintaining a symmetric position to the field lines means to be in some degree of harmony with the environment and thus might increase the animals' well-being, maybe even resulting in slightly more favorable conditions for some physiological processes (e.g. G. Becker and Speck 1964). Magnetic alignments thus might be related to the tendency of many animals to avoid local disturbances of the magnetic field and unnatural intensities (see Sect. 7.2).

The significance of the cardinal directions to bees dancing on a horizontal comb in the absence of visual cues is also difficult to understand. There is no great benefit in this oriented dancing, because the purpose of waggle dances, namely, a transfer of directional information, cannot be achieved in this way; only the distance to the food source is still indicated by the frequency of waggling. The bees, in their urge to dance, seem to fall back on the only external cue available. The special situation when the dances are symmetric to the field lines is also underlined by the observation that, on the vertical comb, the misdirection disappears when the direction of the dance and the inclination coincide, even if the reasons for this are not yet clear. A related phenomenon might be involved in the orientation of sandhoppers: *Talorchestia martensii* (Crustacea: Amphipoda: Talitridae) uses a magnetic compass for orientation (Pardi et al. 1988; see Sect. 4.3.2.2); however, their orientation was more concentrated when the compass courses were close to the N-S axis of the magnetic field.

With the building activity of termites, the significance is evident. Many termites are blind and live under conditions in which other directing factors are reduced, while the magnetic field is readily available. They have to build galleries while searching for food. In the absence of directing olfactory cues, etc., choosing the cardinal magnetic directions means that the actions of hundreds of individuals are united in a common task. The cardinal axes of the magnetic field as basic directions in the absence of other meaningful cues might facilitate the organization within the colony, at the same time leading to an optimal coverage of the space around the nest.

3.6 Summary

A case of magnetotaxis manifesting itself in a 2:1 ratio of larvae in the north and in the south was described for the miracidia of an eyefluke. The larvae infect bentic snails; the magnetotactic response augments a positive geotaxis and thus helps the larvae to reach the habitat of their hosts.

Alignment responses have been described for a number of insects; the best-documented cases relate to termites, flies and Honeybees. In vertebrates, alignment behavior has been reported for goldfish and eels. As the animals position themselves parallel, antiparallel or perpendicular to the magnetic vector, alignments usually lead to quadrimodal orientation. The ratio between animals in the 45 ° sectors centered around the cardinal direction and the intermediate sectors is mostly in the range 2:1. In the vertical plane, alignment takes place along the angle of inclination and its projections; on vertical walls along a N-S axis, it leads to deviations from the vertical. Most often, resting behavior is involved, but alignments are also found in building activity of termites and the waggle dance of Honeybees. In the latter case, the axis of inclination in the vertical and the cardinal axes of the magnetic field in the horizontal are prominent and/or are the preferred directions.

The biological significance of alignments with respect to the magnetic vector is not always clear. They may simply represent some kind of basic orientation which some animals follow when their behavior is not controlled by other factors. In termites building galleries in the cardinal compass direction, an alignment symmetrical to the magnetic field may help to coordinate the activities of numerous workers in a meaningful way.

Chapter 4

Compass Orientation

A compass is a means to locate directions, i.e. it indicates where north, south, east and west lie. The crucial difference between compass orientation and alignment behavior is that there are no prominent directions – for a compass, all courses are of equal relevance. Animals using a magnetic compass can select any angle with respect to the magnetic field. Compass orientation represents *Menotaxis* as defined by KÜHN (1919).

Many animals make use of a compass in a variety of situations. The set direction, i.e. the specific course in a given situation, depends on the geographic relationship between the animal's present location and its goal; it varies according to the behavioral context. An example given in Figure 4.1 may illustrate this point: Bank Swallows, *Riparia riparia* (Hirundinidae), when leaving their colony to collect food, fly in directions leading them to the next lake which provides good hunting for aerial plankton. They know the respective courses from

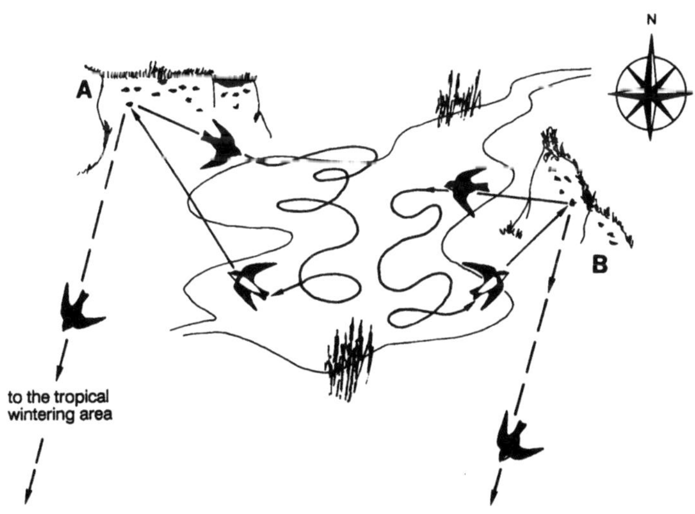

Fig. 4.1. A compass can be used to locate various types of courses: Bank Swallows from different colonies fly different courses to reach a nearby lake – an easterly course from colony A and a westerly from colony B. Their return courses depend on the location where they decide to head back. For autumn migration, swallows from both colonies fly an innate course, the population-specific southerly migratory direction (*dashed line*) towards their tropical winter quarters (see text for more details)

previous flights. After successful foraging, the birds return by flying courses which lead them home; these courses are determined by navigational processes. In these behavioral situations, the set directions are variable, and vary between different colonies. During autumn migration, however, all Bank Swallows fly constant southerly courses, their migratory direction, in order to reach their winter quarters in Africa or South America. These directions, which depend on the geographic relation between their northern breeding areas and tropical destinations, were established as innate courses during evolution, i.e. they are based on genetically encoded information (HELBIG 1991a). Although the set directions mentioned above are of different origin, the same compass mechanism can be used to locate them.

The compass use of the magnetic field is the one best understood so far. Considerable research efforts during recent years have resulted in a detailed analysis of magnetic compass orientation in a variety of animals and in a variety of situations. These findings are presented in Chapter 4, with emphasis on the behavioral situations in which magnetic compass orientation is observed. The origin of the set direction will be mentioned whenever it is known. Test apparatus and procedures will also receive due attention, in particular in those cases where it was found to affect the outcome of experiments.

4.1 The Magnetic Compass of Birds

The magnetic compass of birds represents the most thoroughly analyzed case of magnetic compass orientation so far. It was first described in connection with migration behavior. During migration, birds fly in a more or less constant direction for several weeks until they reach their wintering grounds. It is long known that caged migrants become restless at that time of the year when their free conspecifics undertake their migratory flights; KRAMER (1949) pointed out that they also show directional tendencies which correspond to the headings of their free-flying conspecifics.

The spontaneous tendencies of caged migrants provided a useful baseline for the analysis of orientation, as it allowed controlled studies in a limited space where conditions could be easily manipulated. The sun compass and a star compass were described as mechanisms for directional orientation in birds (see Sect. 5.1.2). Migrants were also able to find their migratory direction in the absence of celestial cues, however, which strongly suggested the existence of another, 'non-visual' compass mechanism. The magnetic field became a most likely candidate when the orientation was found to break down in a steel vault where, among other factors, the geomagnetic field was markedly reduced (FROMME 1961). Yet first attempts to affect the birds' orientation with artificial magnetic fields were inconclusive (KRAMER 1949; FROMME 1961).

4.1.1 Demonstrating Magnetic Orientation in Caged Migrants

Magnetic orientation was first demonstrated in European Robins, *Erithacus rubecula* (Turdidae), a small passerine species migrating at night. The test cage (1 m diameter, 35 cm high) was equipped with eight radially positioned perches and allowed the bird to move freely. The frequency of activation of the perches was recorded, i.e. the positions of the bird within the cage served as criterion for orientation.

The behavior of robins was compared in three test situations which differed in the direction of magnetic North, while total intensity and inclination were similar and all nonmagnetic parameters were identical. The birds responded to the shift of magnetic North with a corresponding deflection of their directional tendencies (Fig. 4.2), indicating that they used the magnetic field as a compass to find their migratory direction (W. WILTSCHKO and MERKEL 1966; W. WILTSCHKO 1968).

These results were first received with great scepticism, mainly because several independent groups of scientists had been unable to confirm the basic finding, namely, that orientation occurred in the absence of celestial cues (for summary, see WALLRAFF 1972). The reasons for these negative results lay in a variety of methodical and technical problems which have been discussed in some detail by ABLE (1994).

One reason why many authors hesitated to accept magnetic compass orientation lies in the fact that the directional tendencies look much more clear-cut when celestial cues are available. In homogeneous dim light, the differences in activity in the various sectors of a cage are often rather small. Sometimes, they were so small that authors argued, since the behavior was random, any

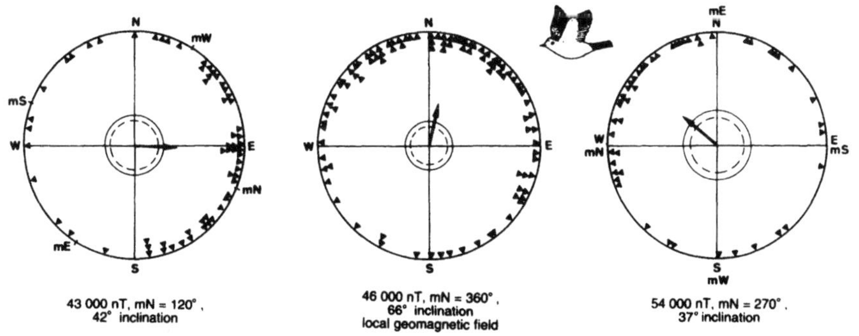

Fig. 4.2. Demonstrating magnetic compass orientation in European Robins, *Erithacus rubecula*, small night-migrating passerines: orientation behavior recorded in cages in the local geomagnetic field (*center*), and in experimental magnetic fields with magnetic North turned clockwise to ESE (*left*) and counterclockwise to W (*right*). Symbols *at the periphery of the circle* indicate the headings of single nightly recordings; *arrows* represent the mean vectors with the length proportional to the radius of the circle = 1. The *two inner circles* mark the 5% (*dashed*) and the 1% significance border of the RAYLEIGH test. (After W. WILTSCHKO 1968)

heading based on these data had to be meaningless. This view is based on a misconception of cage data, however (see WALLRAFF 1972; ABLE 1994 for detailed discussions). Birds perform only one independent directional selection per night (R. WILTSCHKO and WILTSCHKO 1978a). This is represented by the nightly heading; consequently, only the nightly headings can enter statistical analysis as independent data points. This means that the decision whether or not birds were oriented was shifted to the level where more headings were available, which earned magnetic orientation the reputation of requiring an extensive statistical analysis. The distribution of activity, on the other hand, reflects how well the respective heading was maintained; here, visual marks have a strong concentrating effect (see Sect. 5.2.3).

Cage design also proved to be of crucial importance. WALLRAFF (1972) repeated his own negative experiments (WALLRAFF 1966a) using an original cage from WILTSCHKO and MERKEL and also obtained positive results (WALLRAFF 1972). This led to the belief that magnetic orientation was only possible in this specific type of cage (e.g. WALLRAFF 1978). Later experiments, however, produced well-oriented behavior in the absence of visual cues also in cages of different design, like the funnel-shaped cage introduced by EMLEN and EMLEN (1966). It became obvious that there are considerable differences between species and how well a given type of cage reflects their orientation. Under identical test conditions, some species were much better oriented in one type of cage than in another (e.g. EMLEN et al. 1976; BECK and WILTSCHKO 1983). Hence, failure to find oriented behavior often indicated inadequate cage design rather than the birds' inability to orient.

4.1.2 Functional Characteristics

The magnetic compass of birds was analyzed by testing European Robins in various experimental magnetic fields. The results showed that it differed from the technical magnetic compass used by humans in some remarkable aspects.

4.1.2.1 Functional Range

The birds' magnetic compass is narrowly tuned to the total intensity of the ambient magnetic field (W. WILTSCHKO 1978). At the test site in Frankfurt a.M., Germany (40 °08'N, 8 °40'E), the local intensity is about 46 000 nT. The robins would tolerate fields in the range of 43 000 to 56 000 nT, but they were no longer oriented when the total intensity was reduced by 25 % to 34 000 nT or, even more surprising, when it was increased by ca. 30 % to 60 000 nT (see Table 4.1)[1].

[1] This latter finding provides an explanation why some earlier attempts to demonstrate magnetic orientation had failed: the test fields had been of unphysiological strengths (e.g. KRAMER 1949; FROMME 1961).

These data suggested a functional window which is narrower than the range of intensities most birds will encounter on their annual migration. So it was not unexpected that the magnetic compass can be adapted to fields outside this range. Robins housed for at least 3 days at 16 000 nT, which is lower than any field found on earth, were able to orient at this intensity. Corresponding findings were obtained with birds kept in a strong field of 150 000 nT (Table 4.1). Apparently the potential window of the magnetic compass is much wider than the actual functional range (W. Wiltschko 1968, 1978).

Table 4.1. Orientation of European Robins, *Erithacus rubecula*, in magnetic fields of various intensities; spring experiments

Intensity (nT)	Inclination	Tests	Vector length	Significant direction
1. Robins caught and kept in the local geomagnetic field of 46 000 nT				
16 000	31 °	79	0.11	--
32 000	60 °	51	0.04	--
34 000	57 °	104	0.07	--
43 000	42 °	62	0.39	mNNW[a]
46 000	66 °	96	0.39	NNE[a]
46 000	62 °	25	0.40	mENE
46 000	57 °	32	0.43	mNNE
46 000	48 °	73	0.35	NE
54 000	37 °	49	0.42	mNE[a]
60 000	45 °	31	0.12	--
68 000	25 °	68	0.09	--
81 000	8 °	42	0.11	--
82 000	30 °	28	0.06	--
105 000	23 °	50	0.27	--
2. Robins caught at 46 000 nT and kept at 16 000 nT				
16 000	31 °	25	0.35	N
46 000	66 °	25	0.51	NE
3. Robins caught at 46 000 nT and kept at 150 000 nT				
46 000	66 °	33	0.38	NNE
81 000	20 °	57	0.11	
150 000	45 °	56	0.38	ENE

m designates directions given with respect to magnetic North of artificial magnetic fields. (Data from W. Wiltschko 1978; W. Wiltschko and Wiltschko 1972) [a] See Fig. 4.2.

The nature of the adaptation process is still unclear – it is neither a shift nor a simple enlargement of the functional range. Robins kept at 16 000 nT continued to be oriented in the local geomagnetic field of 46 000 nT. The same was true for the birds housed at 150 000 nT; these birds, however, were unable to orient in an intermediate field of 81 000 nT. This suggests that experiencing a given intensity for a certain period of time will form a new window at the respective intensity (W. Wiltschko 1978). For migrating birds, it means that during natural migration, they can gradually adapt to the slowly decreasing intensities they meet en route as they enter lower latitudes. This will lead to a considerable enlargement of the functional range. The same flexibility may have enabled the magnetic compass to cope with variations in intensity as they may have occurred in the geological past.

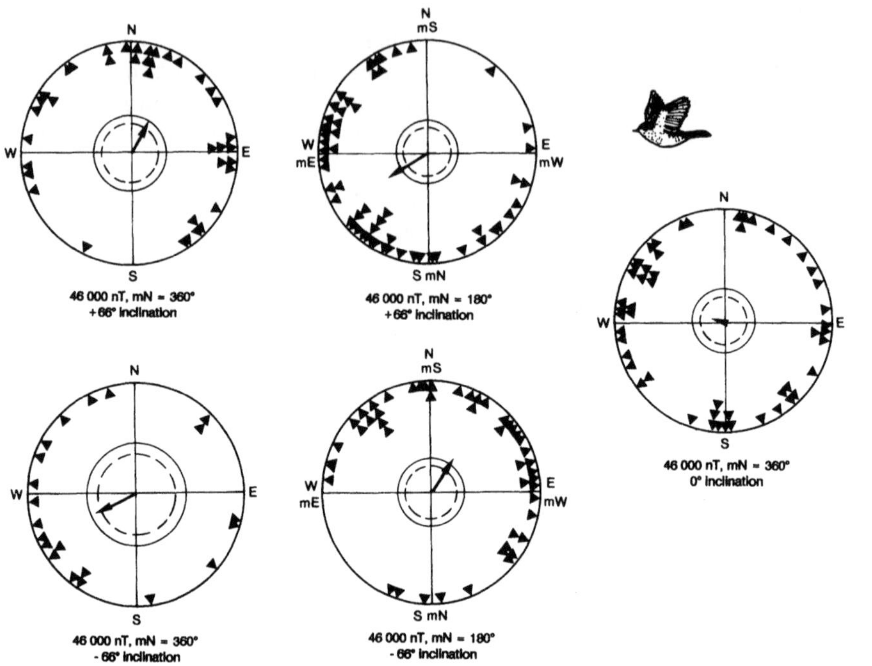

Fig. 4.3. Orientation behavior of European Robins tested in magnetic fields of equal intensity, but with various declinations and inclinations. *Symbols* as in Fig. 4.2. (After W. WILTSCHKO and WILTSCHKO 1972)

4.1.2.2 'Inclination Compass'

Unlike our technical compass, which makes use of the field's polarity, the magnetic compass of birds is based on the axial course of the field lines and their inclination in space. The results in experimental fields with various (positive) dip angles (see Table 4.1, tests at 46 000 nT) suggested that the birds do not perceive separate components, but the magnetic vector as a whole. The polarity of the vector, however, is irrelevant: reversing the vertical component had the same effect as reversing the horizontal component, and a reversal of both components, which means an inversion of polarity while maintaining the course of the field lines, did not alter the birds' behavior (Fig. 4.3; W. WILTSCHKO and WILTSCHKO 1972). In a horizontal field, the birds appeared to be disoriented.

As the course of the field lines alone is bimodal and provides only a north-south axis, the birds must interpret inclination to obtain unambiguous information on directions. Figure 4.4 demonstrates how the *inclination compass* of birds works: it does not distinguish between 'north' and 'south', but instead between 'poleward', i.e. the field lines point to the ground, and 'equatorward', i.e. the field lines point upward. This mechanism has one important advantage when the geological past is considered: the frequent reversals of the geomag-

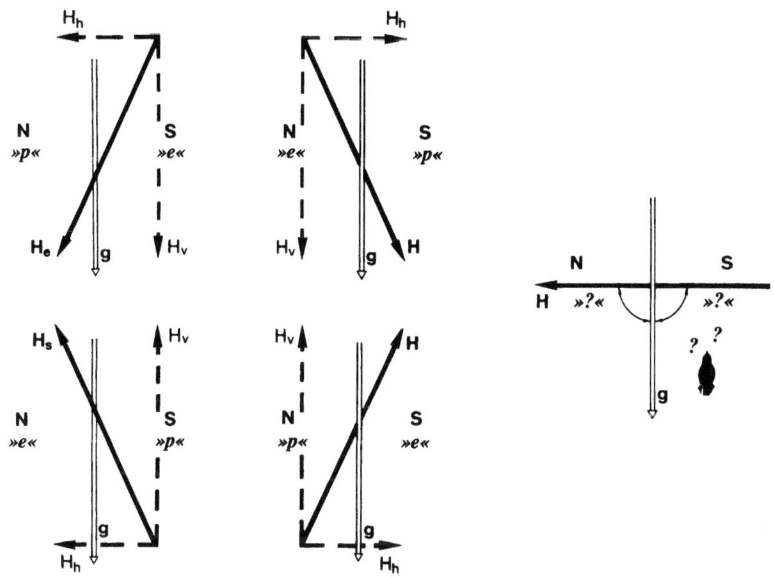

Fig. 4.4. Vertical section through the test fields of the experiments given in Fig. 4.3, indicating the mechanisms of the magnetic compass of birds, which is an 'inclination compass' based on the axial course of the field lines and their inclination. *Upper left diagram* Local geomagnetic field on the northern hemisphere; *upper central diagram* horizontal component reversed; *lower left diagram* vertical component reversed, which corresponds to the geomagnetic field on the southern hemisphere; *lower central diagram* both components reversed; *right diagram* horizontal field lines pointing northward, which corresponds to the situation at the magnetic equator. *H* magnetic vector; H_e vector of the local magnetic field at the test location; H_S vector corresponding to a southern hemisphere field; H_h, H_v horizontal and vertical component of the magnetic field; *g* gravity vector. *N*, *S* geographic North and South; » *p* «, » *e* « 'poleward' and 'equatorward', readings of the birds' inclination compass which distinguishes between 'poleward' and 'equatorward' instead of between magnetic North and South. (After W. Wiltschko and Wiltschko 1972)

netic field changed polarity, but they did not alter the general course of the field lines – because of this, they did not affect the inclination compass.

4.1.2.3 Constant Magnetic Intensity Required

Birds can use the magnetic field for orientation only if it remains at a constant intensity level for a brief interval. Because of the nature of the inclination compass, it was possible to test birds in alternating fields of various impulse shapes and frequencies. Oriented behavior was found when the field lines varied along a fixed axis as long as there were certain intervals with constant intensity (Fig. 4.5).

A saw-shaped impulse of 0.8 Hz, which stayed in the functional range for about 0.25 s twice per period, being 0 in between, was as effective as a rectangular impulse of 1 Hz, varying between 46 000 nT north down and 46 000 nT

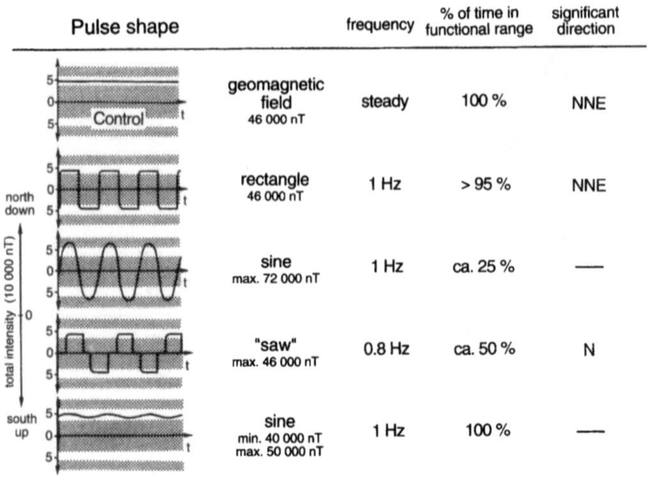

Fig. 4.5. Pulse shapes of various alternating fields and data on the orientation behavior of European Robins. The local geomagnetic field served as control; the vector of the test fields varied along the axis of inclination. The *open spaces between shading* mark the functional range of the magnetic compass, which was assumed to range from ≈ 38 000 nT to ≈ 60 000 nT, cf. Table 4.1. (Data from W. WILTSCHKO 1978)

south up. Fields of 1 Hz with continuously changing intensities, e.g. sine-shaped pulses, did not allow orientation, even if they varied between 40 000 and 50 000 nT and thus did not leave the functional window (W. WILTSCHKO 1978).

4.1.2.4 Accuracy

The question about the accuracy of the magnetic compass cannot be answered properly yet. The behavior of caged migrants, with vector lengths in the range of 0.4 to 0.5, does not seem to suggest a very high accuracy. However, in most cases, the data of numerous individuals have been pooled, and interindividual differences in their set directions have contributed to the general variance. Data on the performance of individual birds are scarce and, in most cases, the number of tests per bird is small. In a recent study, the vector lengths of 12 European Robins, contributing between 4 and 9 headings each, ranged from 0.15 to 0.99. The median of 0.77 suggests an average accuracy of about ±40° (W. WILTSCHKO and WILTSCHKO 1995). In two other species, the vector lengths of individual birds were similar (W. WILTSCHKO et al. 1993b; WEINDLER et al. 1995).

The true accuracy of the individual birds is even better, however, because only a certain part of the variance may be attributed to an inaccuracy of the birds' capacity to detect directions. The other is due to an imperfect reflection of the behavior by the rather artificial test situation in a cage. This latter part may be considerable. In addition, marked differences between individuals, as indicated by the wide range of individual vector lengths, must be assumed. The behavior of free-flying birds indeed suggests a rather high accuracy. Pigeons released under overcast skies, i.e. in a situation in which they are assumed to rely on the magnetic compass (see Sect. 4.1.4.1), may produce vectors above 0.9, which means that in this case interindividual variances *and* individual inaccuracy in the navigation process *and* compass use add up to an angular deviation of less than 25°. If pigeons are used to flying under overcast skies, their orientation under sunny and under overcast skies does not differ (KEETON 1969, 1974; R. WILTSCHKO 1992; see. Sect. 5.2.1.1), which indicates that the accuracy of the magnetic compass equals that of the sun compass.

4.1.3 The Magnetic Compass in Migratory Orientation

Meanwhile, a number of other migrant species have been tested in fields in which magnetic North had been shifted. Their responses, summarized by ABLE (1994), are in excellent agreement with predictions. Table 4.2 lists the species of passerines that have so far been found to use a magnetic compass.

Table 4.2. Migratory birds that have been shown to use a magnetic compass

Bird species	Reference
European Robin, *Erithacus rubecula* (Turdidae)[a]	W. WILTSCHKO and MERKEL (1966) W. WILTSCHKO and WILTSCHKO (1972)[a]
Wheatear, *Oenanthe oenanthe* (Turdidae)	SANDBERG et al. (1991a)
Pied Flycatcher, *Ficedula hypoleuca* (Muscicapidae)[a]	BECK and WILTSCHKO (1981)[a]
Common Whitethroat, *Sylvia communis* (Sylviidae)	W. WILTSCHKO and MERKEL (1971)
Garden Warbler, *Sylvia borin* (Sylviidae)[a]	W. WILTSCHKO (1974)[a]
Subalpine Warbler, *Sylvia cantillans* (Sylviidae)	W. WILTSCHKO and WILTSCHKO (1975a)
Blackcap, *Sylvia atricapilla* (Sylviidae)[a]	VIEHMANN (1979)[a]
Goldcrested Kinglet, *Regulus regulus* (Sylviidae)[a]	WEINDLER (1994)[a]
Dunnock, *Prunella modularis* (Prunellidae)	BINGMAN and WILTSCHKO (1988)
Silvereye, *Zosterops lateralis* (Zosteropidae)[a]	W. WILTSCHKO et al. (1993a)[a]
Yellow-faced Honeyeater, *Lichenostomus chrysops* (Meliphagidae)[a]	MUNRO and WILTSCHKO (1993b)[a]
Red-eyed Vireo, *Vireo olivaceus* (Vireonidae)	SANDBERG and MOORE (in prep.)
Scarlet Grossbeak, *Carpodactus erythrinus* (Fringillidae)	SHUMAKOV and VINOGRADOVA (1992)
Chaffinch, *Fringilla coelebs* (Fringillidae)	BAECKMAN et al. (in prep.)
Indigo Bunting, *Passerina cyanea* (Emberizidae)	EMLEN et al. (1976)
Savannah Sparrow, *Passerculus sandwichensis* (Emberizidae)	BINGMAN (1981); ABLE and ABLE (1993a)
Snow Bunting, *Plectrophenax nivalis* (Emberizidae)	SANDBERG and PETTERSSON (in prep.)
Bobolink, *Dolichonyx oryzivorus* (Icteridae)[a]	BEASON and NICHOLS (1984); BEASON (1989a)[a]

[a] Species that were shown to use an inclination compass, and the respective reference.

4.1.3.1 A Widespread Mechanism Among Migrants

Most of the birds listed in Table 4.2 breed in the northern hemisphere where the inclination is positive, pointing downward; only two species, the Yellow-faced Honeyeater, *Lichenostomus chrysops*, and the Australian Silvereye, *Zosterops lateralis*, are southern hemisphere migrants. In both cases, their magnetic compass proved to be an inclination compass. Especially the data on the honeyeater (MUNRO and WILTSCHKO 1993b) are interesting, as the family Meliphagidae is endemic to the Australian faunal region and not closely related to any other species studied so far. This suggests that the inclination compass is a rather widespread mechanism among birds.

Orientation in the absence of celestial cues has been reported for several more species of passerine migrants. Meadow Pipits, *Anthus pratensis* (Motacillidae), and European Starlings, *Sturnus vulgaris*, oriented in their migratory direction even when tested under total overcast (ORTH and WILTSCHKO 1981; R. WILTSCHKO 1981). Magnetic compass orientation is also indicated in a non-passerine migrant, the Ring-billed Gull, *Larus delawarensis* (Charadriiformes). When young gulls of this species were released away from their colony, they departed in directions that corresponded to their future migratory direction. Oriented behavior was observed under sun as well as under overcast sky. Experimental birds wearing a small magnet on their heads (intensity ca. 98 000 nT) no longer showed such a preference, suggesting that their directional orientation was based on magnetic cues (SOUTHERN 1972a). Similar tests with young chicks of Herring Gulls, *Larus argentatus*, produced inconclusive results (F. MOORE 1975), whereas chicks of Black-headed Gulls, *Larus ridibundus*, failed to show consistent directional tendencies (VIKSNE et al. 1978).

4.1.3.2 Transequatorial Migrants

The list in Table 4.2 includes three species of long distance migrants, the Garden Warbler, *Sylvia borin*, the Pied Flycatcher, *Ficedula hypoleuca*, and the Bobolink, *Dolichonyx oryzivorus*. All three have been shown to use an inclination compass, as described for European Robins. Robins migrate short distances only and never reach the magnetic equator; transequatorial migrants, however, are faced with two problems: (1) the inclination compass becomes ambiguous when the field lines are horizontal, and (2) beyond the magnetic equator, the birds must reverse their migratory direction with respect to the inclination compass in order to continue in the same (geographic) direction.

Birds that cross the magnetic equator twice a year must have developed means to overcome this situation. Experiments with Garden Warblers indicate how they master the problem of changing their migratory direction from 'equatorward' to 'poleward'. After being kept in a horizontal magnetic field for 2 days, the test birds reversed their directional tendencies (Fig. 4.6), whereas untreated controls continued southward, i.e. equatorward, until autumn migration ended (W. WILTSCHKO and WILTSCHKO 1992). Apparently, the horizontal

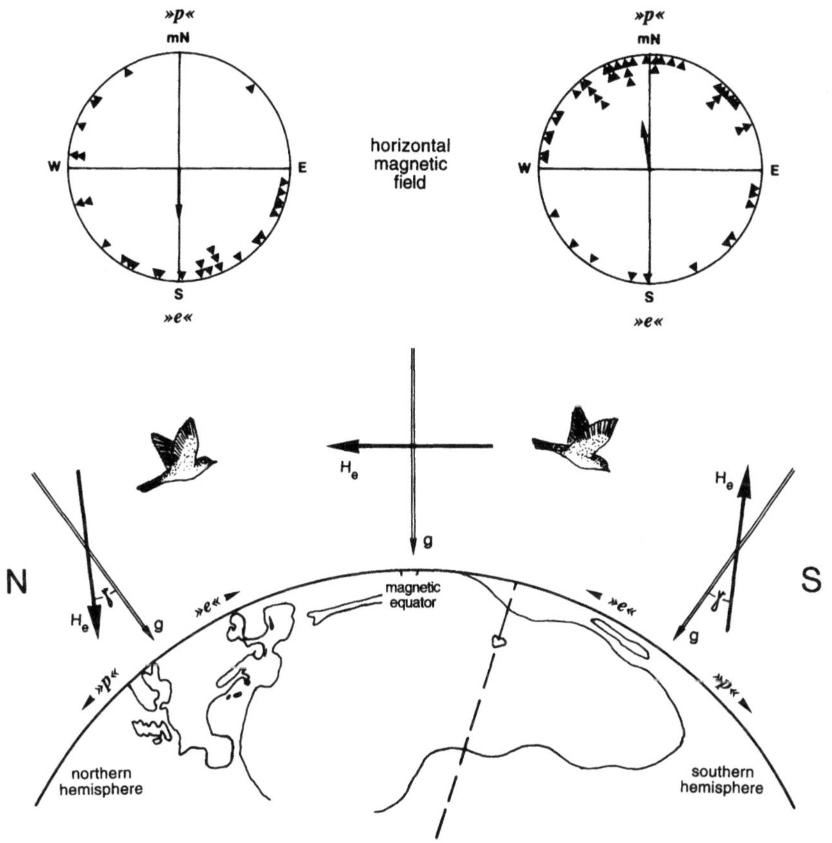

Fig. 4.6. A long-distance migrant crossing the magnetic equator during autumn migration: the birds start 'equatorward' and beyond the magnetic equator continue 'poleward'. Circular diagrams *above* show the orientation of Garden Warblers, *Sylvia borin*, recorded in the local geomagnetic field of Frankfurt a.M. *Left* At the beginning of the migratory season; *right* during the later part of the migration season, after the birds had stayed for 2 days in a horizontal magnetic field. *Symbols* and *abbreviations* as in Fig. 4.2. *Below* Migration flight of Garden Warblers from Europe to Africa. (Data from W. WILTSCHKO and WILTSCHKO 1992)

field itself serves as a trigger and tells the birds to begin flying 'poleward' (see also Sect. 6.3.1).

This leaves the question of how transequatorial migrants cope with the situation at the magnetic equator itself. In laboratory tests, Garden Warblers, like robins, were disoriented when a horizontal magnetic field was the only available cue (W. WILTSCHKO 1974). This suggests that in nature, additional factors are involved; their interaction with magnetic information will be discussed in Section 5.2.2.3.

4.1.3.3 Innate Information on the Migratory Direction

In migratory birds, the magnetic field also serves as external reference for innate information on the migratory course. This is shown by experiments with several species of young migrants that had been hand-raised and kept in the local geomagnetic field under conditions which deprived them of all celestial cues. Tested under the same conditions, these birds headed in their normal migratory direction (W. WILTSCHKO and GWINNER 1974) and responded to shifting magnetic North in the expected way (BECK and WILTSCHKO 1982; BINGMAN 1983a; ABLE and ABLE 1990b; see Fig. 5.4). Thus, the magnetic field alone provides sufficient information to allow proper migratory orientation, which suggests that the migratory direction is genetically encoded as a course relative to the magnetic field.

This has an interesting consequence for birds living in the different hemispheres: northern migrants fly southerly courses in autumn, whereas southern migrants fly northerly courses in southern autumn, i.e. their migratory movements are mirror images. In terms of the magnetic inclination compass, however, both directions mean 'equatorward' (cf. Fig. 4.6). Thus, because of the differences in inclination, migrants in the northern and southern hemisphere may have a common migratory program which tells them: "*fly 'equatorward' when the day length decreases in autumn*" (see W. WILTSCHKO et al. 1993a for discussion).

4.1.4 Magnetic Compass Orientation in Homing

Carrier Pigeons, descendants of the Mediterranean Rock Dove, *Columba livia* (Columbiformes), are the only non-migrants, and at the same time the only non-passerines besides the gull species, whose magnetic orientation has been studied. Pigeons use the magnetic compass in a different behavioral situation, namely, in *homing*, where the courses are not fixed, but determined by a navigational process (cf. Fig. 4.1). Any analysis of pigeon orientation is handicapped by the fact that the birds have to be released in order to record their directional tendencies. This makes manipulations of the ambient magnetic conditions extremely difficult. Attempts to record directional tendencies in cages, as a parallel to the technique applied for migrants, were mostly unsuccessful. This left the possibility to disrupt magnetic information by various means.

4.1.4.1 Locating the Home Course

Early efforts to affect pigeons as well as other birds by applying magnets to their backs, heads, or wings were unsuccessful (see Table 6.2) so that the magnetic field was not considered to play a role in homing. This changed when KEETON (1971) reported that magnets caused disorientation in pigeons that were released unter total overcast (Table 4.3). Concluding that the magnetic field was only used when the sun was not visible, he suggested that the magnetic compass might represent an alternative to the sun compass (see Sect. 5.2.1).

Table 4.3. Homing pigeons: effect of magnets under overcast sky

N	Controls wearing brass sign.	median r	Birds wearing magnets sign.	median r	Significant difference	Region	Reference
11	73%	0.78	37%	0.45	**	New York	KEETON (1971)
20	30%	0.47	35%	0.49	n.s.	New York	B. MOORE (1988)
4	100%	0.78	25%	0.41	*	Italy	IOALÈ (1984)

N = number of samples; sign. = percentage of samples which are significantly oriented (at least $p < 0.05$, Rayleigh Test). Significant difference (Mann Whitney Test): ** = $p < 0.01$, * = $p < 0.05$, n.s. = not significant.

KEETON's findings provided first indirect evidence for the use of a magnetic compass in homing. His findings were later confirmed independently by IOALE (1984) in Italy (Table 4.3). Unfortunately, KEETON's own later data (published by B. MOORE 1988) were inconclusive: control pigeons, well oriented in the early tests, for reasons unknown failed to show oriented behavior under overcast, thus depriving KEETON of a baseline for further studies. The same is true for similar experiments by LAMOTTE (1974).

More direct evidence for magnetic compass orientation in homing was reported by WALCOTT and GREEN (1974), who equipped their birds with pairs of small, battery-operated coils that changed the field around the head. Under sun, these coils did not affect orientation; under total overcast, the effect depended on the direction of current. When magnetic North of the induced field pointed downward, the pigeons' vanishing bearings were homeward oriented, whereas when north was pointing up so that the resultant field roughly corresponded to an inversion of the vertical component, the test birds showed a tendency to fly away from home (Fig. 4.7). Corresponding experiments by VISALBERGHI and ALLEVA (1979) produced similar results. These findings also suggest that the magnetic compass of pigeons is an inclination compass.

RANVAUD and colleagues studied the orientation of homing pigeons at the magnetic equator where the magnetic compass is ambiguous, at local noon when the sun passes through in zenith. The birds headed towards magnetic North. This tendency differed from the orientation observed in the morning or in the afternoon, when the sun compass was available. First data seemed to suggest that the behavior at noon was disrupted by magnets and thus might be based on magnetic cues (RANVAUD et al. 1986); in later tests, however, the effect was no longer observed (RANVAUD et al. 1991). The interpretation of these findings remains unclear.

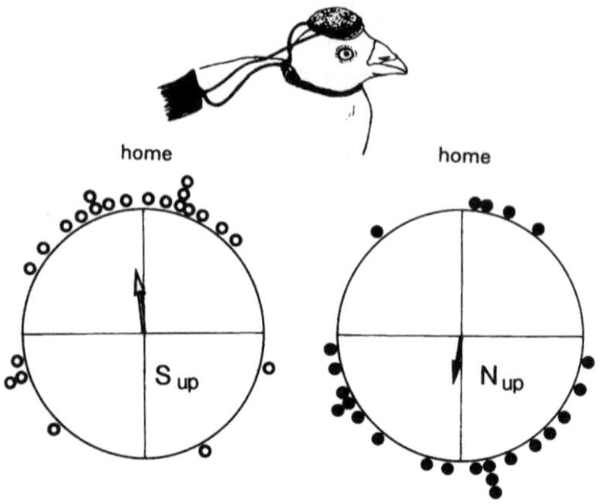

Fig. 4.7. Under overcast, small coils around the head may affect the orientation of homing pigeons. *Upper* diagram Arrangements of coils around the pigeon's head; coils and wiring are marked *black*; *lower diagrams* pigeons were homeward oriented when the polarity of the induced magnetic field pointed downward (South up = *S up*); they reversed their flight direction when the field pointed upward (North up = *N up*). *Symbols at the periphery* indicate the vanishing bearings of individual birds. *Arrows* represent the mean vectors with respect to the home direction = upward, drawn proportional to the radius of the circle. (After C. WALCOTT and GREEN 1974)

4.1.4.2 A Reference for Route Reversal

The findings on homing mentioned so far all refer to locating the home direction after the home course has been determined. However, the magnetic compass may also be involved in the navigation process itself, in particular, when the determination of the home course is based on information obtained during the outward journey. This is indicated by a series of experiments with young, inexperienced pigeons that were transported to the release site in a distorted magnetic field. These birds were disoriented, whereas a second group that experienced the same distorted magnetic field for an equal length of time while waiting at the release site were homeward oriented, like untreated controls (Fig. 4.8). This clearly shows that not the distorted field per se, but *being transported* in the distorted field was crucial (R. WILTSCHKO and WILTSCHKO 1978a). The usual interpretation of these findings is that pigeons record the direction of the outward journey with their magnetic compass and reverse it to obtain the home course, which was no longer possible when the disturbance of the magnetic field during displacement prevented the birds from collecting the necessary information.

This navigational strategy, *route reversal* (SCHMIDT-KOENIG 1970), makes use of the magnetic field as a reference system for storing information on the course of the outward journey and the home course. It seems to be a first simple strategy used by very young birds before they have obtained sufficient experience to make use of other navigational means (R. WILTSCHKO and WILTSCHKO

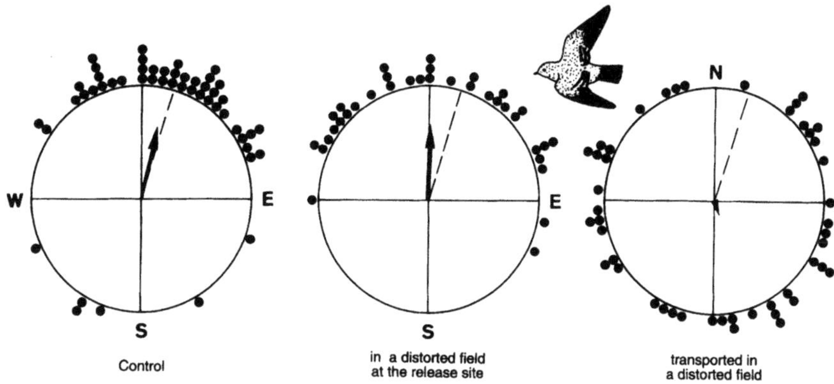

Fig. 4.8. Transportation in a distorted magnetic field disrupts the initial orientation of very young, untrained pigeons, whereas the same field does not interfere with orientation when presented at the release site. The home direction of 16 ° is indicated by a *dashed radius*; other *symbols* as in Fig. 4.7. (After R. WILTSCHKO and WILTSCHKO 1978b)

1985). The navigational strategy of adult birds and possible interactions of the magnetic compass with other cues will be discussed in detail in Section 5.2.1.

4.2 Magnetic Compass Orientation in Other Vertebrates

Orientation responses to natural and artificial magnetic fields have also been reported for a number of other vertebrates. For a long time, experimental evidence was rather scarce and seldom as clear-cut as in birds. The main reasons were (1) problems in finding behaviors where the animals reliably showed a consistent directional preference, and (2) problems in providing an adequate test situation where such directional preferences could be readily recorded as a baseline for experimental manipulations. Recently, however, some of these problems have been overcome. Table 4.4 lists the species of vertebrates that responded to changes in magnetic North with a corresponding shift in their directional tendencies.

For several species not listed in Table 4.4, findings suggesting magnetic orientation have been reported. In these cases, a direct demonstration of a magnetic compass is still lacking, but a breakdown in orientation in the absence of meaningful magnetic information has been observed. In some papers, the use of a magnetic compass is discussed as the most probable possibility considering the respective situation as a whole.

Table 4.4. Vertebrates that have been shown to use a magnetic compass

Species	Reference
Chondrichtyes, Elasmobranchia	
Round Stingray, *Urolophus halleri* (Urolophidae)	Kalmijn (1978a)
Osteichthyes, Teleostei	
European Eel, *Anguilla anguilla* (Anguillidae)	Karlsson (1985)
American Eel, *Anguilla rostrata* (Anguillidae)	Souza et al. (1988)
Sockeye Salmon, *Oncorhynchus nerka* (Salmonidae)	Quinn (1980)
Chinook Salmon, *Oncorhynchus tschawytscha* (Salmonidae)	Taylor (1986)
Amphibia, Urodela	
Red-spotted Newt, *Notophthalmus viridescens* (Salamandridae)	Phillips (1986a, b)
Cave Salamander, *Eurycea lucifuga* (Plethodontidae)	Phillips (1977)
Reptilia, Testudines	
Loggerhead Turtle, *Caretta caretta* (Cheloniidae)	Lohmann (1991)
Leatherback Turtle, *Dermochelys coriacea* (Dermochelidae)	Lohmann and Lohmann (1993)
Aves	
Pigeon, *Columba livia* (Columbidae)	Walcott and Green (1974)
Passeriformes: 15 species (9 families)	cf. Table 4.2
Mammalia, Eutheria	
African Mole Rat, *Cryptomys* sp. (Bathyergidae)	Burda et al. (1990)
Man, *Homo sapiens* (Hominidae)	Baker (1989b); Murphy (1989)

4.2.1 Fishes

In fish, orientation research interest has largely focused on the two groups that undertake long distance migrations, namely, eels and salmons. Their migratory behavior is fundamentally different. Eels are catadromous, hatching in the ocean and migrating to fresh or brackish water, where they spend several years growing. When mature, they return to the sea and migrate to their natal area to spawn. Salmons, in contrast, are anadromous; they hatch in streams or the upper courses of rivers and migrate to the ocean. Here, they stay one or more years until they mature and return to fresh water to their natal river to spawn. – These species are assumed to possess well-developed orientation abilities, and hence, they were prime candidates in the search for magnetic effects. The complex problems of orientation in the course of their extended migrations will be discussed in Section 5.3; here, we focus on data demonstrating or suggesting the use of a magnetic compass.

4.2.1.1 Eels (Anguillidae)

First indications that fish are able to use the magnetic field for directional orientation came from experiments with yellow eels, the stationary freshwater form. The test apparatus was a large maze-type structure; the eels were introduced singly and could choose between six tubes leaving a central hexagon every 60 °, at 30 °, 90 °, 150 °, etc. The behavior was bimodal. At two test sites in eastern Europe, European Eels, *Anguilla anguilla*, spontaneously preferred the E-W axis with 24 and 31 % more choices than expected by chance. At a third site, a less pronounced preference of the 30 °-210 ° axis was observed. When the magnetic field was compensated, these tendencies disappeared, which suggests that they were based on magnetic information (Branover et al. 1971). Continued studies in the same maze yielded similar results (Gleizer and Khodorkovsky 1971; Ovchinnikov et al. 1973). However, when Zimmermann and McCleave (1975) tried to repeat these tests with elvers of the American Eel, *Anguilla rostrata*, using the same type of test apparatus, they did not find any directional preference, which deprived them of a baseline for further magnetic tests.

Souza et al. (1988) tested American yellow eels in an octagonal tank. The fish were released in a central vestibule from where they could enter any of eight chambers at the periphery. In the geomagnetic field and in a field of doubled intensity (it is unclear whether this applies to the total intensity or the horizontal component only), most eels were found in the northern chambers, whereas under two test conditions with magnetic north reversed, the most frequently visited chambers were southern ones, thus indicating the use of a magnetic compass. Tests with European yellow eels in two different types of apparatus produced unclear results (Tesch et al. 1992).

The remaining studies used maturing silver eels at the beginning of their migratory phase. American silver eels caught at the river mouths in Nova Scotia showed pronounced directional tendencies which mostly corresponded well with the presumed migratory direction toward the spawning area in the Sargasso Sea (Miles 1968). The behavior persisted in the absence of celestial cues, which suggested a non-visual orientation mechanism, but its nature was never analyzed. European silver eels from the northern German population, in contrast, showed only weak bimodal preferences for various axes. Reversing the horizontal component had little effect (Tesch 1974a; Tesch et al. 1992). When the horizontal component was compensated, however, the eels headed unimodally toward the east (Tesch 1974a). These findings are hard to explain, and an effect of some local cue that gained importance in the absence of magnetic information cannot be excluded.

Karlsson (1985) tested silver eels of the Baltic population in a tank one at a time. He did not find a common directional tendency; however, some individuals showed a fairly constant unimodal preference for an individual mean, which changed when magnetic North was reversed (Fig. 4.9), but other individuals were less well oriented and seemed to be less affected by changes in magnetic North.

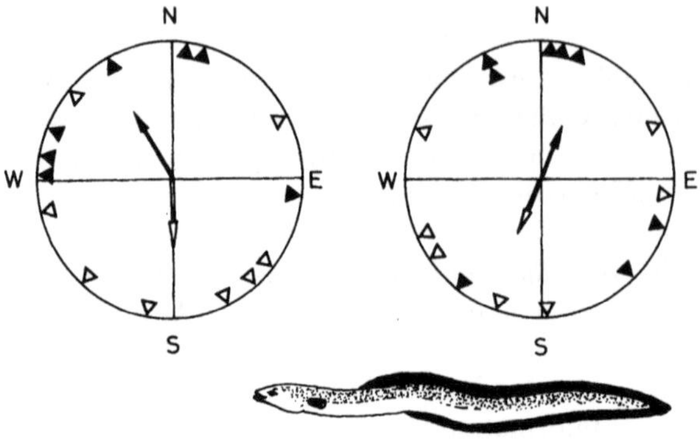

Fig. 4.9. Directions preferred by two individual European silver eels, *Anguilla anguilla*, tested in an octagonal arena. *Symbols at the periphery of the circles* indicate the orientation of the body axis; *open symbols* in the local geomagnetic field; *solid symbols* when the horizontal component was reversed. The *arrows* represent the respective mean vectors. (After KARLSSON 1985)

Taken together, the available data suggest that eels can use the magnetic field for directional orientation, but the evidence is not as clear-cut as one might wish. The fundamental problem lies in difficulties to obtain stable directional tendencies as a baseline for experimental manipulations. The orientation was often bimodal, largely scattered; sometimes it even varied during the course of the experiments (e.g. SOUZA et al. 1988).

The technical difficulties were similar to the situation of the first studies with birds, i.e. inadequate design of the test apparatus, unsolved problems with criteria for orientation and with data processing were a serious handicap. None of the tanks used were really satisfactory. In addition, KARLSSON (1985) pointed out that being confined in an enclosure over a longer period of time might per se affect orientation. In nature, eels must change direction when they hit an obstacle, a behavioral constraint which might lead to total confusion in a tank. The important role of criteria for orientation is indicated by a study of KARLSSON (1985) who compared two criteria and found marked differences between individuals. In many other studies, individuals were not identified, and the data from repeated counts over a certain period of time were pooled, counting the same individuals repeatedly, which is problematic in itself. Additional difficulties arose from the fact that the behavior depended to some extent on age and origin of the test animals and was often susceptible to a number of outside factors, like changes in water temperature, salinity, etc. (OVCHINNIKOV et al. 1973; TESCH 1974a).

The most serious problem, however, appears to be motivation. Eels are expected to be able to orient well because they undertake long migrations – yet apparently, it is impossible to study the marine migratory forms of eels, e.g. larvae or mature silver eels. The test animals were mostly in the stationary yellow form, which cannot necessarily be expected to show directional preferences associated with migration. Indeed, the most clear-cut orientation found in yellow eels appears to be an alignment (TESCH and LELEK 1973a; cf. Fig. 3.10).

Moreover, most silver eels used in the studies can hardly be expected to show migratory orientation. Normally, silver eels start their active migration only after they have reached saline waters. In fresh or brackish water, where the majority of test animals were caught, they more or less passively drift in the current down the rivers (Tesch 1974a). The biological significance of directional tendencies other than migratory behavior is unclear so that it is unknown under what conditions they might be replicable.

4.2.1.2 Salmons (Salmonidae)

In salmons, the test situation is more favorable, as animals in the migratory phase are easily obtained in large numbers, when the young salmon fry migrate from the spawning beds to the nursery lakes, or later, when the yearling fish as smolts move downriver towards the ocean (see Sect. 5.3.2).

Quinn (1980) was the first to demonstrate a magnetic compass in the Sockeye Salmon, *Oncorhynchus nerka*, a species whose spawning sites are the rivers of the Pacific coast of North America. Between 20 and 30 fry (= youngs within the first days after hatching) were released in the center of a centrally symmetric tank with four or eight exits through which the fry left the tank to be caught and counted. These numbers were used to calculate a mean vector. The directional tendencies obtained this way were not very pronounced, with vector lengths frequently only little above 0.1; however, since several hundred individuals were tested, they were nevertheless significant. When magnetic North was turned by Ruben's coils to geographic West in the absence of celestial cues, the fry changed their preference accordingly (Table 4.5).

Table 4.5. Magnetic compass of Sockeye Salmons, *Oncorhynchus nerka*

Magnetic field	Celestial cues	Fry tested	Vector length	Direction: degrees	geogr.	magn.
Normal, mN = 22.5 °	visible	1394	0.22***	185 °	S	m S
Normal, mN = 22.5 °	not visible	1289	0.14***	155 °	SSE	m SE
Turned 90 °, mN = 292.5 °	not visible	166	0.13***	77 °	ENE	m SE

*** indicates significance by the Rayleigh test with $p < 0.001$. (Data from Quinn 1982a)

Experiments with smolts of Sockeye Salmon indicate that magnetic compass orientation is also involved in the seaward migration of this species (Quinn and Brannon 1982). In the closely related Chum Salmon, *Oncorhynchus keta*, the response to altering magnetic North is rather unclear: There are indications that the behavior was significantly affected, but the predicted change in orientation was not observed (Quinn and Groot 1983).

Attempting to analyze the nature of the magnetic compass, Quinn et al. (1981) inverted the vertical component of the magnetic field. This did not lead to a reversal of the preferred direction, which indicates that the magnetic compass of salmons, in contrast to that of birds, depends on the polarity of the magnetic field. A similar study by KITAMURA et al. (1985) was inconclusive, because the behavior proved to be extremely variable, with unexplained changes in direction from day to day.

TAYLOR (1986, 1987) used a different approach to study the directional orientation of Chinook Salmon, *Oncorhynchus tschawytscha*, a population introduced to New Zealand. He kept young fish for 18 months in a rectangular tank with water flowing from west to east, and accustomed them to being fed in the west. This induced a preference for the E-W axis, with emphasis on westerly directions. For the critical tests, individual fish were transferred to a circular arena with radial water flow from four symmetrically arranged inlets; visual and other local cues were excluded. In this situation, the salmons preferably oriented their body axis along the E-W axis. When magnetic North was turned by 90 ° to the east, the salmons changed their preferred axis accordingly to N-S. After return to the normal magnetic field, the salmons again preferred the E-W axis (Fig. 4.10). A null magnetic field resulted in random orientation or in directions that differed from fish to fish (TAYLOR 1986). These data clearly show that salmons can use the magnetic field for directional orientation not only during migration, but also in other behavioral situations which require locating acquired directions.

Magnetic orientation has also been suggested for Rainbow Trouts, *Oncorhynchus mykiss* (= *Salmo gairdneri*). In a circular tank, adult as well as juvenile trouts preferred headings which lay slightly east of magnetic North. This orientation was maintained in the absence as well as in the presence of a weak current, but it broke down when the tank was placed in a null magnetic field produced by Mu-metal shielding. CHEW and BROWN (1989) discuss the behavior as an alignment to the magnetic field (cf. Sect. 3.4), but the fact that the preferred direction is unimodal and deviates from magnetic North suggests compass orientation rather than alignment behavior. The origin and the significance of this orientation are unknown, however.

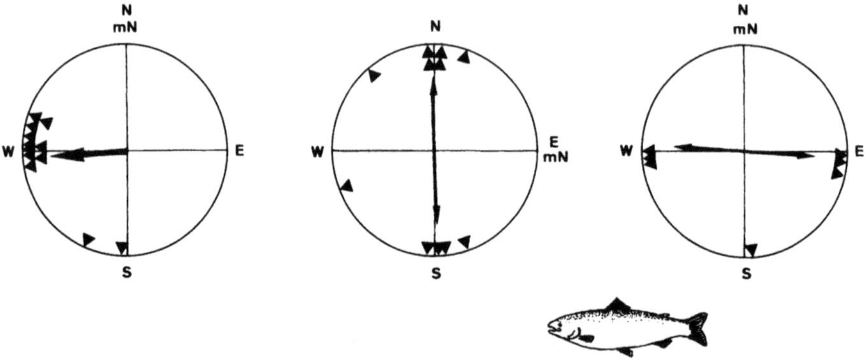

Fig. 4.10. Orientation of Chinook Salmon, *Oncorhynchus tschawytscha*, accustomed to having water flow and food coming from the west. *Left* In the local geomagnetic field; *center* with magnetic North in the east; *right* back in the local geomagnetic field. The *symbols at the periphery of the circle* give the axial directions of individual salmons, marking that side of the axis where the majority of headings lay. The *arrows* represent the mean vector or the mean axis. (After TAYLOR 1986)

4.2.1.3 Elasmobranchs

The only other fish species studied in view of magnetic orientation belong to the elasmobranchs. KALMIJN (1978a) reported that he could successfully train two individuals of Round Stingrays, *Urolophus halleri* (Myliobatiformes), to look for food in the magnetically eastern enclosure and to avoid the western one. In the training phase, the relationship between magnetic North and geographic North was altered from day to day in an irregular order. The animals learned the task and performed 75 and 79% correct, respectively. In the critical tests, magnetic North was reversed in a random order from trial to trial. The animals needed some time to adjust to the frequent changes in the magnetic direction, but later they reached a score of 66% correct, which is highly significant. This is one of the few cases where it was possible to demonstrate magnetic compass orientation by conditioning experiments (see Sect. 8.2.1).

The other case involves a spontaneous tendency in Leopard Sharks, *Triakis semifasciata* (Selachii: Carcharhiniformes). Each morning, a group was found resting along the periphery of their pool slightly off of magnetic North. This tendency persisted when the pool was covered with plastic and did not change when the pool was moved to another site. However, when the magnetic field was largely compensated, the sharks no longer preferred the northern direction (KALMIJN 1978a).

4.2.2 Amphibians

In amphibians, well-developed orientation abilities are implied by the typical lifetime tracks, which include migrations between feeding sites, breeding sites and wintering sites, even if these occur on a much smaller spatial scale (see Sect. 5.3.3). Since amphibians preferably move on moist and rainy nights, it seemed promising to look for magnetic orientation in this group.

4.2.2.1 Salamanders and Newts (Urodela)

First experiments indicating magnetic compass orientation in amphibians used the cave salamander *Eurycea lucifuga* (Plethodontidae). Two groups of 15 salamanders each were housed in a corridor at the end of which they found damp limestone where they liked to rest. The end changed in regular intervals so that the animals got used to moving along the corridor, which for one group corresponded to the magnetic N-S axis, for the other to the magnetic E-W axis. Critical tests were conducted in the geomagnetic field and in a field turned by 90°. Now the two corridors formed a cross-shaped cage, and both groups were released together in the center. The animals moved to the ends, with each group showing a weak preference for that magnetic axis along which it was used to moving (PHILLIPS 1977).

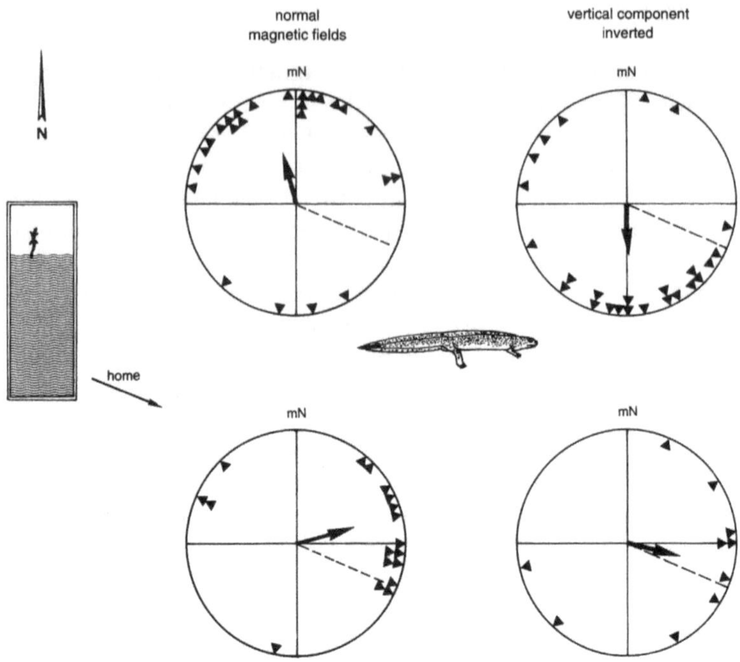

Fig. 4.11. Magnetic compass orientation in the Red-Spotted Newt, *Notophthalmus viridescens*. *Left circular diagrams* Tests in magnetic fields with downward inclination; *right diagrams* tests in magnetic fields with upward inclination. Magnetic North was shifted to all four cardinal compass directions; the headings of the newts are given with respect to magnetic North. *Upper diagrams* Shoreward orientation after increasing the water temperature to 32 °C – the newts preferred a northerly course, which corresponds to the direction towards the shore in their housing tanks (see diagram to the *left*); this direction was reversed when the inclination was pointing upward. *Lower diagrams* Homeward orientation after widely fluctuating water temperatures – the newts prefer a direction that roughly coincides with the course to their home pond, given as a *dashed line*; this orientation was unaffected by a reversal of the vertical component. *Symbols* as in Fig. 4.2. (After PHILLIPS 1986a)

Further experiments were performed with Red-spotted Newts, *Notophthalmus viridescens* (Salamandridae), a species that lives a mostly aquatic life. Adult males from different ponds were kept in housing tanks which had a gravelled bottom and a 'shoreline' either at the north or east end. For testing, the animals were released in the center of a round arena (80 cm diameter). They normally showed a rather weak axial preference for an axis *perpendicular* to the respective artificial shoreline, which corresponds to the 'Y-axis' as defined by FERGUSON (1967), or, in spring, an equally weak preference for an axis whose one end roughly coincided with the home direction (PHILLIPS 1987a). However, when the water temperature, which normally varied between 17 and 27 °C, was raised to 32 °C within 4 hours before the test, the newts showed a strong unimodal orientation towards the respective shoreline, which PHILLIPS (1986a) interpreted as an attempt to leave the pond, since rising water temperatures normally indicate rapidly deteriorating conditions. The direction to the shoreline was located with the magnetic compass, as changes in the magnetic North direction to east, south, and west resulted in a corresponding shift of the preferred direction (Fig. 4.11, upper left diagram). When the vertical component was reversed, now pointing upward instead downward, the animals reversed their directional preference. This indicates that the newts used an inclination compass (Fig. 4.11, upper right diagram; PHILLIPS 1986b).

Whereas rising water temperatures induced orientation towards the landward end of the Y-axis, water temperatures fluctuating between 2–7 °C and 32 °C within 2 days induced unimodal orientation towards a direction that roughly corresponded to the direction of the home ponds, which were 21 to 27 km away (Fig. 4.11, lower left diagram). PHILLIPS (1986b, 1987a) interpreted this as an attempt to leave the present pond altogether and return to the home pond. The direction towards home, too, is located with the help of a magnetic compass, as changes in magnetic North resulted in a corresponding shift of the preferred direction. Reversing the vertical component, however, did *not* result in a reversal of orientation in this case (Fig. 4.11, lower right diagram; PHILLIPS 1986b). This led PHILLIPS (1986b) to postulate *two magnetoreception pathways* in salamanders, namely, an inclination compass for shoreward orientation and a polarity compass for homing.

This is a most puzzling finding. As homing is a two step process (cf. Sect. 2.2.1) and magnetic factors might also be involved in the process determining the home course (see Sect. 6.1.3), it is not entirely clear which component is involved (see also Sect. 9.2.3.2). Only few bearings concerning this question have been published so far, and the data have not been independently confirmed as yet. More data are urgently needed.

4.2.2.2 Toads (Anura: Bufonidae)

Orientation with the help of a magnetic compass is suggested for toads in connection with migration and homing. BUCK (1988) tested European Toads, *Bufo bufo*, caught during their migration to breeding ponds. Recording their behavior in an wooden test arena or in a funnel-shaped apparatus in the laboratory revealed bimodal orientation along the axis of migration. Changes in direction and intensity of the ambient magnetic field caused changes in behavior; but some of the findings are difficult to interpret.

SINSCH (1987, 1990a) tested European Toads and Natterjack Toads, *B. calamitas*, which he caught at their breeding pond, and recorded their orientation after displacements of 150 to 200 m. The toads were released in late afternoon; each carried a tracking device consisting of a sewing maschine bobbin containing 60 m of thread, which was used to measure the extent of the nightly movements. Both species were, in general, well oriented toward their breeding site. Sky cover did not affect orientation. Depriving the animals of magnetic information by gluing bar magnets (7500000 nT at 1 cm distance) on their heads, however, resulted in disorientation (Fig. 4.12); controls carrying brass

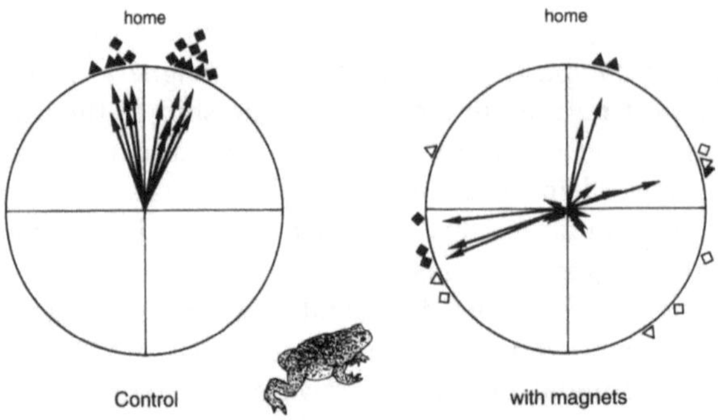

Fig. 4.12. Magnets affect the orientation of displaced European Toads, *Bufo bufo*. *Left* Controls wearing a brass bar; *right* experimental animals wearing a magnet on their head (weight 0.95 g; field strength at 1 cm ca. 7 500 000 nT). The *arrows* indicate the mean vectors of 14 releases of 20 toads each with respect to the home direction; the *symbols at the periphery of the circle* mark the mean direction of each sample (*solid* significant by the Rayleigh test; *open* non-significant samples). *Triangles* indicate releases with the north of the magnet pointing upward, and the respective controls; *diamonds* indicate releases with the south pointing upward. (After Sinsch 1987)

bars of equal size and weight continued to be well oriented (Sinsch 1987, 1992). The amount of locomotory activity and the straightness of the path were not affected by the magnets, indicating that the motivation to move was unaffected. These findings suggest that magnetic information is incorporated in the orientation system of toads. The tests do not allow one to define the role more specifically; the use of a magnetic compass is most likely (Sinsch 1989, 1990b).

The interaction of magnetic orientation with other factors such as acustic cues from a chorus of conspecifics, and olfactory and visual cues are discussed in Section 5.3.3.

4.2.3 Reptiles

The only reptiles whose ability to use the magnetic field for orientation has been studied so far are turtles (order Testudines). The behaviors analyzed include directional training and homing as well as directional tendency associated with migration.

4.2.3.1 Box Turtles (Testudines: Emydae)

First reports on magnetic orientation in reptiles concerned the box turtle *Terrapene carolina*, a species that lives in wetlands but usually prefers to stay on land. The turtles were released in water-filled training tanks and were trained to swim ashore either at the eastern or western end of the tank. In critical tests, when they were released in the center of a round pool (3 m diameter), each group showed a significant preference for the respective direction. These tendencies disappeared when a strong magnet (920 000 nT at 4 cm) was taped to the center of the carapax, suggesting that the turtles located their training directions with the help of a magnetic compass (MATHIS and MOORE 1988).

Displacement experiments indicate that box turtles also use magnetic information in homing. Turtles were transported between 0.7 and 1.2 km in a covered box that blocked all view to outside visual cues. After being released, their behavior was recorded with the help of a tracking device similar to the one described for toads (cf. Sect. 4.2.2.2). Bearings were taken after the turtles had moved 15 m. Controls showed a significant preference for the home direction. Test animals that had been treated with an artificial magnetic field of 61 000 nT (15% increase over the local field of 53 000 nT) during the displacement were disoriented. This suggests that turtles, like young pigeons (cf. Fig. 4.9), determine their homeward course by recording the direction of displacement with a magnetic compass (MATHIS and MOORE 1988).

4.2.3.2 Marine Turtles

More direct evidence for magnetic compass orientation was reported from experiments with hatchlings of the Loggerhead Turtle, *Caretta caretta* (Cheloniidae), and the Leatherback Turtle, *Dermochelys coriacea* (Dermochelidae), marine species distributed worldwide in tropical and subtropical oceans. Breeding is confined to a few sandy beaches only, which may have a long tradition of use by a given population. Directly after hatching, the young turtles crawl rapidly into the sea, probably guided by cues associated with differences in light distribution. They move offshore and begin an entirely marine life; only mature females return to land every 2 to 3 years to deposit their eggs on the beach where they themselves were born.

Experiments were performed in the laboratory with young turtles collected at beaches in Florida, USA, directly after hatching. Tied to an easily movable level arm (see Fig. 4.13), they could swim in a seawater dish in any chosen direction. Their swimming directions were automatically recorded. Hatchlings of Loggerheads as well as Leatherbacks showed significant directional preferences in the natural geomagnetic field, which were reversed when magnetic North was turned to geographic South (Fig. 4.13; LOHMANN 1991; LOHMANN and LOHMANN 1993).

In experiments analyzing the functional properties of their magnetic compass, the responses of Loggerhead Turtles were similar to those previously observed in European Robins (cf. Fig. 4.3): When the vertical component of the field was inverted, the turtles reversed their directional tendencies, while reversing both the vertical *and* the horizontal component did not change the behavior. In a horizontal field, the turtles were disoriented (LIGHT et al. 1993). Thus, the magnetic compass of Loggerhead Turtles is an inclination compass.

An analysis of the origin of the magnetic compass course is currently in progress. On their natal beaches, hatchlings use light cues associated with the

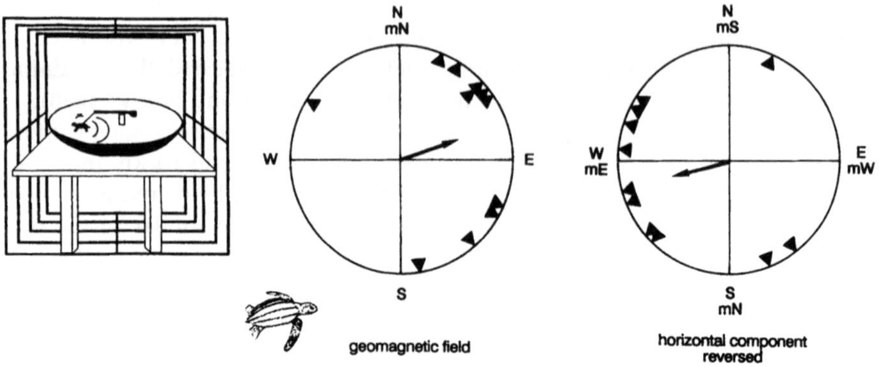

Fig. 4.13. Orientation of hatchlings of the Leatherback Sea Turtle, *Dermochelys coriacea. Left diagram* Test apparatus in the center of Rubens coils; the hatchling was harnessed and tied to an easily movable lever arm while it swam in a round dish of 1 m diameter. An eastward course was induced by a dim light before testing commenced. *Right diagrams* Directional tendencies recorded in the local geomagnetic field and in a field with reversed horizontal component. *Symbols at the periphery of the circle* indicate the mean heading of a hatchling's first oriented swimming period; the *arrows* give the mean vectors. (After LOHMANN and LOHMANN 1993)

seaward horizon, together with the landward silhouette and the beach slope to find their way to the sea (SALMON et al. 1992). LOHMANN and LOHMANN (1994a) could induce easterly or westerly tendencies by exposing young hatchlings to light from the east or the west. The hatchlings would then prefer the respective direction when tested in the dark, whereas hatchlings not exposed to light failed to show consistent orientation. These data show that the set direction can be acquired via a spontaneous response to light cues. In water, young turtles spontaneously head into the oncoming waves (SALMON and LOHMANN 1989; LOHMANN and LOHMANN 1992), a behavior that helps them to leave the beach and enter the open ocean. Recent experiments (SALMON, pers. comm.) indicate that this response can also set a magnetic course for further swimming.

SALMON and WYNEKEN (1994) summarized the available evidence on the interactions of the various orientation cues that enable hatchling sea turtles to reach the open sea, discussing possible mechanisms and implications. Little is known about their orientation in later life, e.g. how they reach suitable nursing areas, etc. Spontaneous responses to the angle of inclination (see Sect. 6.3.2), which seem to modify magnetic courses, may be part of the still unknown migration program (LOHMANN and LOHMANN 1994b).

4.2.4 Mammals

Among vertebrates, mammals show the greatest variation in size as well as in spatial behavior. On the one hand, there are small shrews and rodents whose entire life may be spent within an area of much less than 100 m diameter; on the other hand, there are large species like ungulates that periodically migrate over several hundreds of kilometers, and marine mammals like cetaceans that change regularly between polar/subpolar and subtropical/tropical waters (cf. BAKER 1978). Most of the migrating forms evade being studied in the laboratory because of their sheer size, so that the knowledge on orientation mechanisms of mammals is largely confined to small, non-migrating forms. Many of these species, however, maintain permanent territories and were found to home after displacement (cf. JOSLIN 1977).

4.2.4.1 Rodents (Rodentia)

Small rodents are the most accessible group for experimentation; hence, it is not surprising that most data are available on their use of magnetic information.

Mice. First suggestions for a magnetic compass in small mammals came from homing experiments with woodmice, *Apodemus sylvaticus* (Muridae). The mice were displaced up to 80 m in their natural habitat, and their directional tendencies were recorded in a cross-shaped enclosure, taking the time spent in each arm as a criterion. A significant preference for the direction towards the point of capture became evident, which persisted when the surrounding landmarks were concealed by a screen (MATHER and BAKER 1980, 1981). In further tests, the test animals were displaced in a magnetic field with an inverted horizontal component, which was accomplished by transporting them along a N-S route in a transportation cage equipped with a battery-operated set of coils. This treatment altered orientation of the woodmice significantly when they were tested in the screened cage. Tested without a screen, however, the woodmice continued to orient towards the point of capture like the controls. Together, these findings show that directional information from the magnetic field obtained during the outward journey is involved, but that factors like landmarks, etc. are dominant when available (MATHER and BAKER 1981).

These observations were received with great scepticism, mainly because the homing abilities of small mouse-like mammals were generally questioned. First attempts to replicate the findings failed to find oriented behavior (SAUVÉ 1988). Yet AUGUST et al. (1989) obtained similar results with White-footed Mice, *Peromyscus leucopus* (Cricetidae): at two sites, the controls were homeward oriented, while the test animals having been transported with magnetic North reversed, showed a significant 163° deviation. The behavior at one site was much more variable than at the other.

Hamsters. Findings on the Golden Hamster, *Mesocricetus auratus* (Cricetidae), are inconsistent. Testing displaced animals in a cage in an experimental magnetic field, MATHER (1985) obtained data suggesting magnetic compass orientation. ETIENNE et al. (1986), in contrast, got negative results when studying the return route taken by hamsters after hoarding in an arena (2.2 m diameter) in the absence of visual, olfactory, acoustic, and tactile cues. Also, attempts to train Djungarian Hamsters, *Phodopus sungorus* (Cricetidae), to look for food in the magnetic North of their nests were unsuccessful (MADDEN and PHILLIPS 1987; see Sect. 8.2.1).

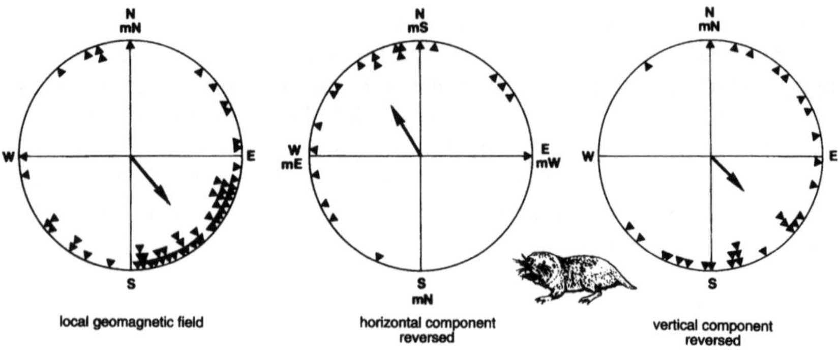

Fig. 4.14. Directions where a family group of mole rats *Cryptomys* sp. of Sambian origin built their nests in a circular arena of ca. 1 m diameter. *Left* Tests in the local geomagnetic field; *center* magnetic North turned to geographic south; *right* inclination pointing upward instead of downward. Symbols as in Fig. 4.2. (Data from MARHOLD et al. 1991)

It is interesting to note that the negative results came from test designs which subjected the animals to frequent changes in the direction of the magnetic field while they were kept in a familiar environment, i.e. in a situation where it cannot necessarily be expected that they pay much attention to changes in magnetic directions. The positive findings, in contrast, were obtained by releasing the animals at a new, mostly unfamiliar site which per se might motivate them to become aware of directional relationships within their surroundings.

Subterranean Mole Rats. The best demonstration of magnetic compass orientation in rodents so far has been based on spontaneous tendencies. African mole rats of the genus *Cryptomys* (Bathyergidae) have a subterranean life; eusocial family groups dig and maintain long, permanent underground burrow systems. The main tunnels are linearly arranged – mostly close to the N-S axis – rather straight, with the nest at one end (ELOFF 1951). The biological significance of this arrangement is unclear as other subterranean mammals have radially arranged tunnel systems. The extension of the tunnels to 200 m and more, however, led to the question how a straight course can be maintained under ground.

Burrowing itself cannot be tested in the laboratory. However, various groups of *Cryptomys* sp. of Zambian origin[2] were found to build their nest again and again against the wall in the SE sector of their large circular arena cage. When the north direction of the ambient magnetic field was altered by means of HELMHOLTZ coils, leaving total intensity and inclination unchanged, the mole rats shifted the sites of their nest accordingly (Fig. 4.14; BURDA et al. 1990). Vector lengths in the range of 0.5 for individual families suggest an accuracy of ± 50 ° to 60 °. When the horizontal component of the magnetic field was cancelled, the nests were found all around the arena. An inversion of the vertical

[2] The test animals belong to two different karyotypes of a group of species that was formerly designated *Cryptomys hottentotus*.

component did not lead to a change in orientation (Fig. 4.14, right). This latter finding shows that the magnetic compass used by *Cryptomys* is a polarity compass (MARHOLD et al. 1991).

Similar tests were performed with the blind mole rat *Spalax ehrenbergi* (Spalacidae), a solitary species that has adapted a similar subterranean lifestyle. The data were much more scattered than in *Cryptomys*, but on the whole, they, too, suggest the involvement of a magnetic compass (BURDA et al. 1991).

4.2.4.2 Horses (Perissodactyla: Equidae)

A study on Domestic Horses, *Equus caballus* (BAKER 1989a), indicated that these animals probably use a magnetic compass to determine the direction of displacement. Six bars were attached to their bridle while they were transported in a horsecar 15–25 km to locations where they had never been before. The controls wore brass bars, the test animals wore bar magnets with a pole strength of 0.016 T; the persons accompanying the horses did not know to which group the horse belonged. After arriving at the release site, the horses were allowed to choose their own route, and their net direction was measured after 15 min. The animals that had worn brass bars showed a significant preference for the direction of their stable. The walking directions of the horses that had worn magnets was much more scattered. Although the sample sizes were rather small, the difference was significant (BAKER 1989a).

4.2.4.3 Man (Primates: Hominidae)

The only other species of large mammals whose orientation behavior has been analyzed with respect to the possible use of magnetic compass information is Man himself, *Homo sapiens*. The test subjects were members of the industrialized societies of Western civilization. BAKER (1980) first reported that people could designate the compass direction to the starting point of the journey after they had been blindfolded and transported by car to an unknown site. The directions named by control subjects showed a significant tendency towards the true home direction, whereas the data of subjects wearing bar magnets (pole strength between 0.014 and 0.030 T) at the back of their heads showed considerably more scatter. These findings encouraged several groups of scientists to turn their interest to spontaneous orientation abilities of humans. Most attempts to replicate BAKER's findings were unsuccessful, while BAKER himself continued his studies with positive results and even claimed that his colleagues often had overlooked positive evidence in their own data (BAKER 1987a). The resulting controversy has been documented in considerable detail by KIRSCHVINK et al. (1985a) and BAKER (1989b); hence, we will give here a brief summary only:

BAKER performed three basic types of experiments:
1. *Walkabouts*: The blindfolded subjects actively walked, being led by a test assistant along unfamiliar routes. The subject was then asked to designate the compass direction and/or point in the direction leading to the starting point.
2. *Bus experiments*: The blindfolded subjects were passively transported to an unknown site; otherwise, the procedures were as described for the 'walkabouts'.
3. *Chair experiments*: The blindfolded subjects sat on a rotating chair which was turned in an irregular fashion. At each of the repeated stops, they were asked to name the direction in which they were facing.
Thus, while the first two approaches involved a navigational task to compute the compass course to the goal and, in a second step, to designate it and/or locate it, the third design involved the last step only.

The magnetic field was altered either by applying magnets (pole strength 0.018 T) to the subjects' heads, mostly at the right temple, or by equipping the subject with PVC helmets which, by means of a built-in coil system, allowed the generation of magnetic vectors along various axes. These artificial magnetic fields moved with the subjects, which in the case of a horizontal field meant that it added to the geomagnetic field in various ways, depending on the direction of movement. The effect of the vertical field was more constant. For the chair experiments, the entire test apparatus was surrounded by a large set of coils that allowed changing the direction of magnetic North.

In walkabouts and bus experiments, BAKER found an overall ability to name the compass courses to the starting point and to point in the respective direction. The degree of accuracy was highly variable and often very low, with vector lengths ranging from below 0.1 to above 0.7. Also, the deviation from the true course was sometimes rather large; in rare cases, even with means in the opposite semicircle. The homeward components (= vector length multiplied by cosine of deviation from home) lay mostly around 0.2 and 0.35. There was no pronounced trend that either naming the compass course or pointing in the direction was more accurate. Altogether, BAKER's data suggest a certain ability of humans to be aware of the direction leading home. Fields generated by the helmets did not seem to affect the orientation in walkabouts or bus experiments; bar magnets tended to lower the home component somewhat, but the effect was marginal.

A number of authors tried to replicate the bus experiments, and, in general, reported negative results; they are listed in BAKER (1987a). In the majority of cases, the authors were unable to find orientation after displacement; hence, the manipulation of the magnetic field could hardly produce any difference. Recalculating their data, BAKER (1987a) pointed out cases of significance. When he combined the probabilities of all independent attempts to replicate his findings, he came to the conclusion that the data, taken as a whole, significantly deviated from random, thus suggesting an ability for non-visual orientation. He also claimed a marginal effect of magnets and magnetic treatment on naming the compass direction, but apparently not on pointing. However, BOVET (1992) warned against some of the statistical procedures applied. A study by WALMSLEY and EPPS (1988) was confirmatory insofar as the authors also found

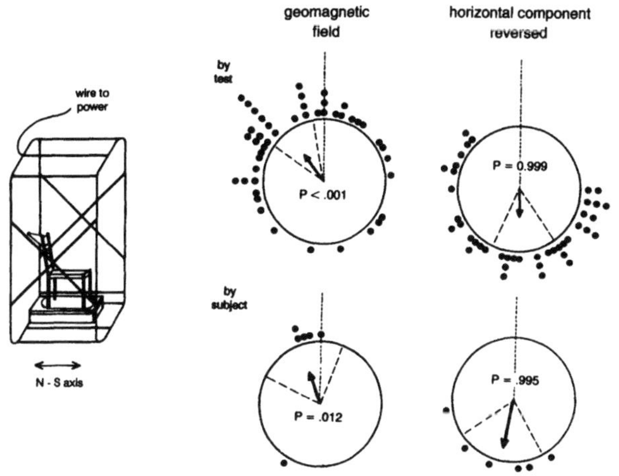

Fig. 4.15. Test apparatus and results of chair experiments with humans, *Homo sapiens*. *Left diagram* Rotating chair in the center of a large set of coils, where the persons were placed for testing. *Right diagrams* Directional estimates of six blindfolded test persons in eight tests each (one person did not complete eight tests because of illness) in the local geomagnetic field and in a field with magnetic North turned to geographic South. In the *upper diagrams, the symbols at the periphery of the circle* give the average error of nine directional estimates per test; in the *lower diagrams*, they represent the mean error of all tests per person. The *line pointing upwards* indicates correct estimates. The *arrows* represent the mean vectors; the *dashed lines* indicate the 95% confidence intervals. The *P*-values refer to the V-Test (a derivation of the RAYLEIGH test) with correct estimates as predicted direction. (After BAKER 1989b)

homeward orientation in blindfolded people; an effect of magnetic treatment was not studied.

The chair experiments performed at Manchester University also showed a weak, yet significant trend to name directions correctly. Magnets reduced the component towards the correct direction from 0.21 to 0.08. A shift in magnetic North around the entire apparatus (an altered field more homogeneous than the changes produced by the other methods) resulted in a corresponding change in directions named by the subjects (Fig. 4. 15). This constitutes the only direct evidence for use of an internal magnetic compass in humans so far (BAKER 1989b).

Attempts to replicate this type of experiment were not so numerous, and most of them also produced negative results (see KIRSCHVINK et al. 1985a). A notable exception is a study by BAKER's student MURPHY (1989) who reported good abilities to name directions and a clear effect of magnets. In her study, the subjects were exclusively females, who, according to BAKER (1989b), possess much better orientation abilities than males.

The controversy still continues, as the various authors analyzing human orientation have not reached agreement. It is not easy to form an opinion. Taken as a whole, the evidence available so far may not be completely random, thus suggesting a vague ability of humans to determine the home direction after displacement and to name it and point in the direction, as BAKER claims. The magnetic nature of this ability is even less clear. The use of a magnetic compass in the chair experiments is better documented. In any case, the orientation expressed in the experiments with humans is very weak, so that it is doubtful whether it can be of any practical use. Perhaps members of industrial civilization, who have hardly ever had to face orientation tasks, were the wrong subjects; the picture might become altogether different if tribesmen and people living a more traditional life were studied. At the moment, spontaneous human orientation abilities, being at the level described by BAKER, appear to be a rather unreliable basis for orientation and navigation.

4.2.4.4 The Technical Compass

However, the weak spontaneous orientation abilities do not pose a real problem, as man has been able to overcome any deficiencies by making use of the technical magnetic compass, an instrument in general use in navigation and aviation today. Its origin and history, however, are somewhat obscure. It was first known in China, where some authors date it back as early as the 12th or even 20th century B.C., but the writings on which this view is based also allow different interpretations. A statement of HAN-FEI-TSE that '*the old rulers made instruments pointing south to distinguish east and west*' in the 3rd century B.C., and HIU-TSCHIN's description of a magnetic stone used to give directionality to a needle in 121 A.D. are among the first unequivocal references to a magnetic compass (HENNIG 1931). The regular use of a magnetic compass in seafaring apparently began sometime later, probably because suitable magnetized iron was found only in northern China, far from the coast. The use of a magnetic compass in maritime navigation is well documented from the 5th century A.D. onward, and a text written about 1122 A.D. by SU-KING describes a ship's compass using a swimming needle. From China, the knowledge about the magnetic compass spread throughout southern Asia and to Arabia. Yet it is difficult to ascertain from what time onward it was generelly used by Arabian seafarers (HENNIG 1931).

In Europe, a magnetic compass was first mentioned about 1190 A.D. by the Englishman ALEXANDER NECKAM, who listed it among the utensils needed for a well-equipped ship. Interestingly, he seemed to imply familiarity with the instument, not giving details on its origin or how it worked, so that it might be assumed that it was more or less in common use in northern Europe at that time. This leads to the question as to how it was introduced to European seafarers. There are two possibilities: it was passed on from China by Arabian sailors, or it was an independent discovery by Norsemen. Definite answers cannot be given. The early European references to the magnetic compass relate to its use in the northern seas rather than in the Mediterranean where Christian and Arabian seafarers had close contacts. This in itself might not be significant, however; the need for a weather-independent means to locate directions would be far greater in the North Atlantic because of climatic conditions. On the other hand, Norsemen might have discovered the useful properties of a magnet on their own, as strongly magnetized iron was obtained in mines in southern Norway. Also, Norsemen named their version of the compass *leidarsteinn*, a term from their own language. Another point possibly suggesting an independent discovery rather than taking over a foreign instrument is the European custom to refer to the magnetic compass as indicating 'north', while in China and Arabia, it is said to indicate 'south' (HENNIG 1931). The two concepts might reflect differing traditions associated with the origin of the compass in different regions. At lower latitudes, the direction South as the sun's culmination would be the most prominent direction, while at the latitudes where the Norsemen lived, with the lower sun's arc, North as indicated by the polar star would be the most significant reference.

The magnetic compass as a technical instrument was passed on by tradition among humans and cannot be compared with the internal magnetic compass of animals. Yet the information it provides seems to be totally equivalent to the one animals obtain using natural perception of the geomagnetic field. It is remarkable how easy this type of information can be used and integrated with other sources of directional and spatial information by man. The magnetic compass fits perfectly into our framework of spatial concepts and orientation, in so-called primitive as well as in technical civilizations. From this, it is tempting to speculate that the technical magnetic compass might represent just a technical enhancement of a spontaneous ability which has become rudimentary in modern man of Western traditions.

4.3 Magnetic Compass Orientation in Invertebrates

A magnetic compass has also been described for a number of invertebrate species of the phyla Mollusca and Arthropoda. These groups are among the most highly developed animals; many species are rather motile and show complex spatial behavior patterns, which include dispersal, nomadism, but also regular migrations. Others have a 'home' or a 'home zone' to which they return after foraging. Especially among arthropods, a number of species show extended movements which seem to require efficient orientation mechanisms.

Unfortunately, the true extent of movements is poorly known, as the natural spatial behavior of only a few species has been studied so far. Attempts to analyze orientation mechanisms are even rarer, being limited to certain groups. Table 4.6 lists the species whose directional tendencies could be predictably altered by changing magnetic North.

Table 4.6. Invertebrates that have been shown to use a magnetic compass

Species	Reference
Mollusca, Gastropoda, Opistobranchia	
Sea slug *Tritonia diomedea* (Nudibranchia: Tritoniidae)	LOHMANN and WILLOWS (1987)
Arthropoda, Crustacea, Malacostraca	
Spiny lobster *Panulirus argus* (Decapoda: Palinuridae)	LOHMANN (1985)
Beachhopper, *Talitrus saltator* (Amphipoda: Talitridae)	ARENDSE (1978)
Sandhopper *Talorchestia martensii* (Amphipoda: Talitridae)	PARDI et al. (1985, 1988)
Riparian talitrid *Orchestia cavimana* (Amphipoda: Talitridae)	ARENDSE and BARENDREGT (1981)
Isopod *Idotea baltica* (Isopoda: Idoteidae)	UGOLINI and PEZZANI (1992)
Arthropoda, Insecta	
Compass termite *Amitermes meridionalis* (Isoptera: Termitidae)	JACKLYN (1990, 1992a)
Harvester termite *Trinervitermes geminatus* (Isoptera: Termitidae)	RICKLI and LEUTHOLD (1988)
Honeybee, *Apis mellifera* (Hymenoptera: Apidae)	DEJONG (1982); SCHMITT and ESCH (1993)
Flour beetle *Tenebrio molitor* (Coleoptera: Tenebrionidae)	ARENDSE and VRINS (1975)
Fruitfly *Drosophila melanogaster* (Diptera: Drosophilidae)	PHILLIPS and SAYEED (1993)
Moth *Noctua pronuba* (Lepidoptera: Noctuidae)	BAKER and MATHER (1982)
Moth *Agrotis exclamationis* (Lepidoptera: Noctuidae)	BAKER (1987b)

4.3.1 Mollusks

Among mollusks, the motile species of gastropods and cephalopode would seem to be the most likely candidates for well-developed orientation mechanisms. Cephalopods, however, have not been studied from this point of view. Studies on snails are also few, and they seldom include a search for possible magnetic effects. Exceptions are those by F. BROWN et al. (e.g. 1959, 1960a) who reported that the directional behavior of common mud snails *Nassarius obsoletus* (Prosobranchia, Mesogastropoda: Nassidae) could be modified by bar magnets of various strengths. The response was highly variable, however, and varied in the course of the day and lunar month (see Sect. 7.3.1). The nature of the effect remains unclear, but it is certainly not a '*compass response*' as defined in Section 2.2.3, although it was called so by F. BROWN et al. (1960a).

TOMLINSON et al. (1980a) observed that several species of chitons (Polyplacophora) on the Californian coast often rested with their head oriented towards the northerly semicircle. Attempts to affect this orientation by magnetic manipulations were successful, but it remains unclear whether or not the responses reflect magnetic compass orientation. The only report suggesting a true magnetic compass in mollusks is based on the spontaneous behavior of the marine slug *Tritonia diomedea* (Gastropoda, Opistobranchia, Nudibranchia: Tritoniidae). LOHMANN and WILLOWS (1987) observed that *Tritonia* showed significant preferences for resting positions in certain directions, which disappeared when the horizontal component of the geomagnetic field was compensated. Further tests involved the use of a Y-shaped maze where the slugs could chose between two directions. A preference for one of the arms could be altered by experimentally deflecting magnetic North; *Tritonia* preferred the formerly neglected arm, when it pointed in the preferred direction with respect to the magnetic field.

These experiments are complicated by the fact that the preferred direction was found to vary with time. LOHMANN and WILLOWS (1987) described a significant correlation of the mean angle with the day of the lunar month, i.e. the preferred direction shifted from north at full moon counterclockwise over south at new moon back to north. The biological significance of the orientation is unclear, since the spontaneous behavior of free-living *Tritonia* is largely unknown. The animals feed on pennatulacean cnidarians; the authors speculate that the slow shift in direction might help *Tritonia* to stay in an area of high food concentration, and be more effective in exploiting food patches.

4.3.2 Crustaceans

Crustaceans show a great variety of spatial behavior, including migration, nocturnal foraging trips followed by the return to a temporary or permanent 'home' or, in the case of beach-dwelling groups, to a preferred zone of humidity. Magnetic compass orientation has been documented, in connection with innate courses, with courses originating from the specific spatial layout of the animals' home range and from directional training.

4.3.2.1 Lobsters (Decapoda: Palinuridae)

The spiny lobster *Panulirus argus* was found to return to its home range after displacement (e.g. HERRNKIND et al. 1975). Also, it was known to migrate regularly over up to 50 km, where thousands of individuals follow one another in a single file. Headings vary at different locations, and it is unclear whether they are controlled by local factors only, as HERRNKIND (1983) tends to assume, or whether a spontaneous directional preference is involved. In a field study, non-migratory *Panulirus* did not respond to magnets attached to their backs, but continued to head into the surge of the waves (WALTON and HERRNKIND 1977).

Directional training, on the other hand, suggested orientation with the help of a magnetic compass. LOHMANN (1985) placed lobsters in a round tank which they could leave by one of six tunnels; they were rewarded when they chose the northern one and negatively reinforced when they entered any other (cf. Sect. 8.2.1). Two out of the five individuals learned the task. When these animals were tested with magnetic North deflected, they shifted their preferences accordingly. Interestingly, the distribution of the incorrect choices, showing a maximum opposite the correct direction, suggests that the lobsters have a certain tendency to respond in an axially bimodal way rather than unimodally to the magnetic field.

4.3.2.2 Sandhoppers (Amphipoda: Talitridae)

Magnetic orientation in amphipods was analyzed in species living on beaches or shores of fresh water bodies, using the animals' tendency to move along an axis perpendicular to the borderline land-sea, which corresponds to the Y-axis described for amphibians (cf. Sect. 4.2.2). Landward or seaward movements occur when the animals are in danger of submersion or desiccation, respectively, and thus can be evoked by appropriate experimental conditions. The orientation was found to be mainly directional, with compass courses being controlled by genetically encoded information as well as by experience (see Pardi and Scapini 1983; Ugolini et al. 1988). More details on the orientation system of sandhoppers are given in Section 5.4.

Research on the magnetic orientation the Beachhopper of *Talitrus saltator* produced conflicting results. Van den Bercken et al. (1967) observed oriented behavior in the absence of visual cues. The preferences were less pronounced than the ones observed under the sun, but nevertheless significant. The local geomagnetic field was considered as a possible orientation cue, but attempts to tests this hypothesis were unsuccessful. Studies by Ercolini and Scapini (1972) in Italy failed to produce conclusive evidence even for non-visual orientation. Continuing the studies on *Talitrus* in the Netherlands, Arendse (1978, 1980; Arendse and Kruyswijk 1981) found an axial preference for the Y-axis rather than unimodal behavior. Compensating the geomagnetic field resulted in disorientation; shifting magnetic North resulted in a corresponding shift of the animals' tendencies, thus indicating the use of a magnetic compass. It was the first case documented in an invertebrate species.

Arendse and Kruyswijk (1981) also offered an explanation for previous negative results. Observing the beachhoppers directly, they distinguished two types of movements, namely, crawling and jumping. Only jumpers were oriented, whereas crawlers were not. Ercolini and Scapini (1972), in contrast, photographed their animals and thus could not make this distinction; Arendse and Kruyswijk argued that the non-oriented behavior of the crawlers might have obscured the orientation of the jumpers. A new study in Italy (Scapini and Quochi 1992), however, again failed to produce evidence for magnetic orientation: only 1 out of 12 samples showed the expected directional preference in the absence of visual cues. Ugolini (1994), on the other hand, reported some response to changes in the ambient field, but the behavior of his beachhoppers was quadrimodal. These differences in results from the Netherlands and from Italy might reflect differences in the relative importance of magnetic cues (see Sect. 5.4.2).

Magnetic compass orientation has been demonstrated in two more species of amphipods. One is *Talorchestia martensii*, a sandhopper found in the tropics along the East African coast. While Scapini and Ercolini (1973) in an early study had been unable to observe meaningful non-visual orientation, Pardi et al. (1985, 1988) reported a pronounced unimodal or axial orientation along the axis land-sea. Shifting magnetic North resulted in a corresponding shift of the animals' preferences (Fig. 4.16). Young, laboratory-born specimen spontane-

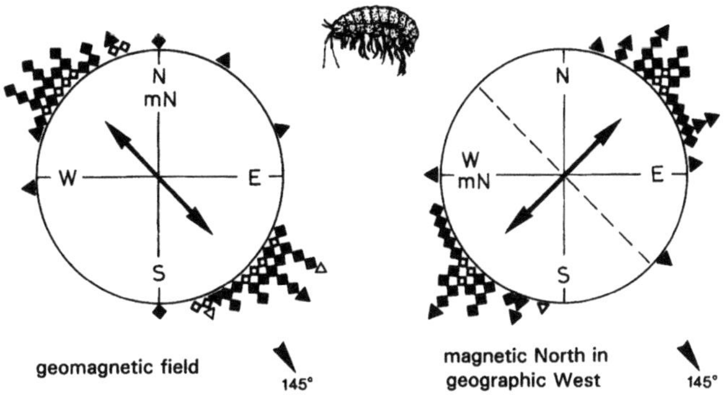

Fig. 4.16. Orientation of the equatorial sandhopper *Talorchestia martensii* in the laboratory in a centrally lit arena. *Left* Tests in the local geomagnetic field; *right* tests with magnetic North shifted 90 ° counterclockwise to geographic west. The theoretical escape direction to the sea (TED) of 145 ° is marked by the *arrowhead outside the circle*; the *symbols at the periphery* indicate the mean headings of individual sandhoppers: *triangles* unimodal behavior; *diamonds* axially bimodal behavior, with both ends of the axis indicated; *solid symbols* samples significant by the Rayleigh test: *open symbols* non-significant samples. The *double-headed arrows* represent the mean axes, the *dashed diameter in the right diagram* marks the axis of the respective controls. (After PARDI et al. 1988)

ously preferred the land-sea axis of their parent population when the geomagnetic field was the only available cue. This indicates that the preference is based on innate information with the magnetic field serving as reference.

The other species is *Orchestia cavimana*, a riparian amphipod found near fresh and oligohaline water (ARENDSE and BARENDREGT 1981). In a centrally lit arena, the animals showed significant unimodal preferences which changed accordingly when magnetic North was turned; compensation of the magnetic field resulted in disorientation (ARENDSE and BARENDREGT 1981). *Orchestia* avoids light as it promotes water loss and desiccation; the directional preference which served as baseline for demonstrating a magnetic compass was the direction pointing away from the light source in the animals' housing container. This direction was apparently remembered as a magnetic course. Changing the position of the lamp in their housing quarters resulted in a corresponding new set direction (ARENDSE and BARENDREGT 1981).

4.3.2.3 Isopods (Isopoda: Idoteidae)

Idotea baltica, a marine isopod found near the coast among algae and *Posidonia*, shows orientation along an axis perpendicular to the borderline land-sea like *Talitrus* and *Talorchestia.* This species was also found to orient with the help of a magnetic compass (UGOLONI and PEZZANI 1992; Ugolini, pers. comm.).

4.3.3 Insects

The number of insect species that have been studied with respect to magnetic orientation is still small. Those showing alignments have already been listed in Chapter 3. Insects showing true compass orientation are mostly species that either show some type of migration, like butterflies, or species that have a 'home', like termites and some hymenopterans. Numerous groups, however, have not been studied at all.

4.3.3.1 Termites (Isoptera)

Mound-Building in Compass Termites. The spectacular mounds of the Australian compass termite *Amitermes meridionalis* (Termitidae), called 'magnetic anthills' by the local population, have often been discussed in connection with magnetic orientation (e.g. G. Becker 1964; von Frisch 1974). These tombstone-like constructions, about 3 m long and very narrow, may reach a height of up to

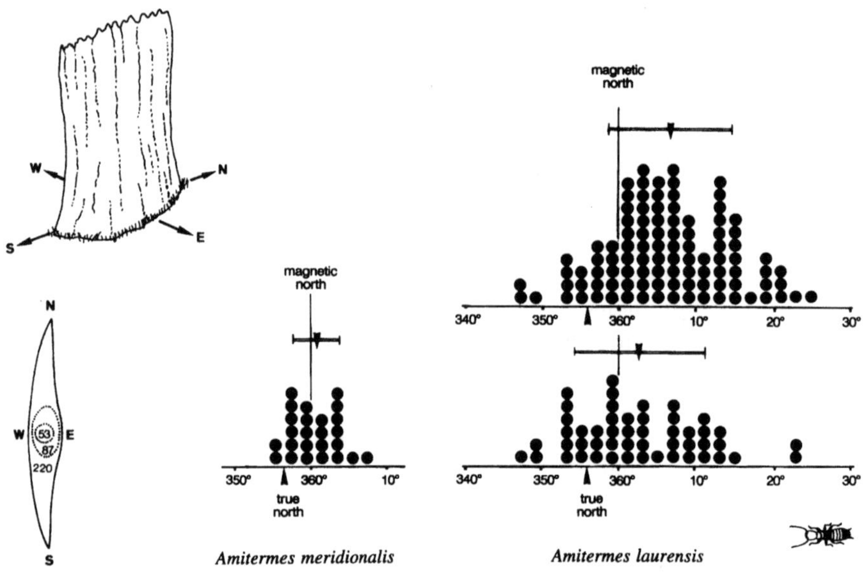

Fig. 4.17. Orientation of magnetic 'anthills' of termites of the genus *Amitermes* as they are found in the Australian Northern Territory. *Left* Side view of a mound of *A. meridionalis*, and outline of the basis showing the elongated shape; the *dotted lines* indicate the changing shapes of growing mounds, with the respective height given in cm. (After Jacklyn 1992a; Duelli and Duelli-Klein 1978) – *Right* Orientation of the longitudinal axes of meridional mounds of *A. meridionalis* in the Darwin area and of *A. laurensis* in Arnhem Land, Australian NT; the latter two samples are from two different fields approx. 3 km apart. As the mounds were measured with a compass, the directions are given with respect to magnetic North = 360 °; true (geographic) North is indicated by an *arrow. Arrows* and *bars* above the data indicate mean directions and standard deviation. (After Duelli and Duelli-Klein 1978; Grigg and Underwood 1977)

4 m. Their longitudinal axis roughly coincides with the N-S direction (Fig. 4.17, left). Building is slow; the mounds start as small, round structures, taking about 7 years to acquire an elongated axis (GRIGG et al. 1988). Such mounds are found only in a limited area in northern Australia, a region with a large amount of rainfall in the wet season and subject to periodic flooding; they usually occur in clusters of 20 to 100 in floodplains and humid areas. Aside from *Amitermes meridionales*, *A. laurensis* and *A. vitiosus* also build meridional mounds when they nest in ill-drained areas, but their mound orientation tends to be less pronounced.

DUELLI and DUELLI-KLEIN (1978) measured the direction of mounds of *A. meridionalis* on the Cox Peninsula near Darwin, Australian Northern Territory, where the local magnetic declination was 3.6 ° E. Mound axes varied in a range of 12 ° only, with the mean deviating less than half a degree from the magnetic N-S axis, while it was significantly different from the geographic N-S axis (Fig. 4.17, right). This seemed to suggest an alignment along the magnetic axis. A more extended study in the same area, however, based on the directions of hundreds of mounds of *A. meridionalis*, showed more scatter and a mean at about 8 °, i.e. ca. 4 ° east of magnetic North (GRIGG et al. 1988). In Arnhem Land (Australian NT), the axes of individual mounds of *A. laurensis* varied between 349 ° and 30 ° (GRIGG and UNDERWOOD 1977). The means of the fields measured were between 7 ° and 11 ° east of north, while in northern Queensland, mound axes of this species deviated mostly to the west of north, up to 17 ° (SPAIN et al. 1983). The orientation of the mound axes is thus typical for a given region. This and the observation that the means of fields often differ significantly from the magnetic N-S axis suggest that the building of meridional mounds might involve compass orientation along defined axes instead of a mere alignment (cf. Sect. 3.2.1).

This raised the question regarding orienting cues. Direct evidence for magnetic orientation is difficult to obtain because the slow growth of the mounds makes experiments almost impossible. Two recent studies, however, strongly suggest that the orientation is indeed based on directional information from the magnetic field. GRIGG et al. (1988) buried strong magnets at the base of four new colonies of *A. laurensis* and *A. vitiosus*, while brass bars were buried at four control colonies. When they checked their test area 7 (!) years later, the control colonies had developed into normal, elongated mounds. None of the colonies treated with magnets had survived, and evidence suggested that they had been given up shortly after the application of magnets. JACKLYN (1990, 1992a) used a different approach to test the magnetic nature of mound orientation more directly. During the dry season, he cut the upper 10 to 15 cm of mounds of *A. meridionalis* and awaited repair. Some of the mounds were supplied with strong permanent magnets in a way that the field at the cut was roughly turned by 45 ° (see Fig. 4.18). In general, the mounds were repaired in the direction given by the base of the mound. However, the magnetic treatment affected the orientation of the elongated cells found in the interior: in normal mounds, they are mostly aligned parallel or perpendicular to the main axis; in the experimental mounds, they were much more scattered, and the perpendicular group was almost totally missing (Fig. 4.18).

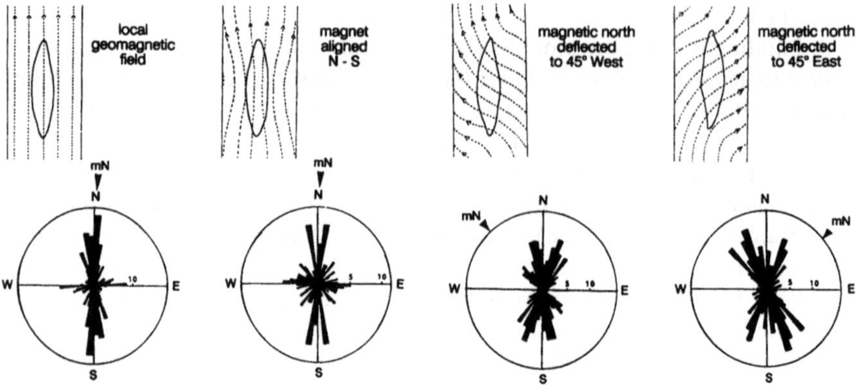

Fig. 4.18. Orientation of elongated cells in mounds of *Amitermes meridionalis*. The respective part of the mound was rebuilt after the top of the mound had been cut off. At some mounds, the ambient magnetic field was deflected with the help of magnets in the ways indicated in the *upper part* of the diagram. The angular distribution of the elongated cells is given *below*; in the deflected field conditions, it differed significantly from the one observed in the geomagnetic field. (After JACKLYN 1990)

The benefit of erecting the mounds roughly along the N-S axis lies in thermoregulation and the facilitation of air ventilation. N-S oriented mounds are warmed up in the morning and in the evening by the low easterly and westerly sun, whereas the hot sun at noon hits only a small part of the surface at a steep angle. The thermoregulatory effectiveness of the mound shape and orientation was experimentally demonstrated by GRIGG (1973). Other termite species construct an extended system of subterranean chambers, thus avoiding large temperature fluctuations; the *Amitermes* species, nesting in periodically flooded habitats which prevents burrowing undergroud, must use other means of thermoregulation. This leaves the question of the origin of the preferred axes and the significance of the observed regional variations. The specific orientation of mounds in a given region may be ultimately determined by local climatic conditions, like predominant wind direction, etc. and has become genetically fixed (JACKLYN 1991, 1992b).

Foraging Harvester Termites. Magnetic orientation in termites in connection with foraging has been reported by RICKLI and LEUTHOLD (1988). The harvester termite *Trinervitermes geminatus* (Termitidae), one of the few species that moves in the open, travels along pheromone trails that form an extended network around the nest. It was observed that returning workers, when coming to a fork, most often chose the track leading directly to the nest. Likewise, they seemed to be able to distinguish the homeward and the outward direction of unbranched trails. To study a possible involvement of magnetic orientation, RICKLI and LEUTHOLD used an elegant technique: returning successful foragers were directed through a tube and were dropped through a trap door into the center of an arena below. Here, eight radially arranged pheromone trails were presented. The termites preferentially chose the one closest to the direction

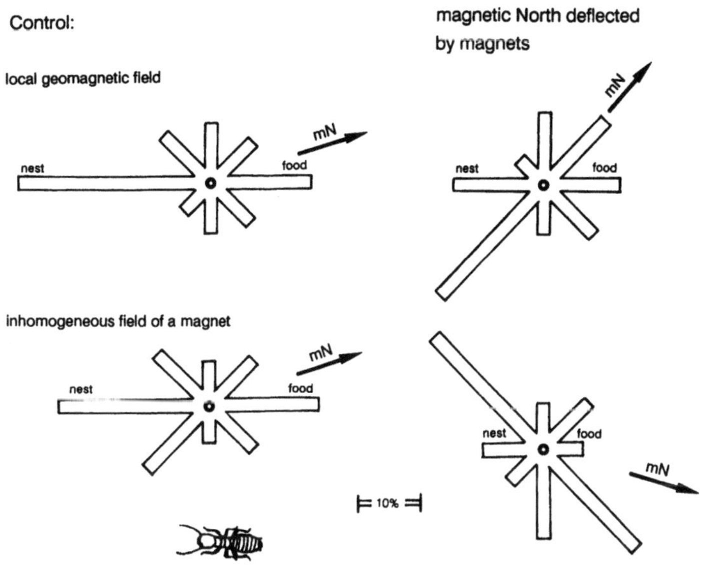

Fig. 4.19. Orientation of homing foragers of the harvester termite *Trinervitermes geminatus* in the local geomagnetic field, and when magnetic north was altered about 33 ° with help of magnets. The *bars* indicate the relative distribution of direction of 200 termites (tests with the magnet aligned along the N-S axis of the geomagnetic field: 143 termites) along 8 radially arranged pheromone trails; the *bar below* gives the length representing 10 %. *nest* and *food* indicate the trails towards the respective goal. The *open circle* in the center indicates the point where the test animals were introduced into the test arena via a trap door. *Arrows* mark ambient magnetic North. (After RICKLI and LEUTHOLD 1988)

leading to the nest. When magnetic North was shifted with the help of a large bar magnet, the preferences shifted accordingly (Fig. 4.19). RICKLI and LEUTHOLD (1988) attributed an increasing bimodality to the inhomogeneity of the field produced by the magnet and to the termites' tendency to reverse direction when they meet an unfamiliar or confusing situation.

The findings show that *Trinervitermes* use directional information from the magnetic field, implying that they record at least roughly the magnetic course of the outward journey during their foraging trips. It must be emphasized, however, that magnetic information is used only in connection with the pheromone trails. Workers that have lost their trail do not continue in the same direction as before, but search until they again find a trail which they can follow.

These findings and the alignments described in Chapter 3 indicate that termites use the magnetic field in various behavioral situations. This is not surprising considering that most termites are blind and live a hidden life, mostly avoiding the open. Many cues that play a prominent role in the orientation of most other animals are not accessible to them, while information from the magnetic field is easily available. Future studies on other species of termites might provide new examples for magnetic orientation.

4.3.3.2 Ants, Wasps, and Bees (Hymenoptera)

The orientation behavior of the social forms of ants, wasps, and bees is fairly well studied. The life of these insects is centered around a nest where the queen lives and the brood is raised. Every day, hundreds of workers swarm out in order to provide food for the colony. The distances involved vary greatly; in flying species like bees and wasps, they are, of course, markedly larger than in ants that have to move on foot. But even here, individuals of larger species may cover distances up to 100 m. As soon as the animals have been successful in obtaining food, they return to the nest on fairly straight routes.

Foraging. These foraging trips have met with great interest for a long time, and attempts to analyze the orientation mechanisms involved have led to the discovery of such important mechanisms like sun compass, use of polarized light, pheromone trails, wind compass, landmark use, etc. (cf. von FRISCH 1968; WEHNER et al. 1983). Surprisingly enough, although compass use is amply documented in the case of the sky compass[3], there was no evidence that bees or ants make use of the magnetic field on their foraging trips. Desert ants of the genus *Cataglyphis* (Formicidae, Formicinae) are disoriented in the absence of directional information from the sky and familiar landmarks (cf. WEHNER 1982), which argues against a non-visual compass. Some forest-dwelling species appeared to be able to forage successfully in the absence of visual and other obvious cues (COSENS and TOUSSANT 1985). Changing magnetic conditions, however, failed to influence the choice of trunk routes of *Formica* species within an arena (ROSENGREN and FORTELIUS 1986). Bees seem to avoid foraging in situations when the sun compass is not available. The search for a backup system allowing bees to orient under cloudy skies did not indicate an involvement of the magnetic compass (DYER and GOULD 1981). – Three recent studies, however, suggest that social Hymenoptera, too, might make use of magnetic information while foraging.

ANDERSON and VANDER MEER (1993) placed nests of the fire ant species *Solenopsis invicta* (Formicidae, Myrmicinae) on a testing table surrounded by coils that, when activated, reversed magnetic North. The ants were allowed to accustom themselves to the situation for 1 h, before a large food item was positioned 22 cm from the nest. This caused the workers to establish a trail between nest and food, which took about 15 min when the magnetic field was constant throughout the experiment. However, when the direction of the magnetic field was altered at the time when the food was introduced, either from normal to reversed or vice versa, the ants needed about twice as long to establish a clearly discernible trail. This suggests that directional information from the magnetic field provides an important cue for orientation.

Honeybees, *Apis mellifera*, seem to make use of a magnetic compass at the start and at the end of their foraging trips. SCHMITT and ESCH (1993) collected bees as they exited their hive and tested them individually in a laboratory. In the

[3] In Hymenoptera, the pattern of polarized light and the sun are integrated, forming a 'sky compass', a functional equivalent to the sun compass (WEHNER et al. 1983).

local geomagnetic field, they preferred easterly directions with a mean at 71 °, which was in rough agreement with the direction in which the opening of the hive was facing. The orientation of the individual bee was rather weak, however, producing vectors in the range of 0.10 to 0.26 only. In a less homogeneous test field with magnetic North deflected 90 ° counterclockwise, the mean direction shifted by 60 ° in the expected direction, i.e. the bees continued to prefer magnetic East. Findings by COLLETT and BARON (1994), on the other hand, suggest that magnetic cues are involved in the final approach to a familiar goal. Bees were found to face landmarks associated with the goal from a standard direction, which may simplify the storage and retrieval of memories. When bees that used to face landmarks in southern directions were trained to a feeding station in the presence of strong experimental fields (ca. 550 000 nT) adjusted in various directions, they were found to face the landmark with respect to magnetic South instead of geographic South. The magnetic field thus appears to provide the directional reference for compass coordinates of landmarks stored in memory.

The last-mentioned finding is surprising insofar as earlier attempts to demonstrate an influence of magnetic cues at the food source had not been very successful. KIRSCHVINK and KIRSCHVINK (1991) tried to train bees to leave the feeder in a constant magnetic direction, but the results were inconclusive (see Sect. 8.2.1). SCHMITT (pers. comm.) tested bees at a distant feeder, applying the technique which he had successfully used on bees leaving the hive. He was not able to find consistent preferences for the home direction, however. This seems to imply that bees use the magnetic compass only when leaving the hive and when arriving at the feeder, but apparently not when starting their return trip. Future experiments may allow one to define the role of magnetic information in the orientation of foraging bees more precisely.

Dancing. Another use of a magnetic compass, similar to the one already described for harvester termites (RICKLI and LEUTHOLD 1988; cf. Sect. 4.3.3.1), was reported by LEUCHT and MARTIN (1990). They found that the magnetic field may be used as directional reference for e-vector orientation when bees perform horizontal dances. In this study, the bees' view to the sky was restricted to a 40 ° or 20 ° sector around zenith, i.e. polarized light was the only visual cue available. The e-vector alone provides bimodal information, yet most dances were directed unimodally towards the food source. Compensation of the geomagnetic field to less than 4% of its total intensity led to a significant increase in bimodal dances up to 50%, which indicates that magnetic compass information is used to decide which end of the axis provided by polarized light is the correct one.

Comb Building. The bees' nest normally consists of several parallel sheets of honeycomb hanging from the ceiling of a cavity. In apiculture, beekeepers usually provide swarming bees with a new hive containing frames which fix the direction of the future combs. If no such artificial help is presented, however, hundreds of workers, facing the task of building new sheets of comb in the darkness of a cavity, must agree upon its orientation. In order to learn about the mechanism used, bees were moved into circular hives without any frames, etc.

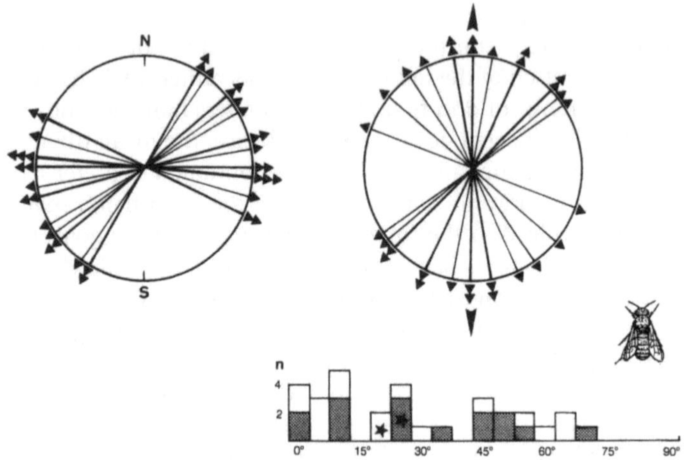

Fig. 4.20. Orientation of combs in the Honeybee, *Apis mellifera. Left* Directions of sheets of comb in the bait hives; *triangles at the periphery of the circle* mark the ends of the axes of each comb which were found to be clustered along an axis slightly counterclockwise of the E-W axis. *Right* Relationship between the axes of newly built combs and the combs in the parent hives. In the *circular diagram*, the axes of the parent combs were set to correspond to the vertical axis and marked by *dark arrows outside the circle*; the ends of the newly built combs are indicated as before. The *lower diagram* gives the angular difference between the axes of the new combs and the comb direction in the parent hives. *Light* Control experiments when both combs were built in the local geomagnetic field; *shaded* experiments in which the magnetic field was altered so that one comb was built in the local magnetic field, the other in a field with magnetic North turned 90 ° to the east; in this case, the angular difference in the magnetic directions is given. *Stars* mark the respective median values. (Data from DEJONG 1982)

They built their combs in directions which almost perfectly agreed with the direction of combs in their parent colony (OEHMKE in LINDAUER and MARTIN 1972). After eliminating possible light cues, magnetic North was deflected around a new cylindrical hive, and the bees were found to change the orientation of their sheets of comb accordingly. Altering the magnetic field by magnets in a non-physiological way led to abnormal combs. These findings, reported by LINDAUER and MARTIN (1972) without giving data or further details, were the first indication for the use of a magnetic compass in bees. IFANTIDIS (1978), studying comb building in relation to the shape of the hive and to various external cues, did not find a relationship between the direction of combs in a new hive and that in the parent hive; he claimed that the magnetic field was of no importance[4]. GOULD (GOULD et al. 1978; GOULD 1980) twice mentions in a note that he was unable to replicate the findings reported by LINDAUER and MARTIN (1972); unfortunately, no data are given.

---

[4] In his Table 1, IFANTIDIS (1978) lists 22 cases in which the comb directions in a new hive and in the preceding hive can be compared. These data do not rule out a relationship between successive comb directions of the same swarm, even if such a tendency does not reach significance ($0.10 > p > 0.05$, with the direction of the preceding hive as predicted direction).

A study by DeJong (1982), however, produced positive results. Feral bee swarms were trapped in cubic bait hives. After being transferred to the laboratory, they were repeatedly forced to build new combs in hives where no other obvious cues were present. Young swarms, which had entered the bait hive just recently and had eggs only, maintained their comb direction fairly well. Older swarms with broods in all stages did not show such a relationship – a finding that might explain the negative outcome of other studies. As a consequence, DeJong (1982) based his tests for magnetic orientation on young swarms only. The bees had to build new combs every 2 days, alternately in the geomagnetic field and in a field of equal intensity with magnetic North turned 90 ° clockwise. The magnetic directions of the new combs were found to be significantly related to the magnetic direction of the previous combs, even if the scatter was considerable (Fig. 4.20, right and lower diagram).

Another interesting phenomenon emerges from the data of DeJong (1982). The orientation of the original sheets of comb in the bait hives showed a significant tendency for an axis running 76 °-256 ° ($r = 0.61$, $n = 18$, $p < 0.01$, Rayleigh test, see Fig. 4.20, left). The origin of this axis and its significance is unclear. It might reflect a general spontaneous preference of the E-W axis in wild-living bees, or it might reflect some local tradition. Maybe the swarms caught in the bait hives originated, generations back, from one common colony whose comb direction was passed on with the magnetic field as reference.

Accuracy. Attempts to estimate the accuracy of magnetic compass orientation in bees yield diverging results. Lindauer and Martin (1972) originally mentioned 2 °, based on their comb-building experiments; however, with about 25 °, the median deviation in DeJong's experiments was markly larger. The data by Ifantidis (1978) and by Schmitt and Esch (1993) seem to suggest even less accurate orientation. It is interesting that the accuracy indicated by experiments on compass orientation is markly lower than that observed in alignment responses, which is ± 4 ° in the zero crossings of misdirection (Martin and Lindauer 1977) and in the range of ± 9 ° when horizontal dances are considered (Towne and Gould 1985; cf. Sect. 3.3.2).

4.3.3.3 Beetles (Coleoptera)

Magnetic orientation was first considered in connection with the migration flights of the cockchafer *Melolontha melolontha* (Scarabaeidae). Their larvae live in meadows and open areas, while the adult beetles feed on the leaves of certain deciduous trees in forests that may be up to 3 km distant. When the young imagines emerge from the ground at dusk, they take off towards higher points of the skyline, i.e. their direction is determined by the visual appearance of the landscape. When they are ready to lay eggs, however, they normally return to the area where they themselves emerged (Schneider 1967). Their orientation is not affected by cloud cover, which suggested that they used a magnetic compass to locate the return course. Schneider (1957) reported that he had been able to affect the directional tendencies of *Melolontha* with bar magnets, but these studies were not continued. A clear demonstration of magnetic compass orientation in *Melolontha* is still missing.

Another species, the Flour Beetle *Tenebrio molitor*, was tested in various magnetic fields produced by HELMHOLTZ coils (see Table 4.7). Similar to the study with the amphipod *Orchestia cavimana*, directional tendencies induced by the light conditions in the housing containers provided a reliable baseline for experimental manipulations. The animals orient on an axis that coincided with the direction towards and away from the light source. Normally, they are negatively phototactic in order to avoid desiccation, but under conditions of high humidity, they reverse their response. When tested in a round arena under isotropic light, they showed clear tendencies which coincided with the direction away from the light in their housing containers. This orientation could no longer be observed when the magnetic field was compensated, and it shifted correspondingly when magnetic North was deflected by 120 ° (Table 4.7). In a field with horizontal field lines, the beetles showed unimodal orientation (Table 4.7, last line), which shows that their magnetic compass is based on polarity rather than inclination. Since total intensity in these tests was reduced to ca. 16 000 nT, i.e. 30 % of the natural local intensity of 47 000 nT, the data also suggest that the magnetic compass of *Tenebrio* is much more robust against intensity changes than the magnetic compass of birds (ARENDSE and VRINS 1975; ARENDSE 1978).

Table 4.7. Magnetic compass orientation of Flour beetles, *Tenebrio molitor*

Magnetic field	dal	n	Direction	Length	Δ dal
1. Beetles from a container with isotropic light					
Geomagnetic field	–	60	(202 °)	0.13 n.s.	
» »	–	100	(9 °)	0.07 n.s.	
2. Beetles from a container with anisotropic light					
Geomagnetic field	285 °	80	313 °	0.30***	+28 °
Magnetic North = 120 ° ESE	»	68	37 °	0.23*	+112 °**
Field compensated	»	120	(181 °)	0.10 n.s.	
Vertical component only	»	120	(83 °)	0.10 n.s.	
Geomagnetic field	105 °	70	78 °	0.48***	– 27 °
Magnetic North = 120 ° ESE	»	80	259 °	0.21*	+154 °**
Horizontal component only	»	70	84 °	0.45***	+21 °

dal = direction away from light in holding container; Δ dal = deviation of preferred direction from dal. Asterisks at vector length indicate significance by the RAYLEIGH test; non-significant directions are given in parentheses. Asterisks at Δ dal indicate significant difference from dal based in the confidence interval: * = $p < 0.05$, ** = $p < 0.01$, *** = $p < 0.001$, n.s. = not significant. (Data from ARENDSE and VRINS 1975)

Tenebrio uses the magnetic compass to locate a course which was derived from the anisotropy of light in the housing containers. Tests with individuals that had been put into a container with isotropic light as *pupae* indicate that the young *imagines* after metamorphosis remembered the light distribution they experienced as *larvae*. These directional tendencies disappeared after staying under isotropic light conditions for a month (Table 4.7, first lines). Another interesting feature of the magnetic compass of *Tenebrio* is that it was not restricted to constant courses. One group of beetles was kept in a container in front of a window so that the direction of light changed with the movement of the sun. When these beetles were tested in the local geomagnetic field, they shifted their preferred directions accordingly in the course of the day, while at night they preferred a constant course corresponding to the direction away from the window (Arendse and Vrins 1975). This suggests that *Tenebrio* can also use the magnetic compass to locate variable directions, even when establishing the respective compass courses involves the internal clock.

The studies of Arendse and Vrins (1975) and Arendse (1978) as a whole provide an excellent example as to how physiological conditions, previous experience and the magnetic field as an external factor interact in a complex way in the control of orientation behavior.

4.3.3.4 Flies (Diptera)

Young imagines of *Drosophila melanogaster* were exposed to incident ultraviolet light from a defined direction, which varied between groups. For tests, the fruitflies were released under isotropic UV light in the center of an eight-arm testing apparatus. In the local geomagnetic field each group preferred the direction which coincided with the direction toward the light source during the pretest exposure; when magnetic North was turned to east, south, and west, the fruitflies preferred the respective magnetic directions. This behavior was only observed in males, however; females did not show a significant directional preference (Phillips and Sayeed 1993).

4.3.3.5 Butterflies and Moths (Lepidoptera)

Many species of butterflies and moths are known to migrate considerable distances. The most spectacular migrations are reported from the North American Monarch Butterfly, *Danaus plexippus*. Species of other families also migrate various distances. Regular cross-country movements have also been observed in several species that are active at night, like members of the families Sphingidae and Noctuidae (cf. Baker 1978). Therefore, it is surprising that only few attempts have been made to analyze their orientation mechanisms.

Danaids. In one of the few studies, Schmidt-Koenig (1985) recorded the migratory directions of Monarch Butterflies at various sites in eastern North America. The observations were made in fall; the butterflies were well oriented in southerly to southwesterly directions. Under overcast, their orientation was just as good as under sun, suggesting orientation mechanisms independent from celestial cues. Schmidt-Koenig explicitly discussed the possibility of magnetic compass orientation. Because of differences in directions observed at the various sites, however, he favored a model that involved no real compass, but restricts the butterflies to certain directions determined by the local magnetic inclination (Kiepenheuer 1984; see Sect. 6.3.3).

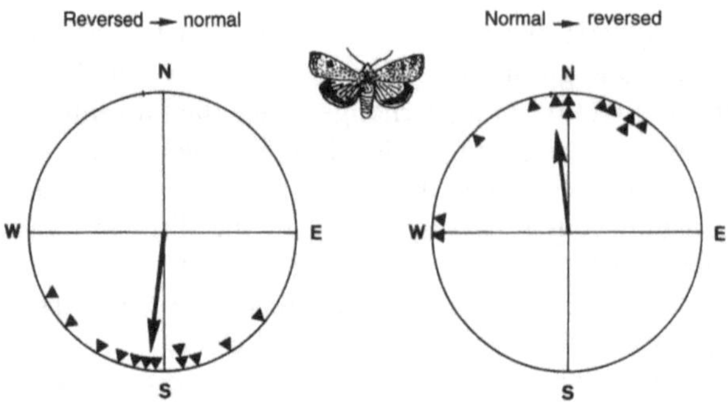

Fig. 4.21. Effect of reversals of the magnetic field on the direction of movements of Large Yellow Underwing Moths, *Noctua pronuba*. Forty moths were introduced into a cross-shaped cage. The horizontal component of the magnetic field was reversed every 10 min, and the moths were counted just before the next change was due. From consecutive counts, a net direction of movement was calculated. The *symbols at the periphery of the circle* indicate the means of these directions for each test night; the *arrows* indicate their mean vectors. (After BAKER and MATHER 1982)

Noctuids. Experiments testing for magnetic orientation used two nocturnal species of noctuids, *Noctua pronuba* and *Agrotis exclamationis* (BAKER and MATHER 1982; BAKER 1987b). For each test, 40 individuals were put in a cross-shaped cage mounted outdoors and surrounded by a pair of HELMHOLTZ coils which could be used to reverse the horizontal component of the geomagnetic field. After a 30-min period of adaptation, magnetic North was reversed in 10-min intervals. The moths were counted each time before the field was changed. The authors recorded the differences in distribution from one count to the next, and used them to compute a net direction of movement. From these directions, a mean vector for both the natural geomagnetic field and the reversed field periods was calculated.

In the local geomagnetic field, *Noctua* was found to show southerly tendencies (Fig. 4.21), while *Agrotis* predominantly moved in directions that were close to the azimuth of the moon. These tendencies are assumed to be associated with directions of migration. It is unclear, however, whether they represent general tendencies or whether they are just individual tendencies of the moths that are caught in the type of light traps used. Reversing the horizontal component resulted in a corresponding reversal of the directional tendencies in both species (Fig. 4.21). This indicates that the moths orient their movements by the magnetic field and that their preferred directions represent magnetic compass courses. There was no obvious difference between overcast and clear nights in the outcome of the experiments, i.e. any other possible cues affected the behavior far less than the magnetic field.

4.4 The Use of the Magnetic Field for Direction Finding

So far, if strict standards are applied, magnetic compass orientation has been demonstrated in 43 species of animals only. They range from 1 marine snail and 12 arthropods to 30 vertebrates from all major groups, 19 of them birds. Hence, although the number of species is still limited, their systematic variety suggests that the magnetic compass is a rather widespread mechanism among animals. Many more species might be shown to use a magnetic compass if it were possible to find behaviors that produce reliable directional tendencies in the laboratory where the ambient magnetic field can be manipulated in a controlled way.

4.4.1 Different Types of Mechanisms

The functional properties of the magnetic compass have been analyzed in very few species only. In birds, a functional window became evident. Orientation was only possible when total intensity did not differ more than about 25% from the ambient intensity; yet when the birds were exposed to higher or lower intensities for an interval of about 3 days, they adjusted their magnetic compass and were able to orient in the new intensity (W. Wiltschko 1978). Systematic tests with other species concerning this question are almost completely lacking. Descriptions of the test fields used suggest that at least the compass mechanism of insects might have a wider functional range. The flour beetle *Tenebrio* was oriented at intensities of only 16 000 nT (Arendse 1978), and bees were oriented in a field of 550 000 nT, roughly ten times the local intensity (Collett and Baron 1994).

The functional mode of the magnetic compass has been analyzed in a few more species. Two types of mechanisms have been described:

1. an *inclination compass*, based on the inclination of the field lines and thus distinguishing between *equatorward* and *poleward*, and
2. a *polarity compass*, based on the polarity of the magnetic vector, which corresponds to our technical compass and leads to a distinction between *magnetic North* and *South*.

The critical test involves inversion of the vertical component of the magnetic field, which alters the course of the field lines without changing polarity. Animals using an inclination compass reverse their directional preferences, whereas animals using a polarity compass are unaffected. The behavior in a horizontal magnetic field is also revealing, as the use of an inclination compass leads to bimodality, often resulting in disorientation, while a polarity compass is not affected.

The inclination compass, which was first described for European Robins, seems to be very widespread among birds. It has been found in all species studied so far from this point of view (cf. Table 4.2), but it is not common to all animals. Table 4.8 shows that two types of magnetic compasses are found among vertebrates (see Sect. 9.4.2). The only invertebrate species whose magnetic compass has already been analyzed for its functional mode is the Flour Beetle *Tenebrio*; these beetles were unimodally oriented in a horizontal field, which indicates a polarity compass (Arendse 1978). On the whole, the present evidence is still too scarce to allow a meaningful estimate of how widespread each mechanism might be.

Table 4.8. Two types of magnetic compass mechanisms among vertebrates

Species	Type	Reference
Oncorhynchus nerka (Osteichtyes: Salmonidae)	pol.	Quinn and Brannon (1982)
Notophthalmus viridescens (Amphibia: Salamandridae)	pol., incl.	Phillips (1986b)
Caretta caretta (Reptilia: Cheloniidae)	incl.	Light et al. (1993)
Dermochelys coriacea (Reptilia: Dermochelidae)	incl.	Lohmann and Lohmann (1993)
Columba livia (Aves: Columbidae)	incl.	Walcott and Green (1974)
Aves, Passeriformes: 8 species of 6 families	incl.	cf. Table 4.2
Cryptomys sp. (Mammalia: Bathyergidae)	pol.	Marhold et al. (1991)
Homo sapiens (Mammalia: Hominidae)	incl.?	Baker (1989a)

Type: pol., compass based on the polarity of the magnetic field; incl., compass based on the inclination of the magnetic field.

The most puzzling case, however, is that of the newt *Nothophthalmus*, where two different mechanisms are indicated in the same species. *Notophthalmus* responded to the inversion of the vertical component when orientation towards the shore was involved, whereas it seemed to ignore the same manipulation when it oriented towards its place of capture, i.e. it seems to use an inclination compass for shoreward orientation and a polarity compass for homing (Phillips 1986b; cf. Fig. 4.11). These findings are difficult to explain. Neither shoreward orientation nor homing involves innate courses. The set direction of shoreward orientation is an acquired course based on experiencing the long-term directional relationship between water and land. The course towards the place of capture is determined by a navigational process, which is based on information also acquired by experience. The two processes do not appear to be so fundamentally different that two separate pathways for magnetoreception for compass information are plausible. Birds, for example, seem to use an inclination compass for migration and homing. However, in the case of homing, the navigational process may also involve magnetic parameters (see Sect. 6.1), so that the response to the inversion of the vertical component may reflect an effect on determining the home direction rather than on the compass itself. Recent experiments by Phillips and Borland (1994) corroborate that important differences exist between shoreward and homeward orientation (see Sect. 9.2.3.2). Further experiments will have to solve this question.

4.4.2 Biological Significance of Magnetic Compass Use

In theory, animals should be able to use directional information from the magnetic field for any kind of spatial orientation. The findings reported in this chapter show that the magnetic compass is indeed used in a wide variety of behaviors and for various purposes. Location of courses and control of the actual directions of movements with the help of the magnetic compass are intrinsically coupled with the function of the magnetic field to provide a reference system for directional information. This information may be innate or acquired and memorized on a short- or long-term basis, depending on the behavioral situation.

4.4.2.1 Courses of Various Origin in Different Behaviors

According to their origin, the set directions located with the magnetic compass (cf. Sect. 2.2.1) may be roughly grouped into the following categories: (1) innate courses, (2) acquired courses established by direct factors indicating the goal, (3) acquired courses derived from the general lay of the land in the home area, and (4) courses determined by navigational processes.

Innate courses. Magnetic orientation was first discovered in connection with the orientation of migrating birds, and it continues to be discussed as a most prominent cue when large-scale migrations are considered. In particular, when inexperienced animals must reach a specific, unknown goal area in a distant region of the world, they must base their orientation on innate information which must be (1) fairly simple in the sense that it does not require too many specific details, and (2) sufficiently reliable to lead the animals safely to their goal. Where stable geographic relationships exist between the starting area and the goal area, selection appears to favor the development of innate programs that fix a specific route by controlling direction and distance (cf. SCAPINI 1988). These programs may differ between populations of the same species, being finely tuned to the animals' specific needs (see Sect. 5.3.2.5). The magnetic field provides a dependable, omnipresent reference system for genetically encoded compass courses, which allow animals to reach their goal area without prior experience. Bird migration, in particular the first flight of young birds to their winter quarters, is an excellent example for extended migrations controlled in this way (see Sect. 5.2.2). The migratory routes of eels and salmons might be determined in a similar way (see Sect. 5.3). For the regular migrations of butterflies such as the Monarch, analogous mechanisms have been discussed.

When such genetically encoded information is passed on from generation to generation, long-term variations in the magnetic field are no longer negligible. In view of innate courses, the most important aspect of secular variations are changes in magnetic declination (cf. Sect. 1.1.3.2), as they tend to shift the respective directions with respect to geographic North. Yet the rate of change is rather slow, and the information on the innate courses is not overly precise so that deviations from the original direction may be compensated by the normal amount of scatter and genetic variability. This variability, on the other hand, also provides a means of modifying innate courses by selection when animals shift their goal area in response to long-term environmental trends (e.g. BERTHOLD et al. 1992).

The magnetic compass mechanisms also appear to be flexible enough to adapt to gradual changes in total intensity in the course of secular variations (see Sect. 1.1.3). Reversals of the geomagnetic field represent more dramatic changes. The inclination compass of birds, newts, and marine turtles works after the reversal just as it did before, since only the polarity of the field lines is reversed, while their general course remains largely unchanged. The situation during the process of reversal itself is unclear, however.

Acquired Courses. Acquired courses of different origin have been described for a variety of behaviors. A situation of particular interest arises when animals have to find a specific habitat, but the location of this habitat cannot be predicted beforehand. In this situation, animals cannot use innate directional information, but must rely on local cues. The orientation of young marine turtles when attempting to reach the open ocean is such a case: the hatchlings find the sea with the help of light cues, and continue to head in the same direction with the help of magnetic cues when they are in the water (LOHMANN and LOHMANN 1994a). Here, local cues indicating the desired habitat set a course which becomes a set direction for magnetic compass orientation and thus can be followed up when the direct factors are no longer available. For the cockchafer beetle *Melolontha*, a similar system has been discussed. When the young *imagines* emerge, they head for the highest point in the skyline in search for a suitable forest to feed on leaves (cf. SCHNEIDER 1967). The respective direction is memorized as a magnetic course; the mature adults fly the reverse course to return to their natal area to lay their eggs. Terrestrial amphibians might also use the magnetic field as reference for memorizing the direction when they leave their natal ponds, thus establishing their individual axis of migration between feeding sites and breeding ponds.

Very often, courses are derived from the general lay of the land in the animals' home range. This allows animals to move fast and efficiently within the home area on remembered routes, minimizing energy expenditure, danger of predation, etc. In the laboratory, this stategy is reflected by courses derived from long-term stable situations in the animals' living quarters. The orientation along the axis land-water, used to demonstrate the magnetic compass in newts, provides a good example for such set directions. They are stable to some extent, but can be easily modified by experience when the animals move or are moved into a new environment. The same is true for the orientation along the axis land-sea in littoral amphipods. Here, the animals are additionally faced with frequent changes in the sandy shoreline caused by wave action, etc. This requires great flexibility and readaptation of the course to the new situation (cf. SCAPINI 1988). Interestingly, in some species of amphipods, innate directional information already allows a rough, adequate response before any learning processes have taken place. The directional preferences of the beetle *Tenebrio* and the amphipod *Orchestia*, and any tendencies induced by directional training are likewise based on stable directional relationships in the animals' home quarters.

Homing. In the above-mentioned cases, the set directions are fixed or at least constant over a considerable period of time. Homing, in contrast, means variable courses depending on the animal's current position; these courses are valid only until the animal has reached 'home'. The magnetic compass in homing is less well documented because of difficulties with the test design. In the termite *Trinervitermes*, the entire homing process could be studied in the laboratory; in vertebrates, however, homing usually involves outdoor experiments where magnetic conditions cannot be altered in a controlled way so easily. Hence, in view of problems to demonstrate magnetic compass orientation directly, indirect evidence, such as deterioration of orientation induced by magnetic treatments, becomes an important indication. Route reversal, i.e. navigation based on directional information collected during the outward journey, offers a fairly easy opportunity for experimental interference, but here, too, the normally available magnetic information is mostly disrupted rather than replaced by different information. When magnetic manipulations during displacement resulted in disorientation, although all factors (including magnetic ones) were available at the release site, most authors concluded that a magnetic compass was used in route reversal. In this way, magnetic orientation is indicated in a number of species not included in Table 4.4, such as box turtles, woodmice, white-footed mice and horses (cf. Sect. 4.2). Route reversal might be a fairly common strategy among animals.

Other Compass Responses. Social insects use the magnetic compass to orient their building activities. The magnetic field determines the axis of construction, and thus helps to coordinate the work of hundreds of individuals (see von FRISCH 1974). In the meridional mounds of *Amitermes*, the direction is assumed to be innate, with differences between populations to suit local climatic conditions (cf. JACKLYN 1992b). In Honeybees, the origin is not entirely clear. In view of fact that a swarming queen is accompanied by a group of workers from her former nest, an acquired direction based on tradition seems to be most likely.

Two cases have been described where the magnetic compass is only used to decide between the two ends of an axis or between a number of choices provided by other cues, namely, when bees orient their dances with the help of the e-vector or harvester termites select one of several odorous tracks. This is a curious phenomenon, because here the animals use magnetic information, but are apparently not able to orient the respective behavior by information from the magnetic field alone.

The advantages of a magnetic compass in the behaviors mentioned so far are evident. In other cases, however, the biological context and significance of the observed behavior are not yet understood. This applies to spontaneous directional tendencies like those of Leopard Sharks and mole rats, as well as for the orientation of the marine snail *Tritonia* which changes counterclockwise in the course of lunar months. These tendencies provided a useful basis for experiments, indicating the use of a magnetic compass, but the ecological interpretation of these observations is unclear.

4.4.2.2 *One Compass for All Tasks?*

The variety of orientation tasks which involve magnetic orientation leads to the question whether the magnetic compass in a given species is used for all kinds of behavior requiring directional information. The findings on birds provide excellent examples for the use of the magnetic compass in different behavioral situations (cf. Fig. 2.3). During migration, the magnetic compass is used to locate an innate course, the migratory direction. In homing, it is used to locate the various home courses established by navigational processes. Young inexperienced pigeons, on the other hand, use the magnetic field also as a reference for recording the net direction of the outward journey in order to determine the home course from this information; in this case, the magnetic compass is part of the navigational process itself.

Both migration and homing involve orientation over considerable distances. It is not known to what extent birds might also use the magnetic compass for orientation on a smaller scale within their home range, since this question largely escapes experimental study. Scrub Jays, *Aphelocoma coerulescens* (Passeriformes: Corvidae), were found to use the sun compass in caching and recovery of seeds (W. Wiltschko and Balda 1989); further studies suggest that the magnetic compass might also be involved (Balda and Wiltschko 1991). So, it would not be surprising if birds also oriented their everyday flights within the home range, between nest, feeding sites, etc. with the help of the magnetic compass, at least on overcast days (see Sect. 5.2.1).

With respect to various possible uses of the magnetic compass, it is unfortunate that studies on the various behaviors carried out with birds involve different species, i.e. migration was studied in small passerines and homing in pigeons. The same is true for termites, where a magnetic compass was found to be used in mound building by *Amitermes* and in homing by *Trinervitermes*. Only very few species have been studied with regard to more than one behavior so far. One of them is the newt *Notophthalmus viridescens* whose shoreward orientation and homing were studied (cf. Sect. 4.2.2.1). A second species is the box turtle *Terrapene carolina*; here, the use of a magnetic compass is suggested in locating training directions as well as in recording the direction of displacement (Mathis and Moore 1988; cf. Sect. 4.2.3.1).

One might tend to assume that animals capable of magnetic compass orientation make use of it whenever directional orientation is required. Finding two apparently different mechanisms involved in shoreward orientation and in homing of the newt *Notophthalmus* (Phillips 1986b) indicates that this is not necessarily true. Different behaviors might require different mechanisms. An interesting case is also provided by Honeybees (cf. Sect. 4.3.3.2). Bees were shown to use the magnetic field to orient their building activities (DeJong 1982), and possibly also when leaving their hive (Schmitt and Esch 1993) and when facing familiar landmarks near the goal (Collett and Baron 1994), but there is no evidence that they use magnetic information to orient their foraging flights. When bees leave the hive, it is unclear whether the set direction was the direction of exit or the direction towards the food source, as both directions coincided in that study. Assuming the former leads to the conclusion that, in bees, the use of the magnetic compass is possibly restricted to orientation tasks associated with long-term stable situations rather than in connection with the frequently changing directions to food sources. The reasons for this are completely unknown, as the findings available so far are still insufficient to assess possible constraints of magnetic compass orientation.

4.5 Summary

Magnetic compass orientation means that animals use the magnetic vector for direction finding. A magnetic compass, first described for birds, has meanwhile been demonstrated in all major groups of vertebrates, ranging from elasmobranch fish to mammals, including man. Among invertebrates, a magnetic compass has been found so far in 1 marine nudibranch snail, and in 12 species of arthropods, which include crustaceans of the orders Decapoda, Amphipoda, and Isopoda and insects such as termites, Hymenoptera, beetles, flies, and moths.

The most important differences between the magnetic compass of birds and our technical compass are:

1. The birds' magnetic compass is closely tuned to the total intensity of the local magnetic field. Its functional range can be adapted to new intensities by continuously exposing the birds to these new intensities for a few days.
2. The birds' magnetic compass functions as an inclination compass, based on the inclination of the field lines without making use of polarity. Hence, birds distinguish between *poleward* and *equatorward* instead of between north and south.

Little is known about the functional characteristics of the magnetic compass in other animals. An inclination compass has also been demonstrated for marine turtles, whereas data on salmon fry and mole rats indicate that these animals use the polarity of the magnetic field. The magnetic compass of the flour beetle *Tenebrio* also seems to be a polarity compass. This suggests that different animals use different types of mechanisms. For the newt *Notophthalmus viridescens*, an inclination compass as well as a compass based on polarity have been described, which are used in different behavioral situations. Evidence is still insufficient to assess how widespread the two mechanisms may be.

Magnetic compass orientation is involved in a wide range of behaviors, where it controls the direction of movement and serves as reference for genetically encoded and memorized directional information. Consequently, the set directions are of various origin: they range from innate courses associated with migration, e.g. in birds, to acquired courses originally established by other cues, e.g. in marine turtles, or those derived from the general lay of the land in the animals' home range or living quarters, e.g. in many amphibians and amphipods. The magnetic compass seems to be also involved in homing, even if in this context direct evidence is rather scarce, and in the building activities of social insects like termites and bees. In other cases, the biological significance of spontaneous tendencies which obviously depend on the direction of the magnetic field is still unclear. Taken together, the evidence available so far indicates that animals may rely on the magnetic compass in a variety of behaviors where directional information is useful, but many questions are still open.

Chapter 5

The Magnetic Compass as a Component of a Multifactorial System

Chapter 4 described experiments that demonstrate magnetic compass orientation in various animal species. Almost all of these studies, however, were laboratory tests performed in the absence of other orientation cues that would be available to the animals in nature. In order to truly understand the role of the magnetic compass, it must not be considered by itself, but in connection with those other factors. Nature provides a multitude of cues, and animals take ample advantage of them. Hardly any behavior analyzed so far has been found to be controlled by one single factor alone.

The present chapter is devoted to interactions of the magnetic compass with other orientation cues. As examples, we selected cases which are fairly well analyzed, like homing and migration in birds, migration in eels and salmons, homing in toads, and orientation of littoral amphipods at the borderline land-sea. At the same time, these examples cover a wide spectrum of behaviors and illustrate the role of the magnetic compass in different orientation tasks and in a variety of environmental situations. They comprise orientation during daytime and at night, the distances involved range from a few meters to thousands of kilometers, they include singular migrations as well as regular, annual migrations, and homing to a specific goal as well as return to a long, narrow zone.

5.1 Orientation Based on Various Factors of Different Nature

Oriented behavior is a highly complex phenomenon. Internal mechanisms like endogenous rhythms and the physiological situation control motivation, and thus determine when a particular behavior occurs. Specific external factors may act as signals for the *onset* or *termination* of movements. They may be releasing factors and triggers, or they may be the desired conditions themselves, e.g. when 'home' is reached after displacement. Here, we do not deal with the general conditions under which orientation takes place, but will focus on the *control of direction*. This, too, involves a multitude of factors.

5.1.1 Different Types of Orientation Cues

For the control of direction, interactions between the magnetic compass and other orientation cues may take place at various levels. They involve the location of courses and the control of the actual direction of movement as well as the establishing and changing of the respective set directions (cf. Fig. 2.3). According to their function, these orientation factors may be classified into different categories:

1. Two alternative compass mechanisms have been described which provide animals with essentially the same type of directional information as the magnetic compass: the *sun compass* and the *star compass*, both based on celestial cues.
2. Directional information from the magnetic field, the sun or the stars can be transferred to factors that are temporally and spatially constant to some extent, like landscape feature, etc. These factors may then temporarily function as secondary compass cues providing directional information.
3. Because orientation is aimed at reaching a location where certain conditions prevail, local cues signaling these conditions (cf. Fig. 2.3) may affect the direction of movements, e.g. at sea shores, animals move away from the high horizon line to approach the water.
4. Factors of a different nature may be involved in the processes of determining the specific courses for compass orientation. Such factors involved in the navigational 'map' indicate positions in homing (see Sect. 6.1) or they may, by being spontaneously attractive, set courses which are then memorized as magnetic courses (cf. Sect. 4.2.3.2).
5. Factors may affect directional orientation by acting as signals or triggers. When the animal's behavior is controlled by a migration program, such a factor may elicit preprogrammed changes in direction and initiate the next step in the program.

The above list includes only external factors affecting directions. Sometimes, internal factors also contribute to the decision as to which specific direction is pursued. Endogenous time programs may control the specific headings during extended migrations (e.g. Gwinner and Wiltschko 1978), or physiological conditions might determine the general strategy in a given situation. For example, there are indications that migratory birds with a large amount of body fat (i.e. energy stores filled) cross high mountain ridges in the course of their route, while lean birds tend to change direction and detour (Bruderer and Jenni 1988). Similar findings were reported from a coastal station: when tested for magnetic orientation, only birds with a high fat score oriented in their migratory direction, whereas lean birds did not (Sandberg 1994). In nature, different kinds of internal and external factors interact to produce the behavioral output, the representations of which scientists attempt to analyze in the laboratory.

5.1.2 Celestial Compass Mechanisms

Most animals have more than one compass at their disposal. The sun compass and the star compass provide two alternative sources of directional information based on celestial cues. Both mechanisms are based on variable cues. The rotation of the earth causes the apparent movement of the sun across the sky in the course of the day and the corresponding movements of the stars across the nocturnal sky; the earth's movement around the sun leads to seasonal changes in the sun's arc and in the appearance of the night sky. Furthermore, the specific position of sun and stars depends on geographic latitude.

Animals utilizing celestial cues for direction finding must have the means to cope with this variability. Different strategies have been developed for the sun and for the stars. Animals using the sun compass compensate for the sun's apparent movements with the help of their internal clock. The classic method demonstrating sun compass orientation is as follows: in *clock-shift experiments*, the internal clock is reset by subjecting animals to an artificial light regime with the light/dark period phase-shifted with respect to the natural one. As a result, the animals misjudge the time of day, and when they are subsequently tested, they show characteristic deflections from untreated controls (PARDI and GRASSI 1955; SCHMIDT-KOENIG 1958; see Fig. 5.1)[1]. Animals using the star compass, in contrast, appear to derive directions from star patterns similarly to people who are able to find North from the constellation of the Big Dipper, *Ursa major*, regardless of its current position (EMLEN 1967a,b). Fortunately, stars in a planetarium can be substituted for the natural sky, so that for demonstrating star compass orientation, the northern stars are projected on the southern side of the planetarium dome. For many purposes, the complex natural sky could also be replaced by a simple pattern of only 16 artificial 'stars' (e.g. W. WILTSCHKO and WILTSCHKO 1976).

The ontogeny of the celestial compass mechanisms has been systematically studied only in a limited number of species, namely, bees, several other arthropods, homing pigeons and night-migrating passerines. The mechanisms compensating for the sun's movement and the ability to interpret stellar configurations appear to be based on experience. The respective learning processes take place early in the animals' life. Exceptions to this are some spiders and littoral amphipods that possess innate mechanisms of compensation for the sun's movements (cf. von FRISCH 1968).

The sun compass was first described by KRAMER (1950) for birds and by VON FRISCH (1950) for bees. Meanwhile, it has been demonstrated in numerous species from all major groups of vertebrates and arthropods and thus appears to be a widespread mechanism among day-active animals. It is used in various behavioral situations, and, like the magnetic compass, it is used to locate courses of various origin. The star compass, in contrast, has only been demon-

[1] In the northern hemisphere, where most clock-shift experiments have been performed, these deflections are counterclockwise when the clock is advanced and clockwise when it is delayed. In the southern hemisphere, the animal's response is reversed.

strated in night-migrating birds in connection with the migratory direction; other animals have not been systematically studied.

5.2 Homing and Migration of Birds

Bird orientation has been studied in two behavioral situations, homing and migration, which are fundamentally different in several aspects (cf. Fig. 4.1). Homing means returning to a familiar site, to a location where the bird has been before. Homing takes place within the bird's home range or at least within the home region, and it involves flights in directions which vary according to the bird's current position with respect to 'home'. Migration, in contrast, means a periodic transfer of the bird's home range to a distant region of the world. The distances involved are much greater; it normally takes several weeks to months to fly there. The migratory direction is more or less fixed, determined by an endogenous program (BERTHOLD 1988), with most migrants moving equatorwards in autumn and returning to their breeding area in spring.

As already mentioned, the magnetic compass is used for direction finding in homing as well as in migration. However, many more factors are involved; they interact with the magnetic compass in various ways, complementing, but also replacing magnetic information (see R. WILTSCHKO and WILTSCHKO 1994a). These manifold interrelations between factors of various nature in bird orientation will be described here, with emphasis on the role of the magnetic compass and its relative importance compared with other cues.

5.2.1 Interaction of Orientation Mechanisms in Homing

Homing involves determining the direction to a distant goal. In homing, the two step-nature of indirect orientation processes (cf. Sect. 2.2.1) was first recognized and described by KRAMER (1961) in his 'map and compass' model: In the first step, the 'map' step, the bird determines its position relative to home and establishes its home direction *as a compass course*; in the second step, the 'compass' step, a compass is used to locate this course. The magnetic compass, involved in both steps, interacts with the sun compass and with the factors constituting the 'map'.

5.2.1.1 Interaction of Sun Compass and Magnetic Compass

In fair weather, birds have the magnetic compass and the sun compass at their disposal. Orientation under sun and under overcast is of similar accuracy (Fig. 5.1), indicating that the sun is not necessary for good orientation. Clock-shift

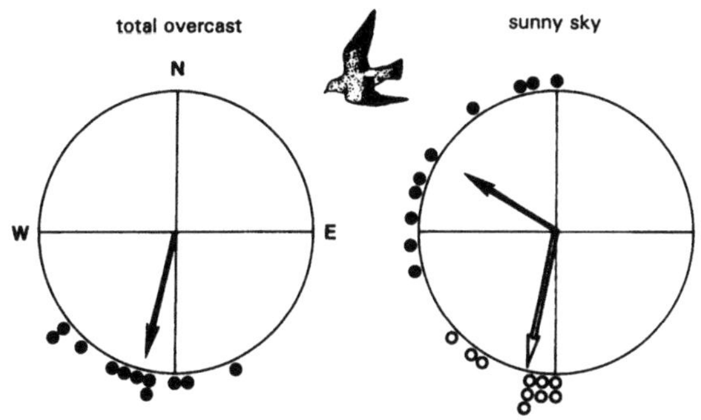

Fig. 5.1. Orientation of homing pigeons under solid overcast and under sunny sky at the same site 40 km north of the home loft, with the home direction 192 °. The *symbols at the periphery of the circle* indicate vanishing bearings of individual pigeons; the *arrows* give the mean vectors proportional to the radius of the circle. In the *right diagram* bearings of pigeons whose internal clock was shifted 6 h slow (*solid symbols*) show a characteristic deflection from untreated controls (*open symbols*), demonstrating the use of the sun compass

experiments, however, demonstrate its dominant role when available: pigeons show the typical deviation (Fig. 5.1, right), although their magnetic compass could have given them correct information. The relationship between sun compass and magnetic compass thus may be summarized in the following way: the sun compass is the preferred system, but it can be replaced without loss when the sun is not available. The mechanisms of directional orientation are thus partly redundant (cf. R. WILTSCHKO and WILTSCHKO 1994a).

The findings mentioned above seem to suggest that the magnetic compass is just a subsidiary mechanism for overcast days. This is not true, however. The magnetic compass plays a very important role during the development of the orientational system, and continues to exert some control on directional orientation even in adult birds.

The Role of the Magnetic Compass During Ontogeny. Experiments with very young homing pigeons suggest that during an early period when they just begin to fly, the sun compass is not yet available to them. During that period, they do not respond to shifting of their internal clock (R. WILTSCHKO and WILTSCHKO 1981); at the same time, their orientation is disrupted with magnets even when the sun is visible (KEETON 1971). This latter finding suggests that young birds rely on the magnetic compass before the sun compass is established (see also W. WILTSCHKO et al. 1987a).

A series of experiments analyzing the development of the sun compass indicates learning processes which take place as soon as the young pigeons gain some flying experience. Normally, the sun compass is established during the third month after hatching, with considerable variation between individuals. Good opportunities to observe the sun proved to be important. Pigeons allowed to see the sun only in the afternoon did not rely on the sun compass in the morning; instead they used the magnetic compass (R. WILTSCHKO et al. 1981). Apparently, young pigeons must observe major parts of the sun's arc at different

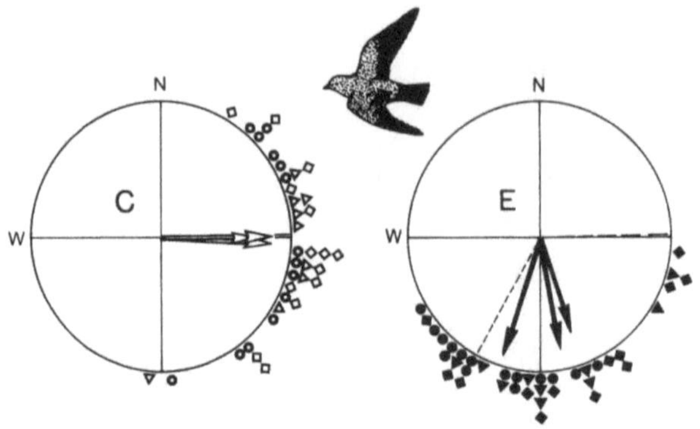

Fig. 5.2. Orientation of young pigeons. During the time when they establish their sun compass, the birds had been kept for 10 sunny days in an altered magnetic field with magnetic North deflected 120 ° clockwise to ESE; immediately after this, they were released under sun. The pretreatment induced a deflection of their bearings. *Solid symbols* mark the bearings of experimental pigeons kept in the shifted magnetic field; *open symbols* give the bearings of control birds kept in the natural geomagnetic field. The data of three releases are indicated by different symbols, with the mean vectors given separately. The home direction 88 ° is marked by a *dashed radius*, another *dashed radius* indicates the expected direction of the experimentals based on the deflection of magnetic North. (After R. WILTSCHKO and WILTSCHKO 1990)

times of the day in order to associate sun azimuth, time and geographic direction (cf. R. WILTSCHKO 1983).

This led to the question of how pigeons estimate the changes in sun azimuth. Experiments exposing groups of young pigeons to the sun in an artificial magnetic field indicate that the magnetic compass provides the reference system. A small loft was equipped with a large set of coils which turned magnetic North 120 ° clockwise to ESE; the inhabitants were able to observe the sun from an aviary on the roof in an abnormal relationship to the magnetic field. When young birds raised in this loft were released under sun, they deviated from untreated controls in a way which roughly corresponded to the deflection of the magnetic field they had experienced. The same was true for young birds kept in the loft for about 10 sunny days during the period when the sun compass develops (Fig. 5.2). This indicates that they had developed a sun compass according to the experimental situation. It changed to normal, however, after several flights under sun (W. WILTSCHKO et al. 1983; R. WILTSCHKO and WILTSCHKO 1990). Old, experienced pigeons, on the other hand, were not affected by a stay in the altered magnetic field. This difference between young and old birds indicates that the development of the sun compass involves a sensitive period. During this sensitive period, young pigeons spontaneously pay particular attention to the directional relationship between sun and magnetic field (R. WILTSCHKO and WILTSCHKO 1990).

The Role of the Magnetic Compass in Experienced Pigeons. The magnetic compass continues to play a role when the sun compass has been established. An analysis of more than 100 clock-shift experiments revealed that the observed deflection[2] is often considerably smaller than expected if the sun compass alone was used (R. WILTSCHKO et al. 1994). First tests indicate that magnets might increase the deflections almost to the expected size (R. WILTSCHKO and WILTSCHKO 1994b). These findings suggest that information from the magnetic compass is also involved under sun. Sun compass and magnetic compass appear to control the direction of flight together.

In addition, the magnetic compass might continue to control the sun compass and adjust it to changing conditions. The sun compass can be dramatically altered by subjecting birds for a longer period of time to a shifted photoperiod. When such pigeons are allowed to fly freely during the overlap time between natural and artificial day, their sun compass adjusts to the experimental situation (W. WILTSCHKO et al. 1984). This means that an entirely new relationship between sun azimuth, time, and geographic direction has to be formed.

The possibility to recalibrate the sun compass is of great importance regarding two aspects: (1) it may provide a means to adapt the sun compass to seasonal changes in the sun's arc, which are no longer negligible to birds living at lower latitudes. (2) It may allow long distance migrants to establish a new sun compass adapted to the local situation when they have reached their winter home. For example, northern migrants wintering in the southern hemisphere need a sun compass to compensate for clockwise movements of the sun in their breeding areas, and one to compensate for counterclockwise movements in their wintering areas (R. WILTSCHKO and WILTSCHKO 1990).

It has not yet been analyzed how the birds proceed when they adapt their sun compass to altered conditions. The process is equivalent to establishing a new sun compass which suggests similar mechanisms, i.e. the magnetic compass may again provide the directional reference system.

5.2.1.2 Mechanisms Determining the Home Course

Displaced birds may use two basic strategies in order to determine their home course: they may rely on information obtained during the outward journey, or they may use local information obtained at the starting point of the return trip (cf. SCHMIDT-KOENIG 1970; W. WILTSCHKO and WILTSCHKO 1982). Experimental evidence suggests that both strategies are involved.

Route reversal based on magnetic information has already been described in Section 4.1.4.2. Birds record the direction of the outward journey with their magnetic compass (during active flight or during passive displacement) and reverse this direction to obtain the home course (cf. Fig. 4.8); i.e. information collected en route is processed shortly thereafter. This simple strategy, which does not require any foreknowledge, plays a role only during a very short

[2] Normally, the pigeons' internal clock is shifted by 6 h. At temperate latitudes, depending on the time of day and season, the expected deflection lies in a range between 90 ° and 135 ° (cf. R. WILTSCHKO et al. 1994).

period after the young pigeons begin to fly. Displacement in a distorted magnetic field soon ceases to affect orientation. In old, experienced pigeons, the effect is very small or non-existent (KIEPENHEUER 1978; WALLRAFF et al. 1980; R. WILTSCHKO and WILTSCHKO 1985)[3], indicating that route reversal based on magnetic information is given up in favor of information obtained at the release site.

Orientation by cues of the release site requires an interpretation of the local 'map' information with respect to the known spatial distribution of navigational factors. In contrast to route reversal, this navigational strategy is based on previously acquired knowledge which must be memorized on a long-term basis. The present concept of navigation by site-specific information is described in detail by WALLRAFF (1974) and W. WILTSCHKO and WILTSCHKO (1982, 1987). The birds' knowledge of the distribution of navigational factors, the navigational 'map', is assumed to represent a *directionally oriented mental image* of their spatial pattern (see Fig. 6.1). The nature of these factors is still open; odors, magnetic parameters (see Sect. 6.1), infrasound and gravity are currently discussed (cf. PAPI 1986; R. WILTSCHKO and WILTSCHKO 1994a). The learning processes establishing the 'map', like those establishing the sun compass, most probably take place as soon as the young pigeons obtain some flying experience. In principle, they involve matching local 'map' information with compass information, i.e. by experience, the young pigeons find out how specific factors change in a given direction (W. WILTSCHKO and WILTSCHKO 1987). Here, too, the magnetic compass might provide the reference system. Experimental evidence is not available, however. These learning processes take place mainly during spontaneous flights and thus escape analysis.

5.2.1.3 A Model For Homing

Figure 5.3 summarizes the interaction of the various orientation factors in the navigation system of homing pigeons as we see them today (W. WILTSCHKO et al. 1991). The magnetic compass plays a crucial role during the development of the navigational system, providing a reference system for route reversal and for the processes establishing the more complex learned mechanisms such as the sun compass and the navigational 'map'. This is indicated by the open arrows in Fig. 5.3. In experienced birds, the magnetic compass provides a backup system for situations when the learned mechanisms cannot be used, but it might also take part in locating directions under sun. At the same time, it probably continues to serve as a reference system for seasonal adaptations of the sun compass and for updating the 'map' as experience increases.

[3] This is true only as long as magnetic fields in the physiological range are used. Effects of displacement in extreme fields (e.g. BALDACCINI et al. 1979; BENVENUTI et al. 1982) are discussed in Section 6.1.1.3.

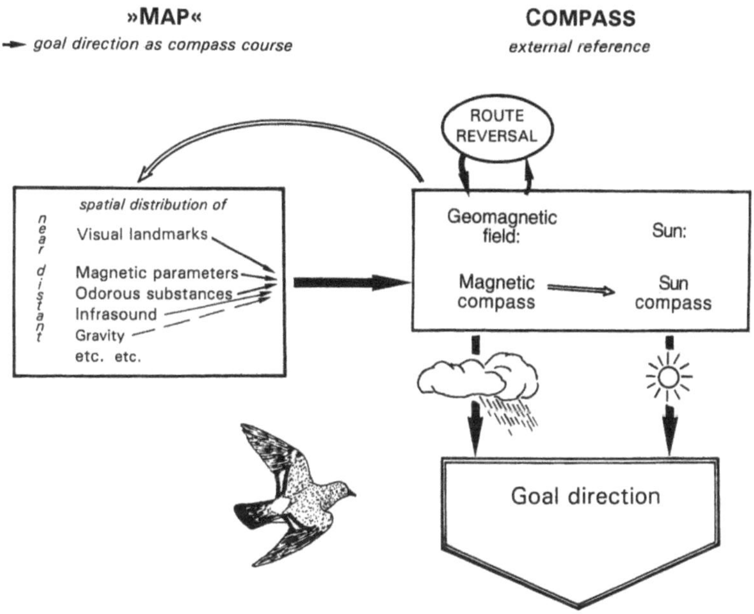

Fig. 5.3. Model summarizing the interaction of various cues in avian navigation and homing. *Open arrows* mark learning processes during the development of the navigational system, *solid arrows* represent the flow of information during the navigational process. For further explanations, see text. (After W. WILTSCHKO et al. 1991)

5.2.2 Interaction of Various Cues in Migratory Orientation

On migration, birds leave their home region in order to move to distant regions of the world. Young birds migrating for the first time must find the wintering area of their species, still unknown to them. Large-scale displacement experiments involving the translocation of thousands of birds perpendicular to their normal migration route (e.g. PERDECK 1958) revealed that the young migrants possess innate information regarding the location of their wintering area. This is given in 'polar coordinates', namely, as direction and distance to be traveled. The distance is determined by an endogenous time program controlling the duration of migration (see BERTHOLD 1988 for review). The directional information is given as an innate compass course, i.e. as an angle to an external reference.

For technical reasons, most research on migratory orientation has focused on passerine species that migrate at night. On clear nights, both the magnetic compass and the star compass provide these birds with directional information. Additionally, as most nocturnal migrants begin their flights at dusk, the setting sun and associated factors, such as horizon glow and polarized light, may be used for direction finding. In several species, more than one compass mechanism has been experimentally demonstrated; the European Robin, *Erithacus rubecula*, and the Blackcap, *Sylvia atricapilla*, have been shown to use all three types of cues (cf. W. Wiltschko and Wiltschko 1991b; Able 1993).

5.2.2.1 Realization of the Starting Course

The first step in migratory orientation requires the transformation of genetically encoded information on the migratory direction into a compass course. The magnetic field may serve as external reference for this process (see Sect. 4.1.3.3), but it is not the only such reference. Experiments with birds that were hand-raised under controlled conditions showed that the migratory direction is also genetically encoded with respect to *celestial rotation* and can be transferred to the star compass (e.g. Emlen 1970a). In several species, information on the migratory direction was found to be represented with respect to both reference systems, i.e. young birds were able to locate their southerly migratory direction as 'equatorward' with their magnetic compass and as 'away from the center of rotation' with their star compass (Fig. 5.4).

Most species examined so far can rely on either system to determine their migratory direction in autumn. There are exceptions, however, which involve birds growing up at higher magnetic latitudes. In Latvia, where the local inclination is 73 °, hand-raised young Great Reed Warblers, *Acrocephalus arundineus*, were oriented only when information from the geomagnetic field and a rotating planetarium sky had been available to them during the premigratory period (Liepa, in Katz et al. 1988). Young Pied Flycatchers, *Ficedula hypoleuca*, of the Latvian population oriented bimodally preferring both ends of the migratory

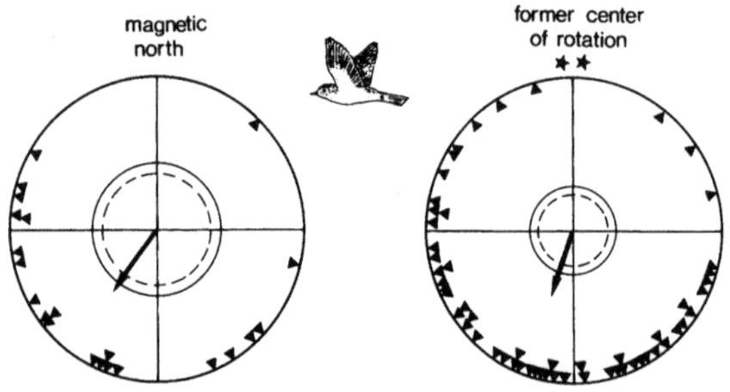

Fig. 5.4. The geomagnetic field and celestial rotation both serve as reference systems for innate information on the migratory direction in Garden Warblers, *Sylvia borin*. *Left* Birds hand-raised and tested in the local geomagnetic field without ever seeing celestial cues. (From W. Wiltschko and Gwinner 1974); *right* birds hand-raised under a rotating artificial sky, tested under the same, now stationary sky in the absence of meaningful magnetic information. (After W. Wiltschko et al. 1987b) – *Symbols at the periphery of the circle* indicate the headings of individual test nights; the *arrows* represent the mean vectors proportional to the radius of the circle. The *two inner circles* are the 5% (*dashed*) and the 1% significance border of the Rayleigh test

axis, when they had to rely on the magnetic field alone. However, when they were exposed to a rotating planetarium sky before migration, they showed unimodal preferences for their autumn migratory direction (WEINDLER et al. 1995). The authors interpret their findings in view of the steep inclination causing possible problems when genetic information has to be transformed into a magnetic course; with the help of celestial rotation, these problems could be overcome.

Even when the magnetic field and celestial rotation alone are sufficient to ensure migratory orientation, they are not really independent, but interact to produce one common course. This was demonstrated by several series of experiments exposing hand-raised birds during their first summer to the natural sky in artificial fields so that magnetic North and rotational North no longer coincided (BINGMAN 1983a; BINGMAN et al. 1985; ABLE and ABLE 1990a). When these birds were later tested without visual cues in the geomagnetic field, they changed their magnetic course. Their new magnetic course corresponded to that which had been their true (geographic) migratory direction during ontogeny (see Fig. 5.5). In Pied Flycatchers, this transfer of directional information was asymmetrical: the magnetic course could be shifted clockwise, but not counterclockwise (Fig. 5.5; PRINZ and WILTSCHKO 1992). ABLE and ABLE (1990b) replaced the natural sky by a simple pattern of artificial 'stars' and obtained corresponding results with Savannah Sparrows, *Passerculus sandwichensis*. Further studies indicate that the daytime sky, too, can alter the magnetic course of autumn migration; here, the pattern of polarized light proved to be crucial (ABLE and ABLE 1990a, 1993b, 1994). Similar findings with adult birds starting on their second or third autumn migration (ABLE and ABLE 1995) indicate that celestial information continues to play an important role when the birds leave their breeding grounds.

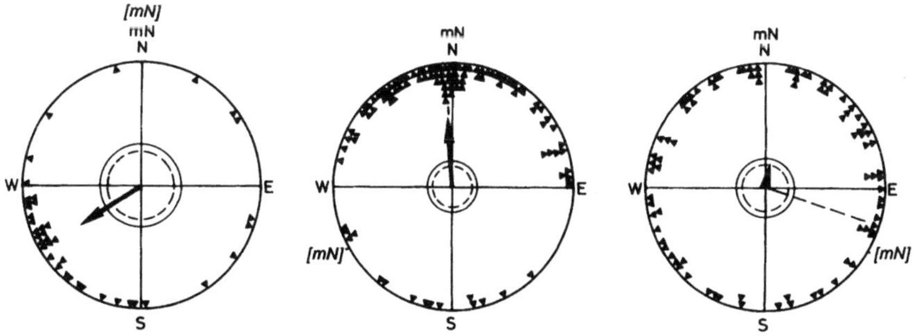

Fig. 5.5. The effect of celestial rotation during ontogeny on the magnetic course during migration: the orientation of handraised Pied Flycatchers, *Ficedula hypoleuca*, tested in the local geomagnetic field in the absence of visual cues. The birds had been exposed to different relationships between magnetic and stellar North during the premigratory period; *[mN]* indicates the former direction of magnetic North. *Left* Controls reared in the local geomagnetic field; *center* birds reared with magnetic North turned to 240 ° WSW; *right:* birds reared with magnetic North turned to 120 ° ESE. The expected directions based on the behavior of the controls and the previous deflection of magnetic North are indicated by a *dashed radius*. *Symbols* as in Fig. 5.4. (PRINZ and WILTSCHKO 1992)

In contrast, an effect of the magnetic field during the premigratory period on the later course with respect to the stars could not be demonstrated. This suggests that the initial calibration of the star compass is independent from the magnetic field (W. Wiltschko et al. 1987b). In summary, when the migratory direction is first established before the onset of migration, innate information with respect to celestial rotation dominates over the corresponding information with respect to the magnetic field (for a summary, see Able 1991b; W. Wiltschko and Wiltschko 1991b). However, the asymmetry observed in flycatchers, which is yet unexplained, indicates that the relationship between celestial rotation and the magnetic field during the premigratory period may not be as simple as it might seem.

The setting sun and associated factors have also been discussed as a reference for innate information on the migratory direction (e.g. Katz 1985; cf. F. Moore 1987). The available data are not entirely clear. Findings by Alerstam and Högstedt (1983), Bingman (1983b) and Able and Able (1990c) suggest that sunset factors obtain directional significance from the magnetic field. Results from Able and Able (1994), on the other hand, indicate that polarized light at sunset and sunrise plays an important role and may affect the course with respect to the magnetic field.

5.2.2.2 Interactions During Migration

The interaction of magnetic field, stars and sunset cues during migration has been examined in a large number of so-called cue-conflict experiments, which have been summarized by Able (1993). The direction of the magnetic field was altered by coil systems, stars were projected on the reverse site of a planetarium dome, and sunset factors were manipulated by changing the position of the sun with a mirror, by changing the e-vector direction with polarizers or by clock-shifting birds.

Interactions of Magnetic Field and Stars. The results on the interaction of stars and magnetic field are largely consistent. When natural stars and an experimentally altered magnetic field gave conflicting information, several species of *Sylvia* warblers and European Robins oriented by the information from the magnetic field. The first responses varied, however. Some species, mainly long distance migrants, followed the magnetic field immediately, whereas others showed a delayed response: they first continued in their previous direction, but changed to the direction according to the experimental field after a few nights of exposition (Fig. 5.6, upper diagrams). In a partly compensated magnetic field not providing meaningful information, the birds oriented by the stars alone. The ones that had been exposed to the altered magnetic field in previous tests, however, now continued to prefer the new, magnetically derived direction (Fig. 5.6, lower diagrams). This suggests that they had recalibrated their star compass according to the experimental magnetic field (W. Wiltschko and Wiltschko 1975a,b; Bingman 1987).

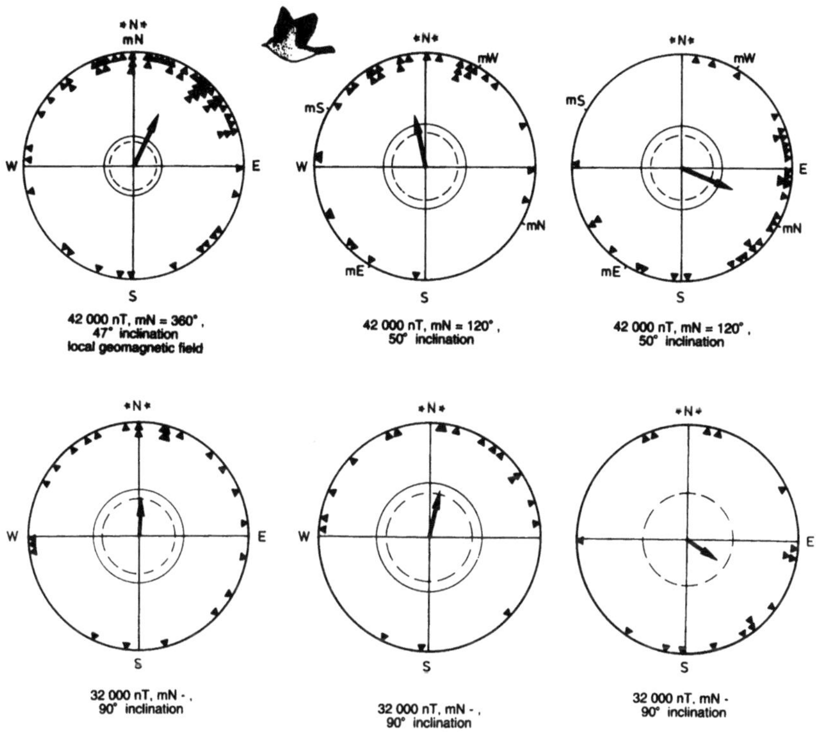

Fig. 5.6. Interactions of magnetic field and stars during migration; orientation of European Robins, *Erithacus rubecula*, under the clear natural sky. One group was tested in the local geomagnetic field (*upper left diagram*), another in a field with magnetic North turned to 120 ° ESE (first two test nights: *upper central diagram*; later test nights: *upper right diagram*). Both groups were also tested in a partially compensated field not providing meaningful information (*lower diagrams*; the data in the respective diagrams correspond to the ones above). *Symbols* as in Fig. 5.4. (After W. Wiltschko and Wiltschko 1975b)

Reverse experiments were performed in a planetarium, where the stars were changed and the magnetic field was kept constant. First tests with Indigo Buntings, *Passerina cyanea*, indicated a dominant role of the stars (Emlen 1967a). However, these birds had been exposed to the test situation for a short time only, so that any delayed response would not have been detected. Later planetarium experiments with Bobolinks, *Dolychonyx oryzivorus*, allowed for this possibility; the birds first changed their directions according to the stars, but eventually they returned to magnetic North (Beason 1987). European Robins ignored a shift of magnetic North in a planetarium as long as the sky looked natural, but responded without delay under an 'unnatural' sky (Lutsyuk and Nazarchuk 1971).

These findings reflect a rather complex interrelationship between the star compass and the magnetic compass during migration. On a long-term basis, the magnetic compass proved to be dominant over the stars and controlled their directional significance. However, the star compass represents, tempo-

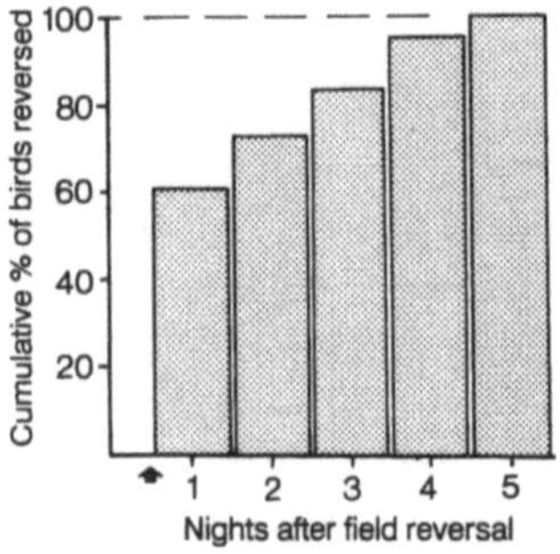

Fig. 5.7. Bobolinks, *Dolichonyx oryzivorus*, tested under a planetarium sky; inverting the vertical component of the ambient magnetic field caused them to reverse their bearings. Individual differences in the response are indicated. (After BEASON 1989a)

rarily at least, a system that allows direction finding independently from the magnetic field. How frequent birds check the stars against the magnetic field seems to vary greatly. Species like the European Robin continued to follow the stars despite conflicting information from the magnetic field for one or two more nights, while *Sylvia* warblers responded immediately to the change in magnetic North. In Bobolinks, BEASON (1987) found considerable differences even between individuals of the same species (Fig. 5.7).

For a star compass, the specific stellar configurations proved irrelevant as long as birds had an opportunity to observe the stars in the presence of meaningful magnetic information. In this way, directional information could also be transferred to a set of artificial 'stars', and these 'stars', which per se had no significance at all, could then be used to locate the migratory direction (W. WILTSCHKO and WILTSCHKO 1976).

Interaction of Cues at Sunset. The role of the setting sun and associated factors like the pattern of polarized light is not entirely clear. Sunset cues enable birds to orient in the absence of magnetic information (e.g. HELBIG 1990, 1991b; PETTERSON et al. 1991). They were first discussed as an independent compass indicating west at the time of sunset; clock-shift experiments, however, suggest that they may be part of the sun compass (e.g. ABLE and CHERRY 1986; HELBIG 1991b).

Table 5.1 lists experiments performed at sunset where sunset cues and magnetic information were in conflict; the results appear rather confusing and do not indicate a clear hierarchy of cues. It was not difficult to alter the birds' orientation by deflecting the view of the setting sun with mirrors. Changing the direction of the e-vector with polarizers resulted in a shift of the birds' behavior; the data of SANDBERG (1988) are the only exception. This seems to suggest that polarized light dominates over all other cues including the magnetic field. Tests which changed magnetic North while the birds could see sunset cues and the natural pattern of polarized light also yielded varying results. Data of Wheatears and European Robins are difficult to interpret.[4] In Dunnocks, infor-

[4] For reasons unknown, the Wheatears did not prefer their migratory direction (SANDBERG et al. 1991a); their orientation, also shown to be affected by magnetic manipulations, is difficult to interpret. In Robins, not all birds were oriented in the seasonally appropriate migratory direction (SANDBERG et al. 1988a,b). A reanalysis of the data revealed that birds with a great amount of body fat were far better oriented than lean birds; these fat birds followed a shift of magnetic North also when the setting sun was visible (SANDBERG 1994).

Table 5.1. Cue conflict experiments performed with night-migrating passerine birds at sunset: magnetic compass versus sunset cues

Species	Dominant cue	Reference
1. Setting sun shifted with mirrors		
Savannah Sparrow, *Passerculus sandwichensis*	Sunset	F. MOORE (1982)
European Robin, *Erithacus rubecula*	Sunset	SANDBERG (1991)
2. e-Vector altered by polarizers		
White-throated Sparrow, *Zonotrichia albicollis*	Polarized light	ABLE (1982, 1989)
Yellow-rumped Warbler, *Dendroica coronata*	Polarized light	F. MOORE and PHILLIPS (1988)
" " " " "	" "	PHILLIPS and MOORE (1992)
European Robin, *Erithacus rubecula*	Magnetic field	SANDBERG (1988)
3. Magnetic North altered by coil systems		
Dunnock, *Prunella modularis*	Magnetic field	BINGMAN and WILTSCHKO (1988)
European Robin, *Erithacus rubecula*	Magnetic field?	SANDBERG et al. (1988a,b)
Wheatear, *Oenanthe oenanthe*	Sunset	SANDBERG et al. (1991a)

mation from the magnetic field was clearly dominant over polarized light cues, as these birds were tested after the disk of the sun was below horizon. The data of BINGMAN and WILTSCHKO (1988) even suggested a similar relationship between magnetic field and the e-vector as it was decribed for magnetic field and stars, i.e. the magnetic field appeared to control the directional significance of polarized light (BINGMAN and WILTSCHKO 1988).

An experiment designed to determine whether manipulations at sunset would have aftereffects, recorded the birds' orientation later at night when the sunset cues were no longer present. After the setting sun had been deflected with mirrors, Savannah Sparrows continued to orient according to the shifted sunset, ignoring contradicting information from polarized light and from the star compass and magnetic compass; F. MOORE (1985) suggested that once the direction was established, it was maintained by other cues, possibly the stars. However, in other experiments, European Robins and *Sylvia* warblers oriented at night with the help of the experimentally altered magnetic field, although they had been exposed to the natural sky from late afternoon onward (W. WILTSCHKO and WILTSCHKO 1975a,b; cf. Fig. 5.6).

Thus, the dominant cue varied between the various experiments. The reasons for these differences are not fully understood. They may reflect species-specific differences, but they may also reflect specific aspects of the test designs (see HELBIG 1990). For example, if the setting sun is indeed part of the sun compass, its dominant role may represent a parallel to the clock-shift experiments (cf. Fig. 5.1). A dominant role of polarized light was indicated when birds were tested under polarizers. Here, however, a cautious interpretation might be advisable. Blackcaps aligned parallel to the axis of polarization and became bimodal, which was significantly different from their orientation to the natural sky (Helbig and Wiltschko 1989). Such bimodal alignments parallel to the e-vector are not uncommon when polarizers are used, suggesting that the responses to the 100% polarized light produced by polarizers may not reflect the true relationship between cues.

5.2.2.3 An Integrated System for Nocturnal Migration

The interrelations between magnetic compass, star compass and sunset factors in the orientation of nocturnal migrants are summarized in Fig. 5.8. During the premigratory period, celestial information dominates over magnetic information, whereas during migration, the magnetic field becomes the most powerful cue. Similar models on migratory orientation, acknowledging a central role of the magnetic compass, have been published by F. Moore (1988) and Able (1991b).

The change in dominance from celestial to magnetic cues as migration proceeds may reflect the respective ecological situations. At higher geographic latitudes, secular variations lead to the largest changes in declination (cf. Sect. 1.1.2.2), which are perhaps no longer negligible when innate courses are con-

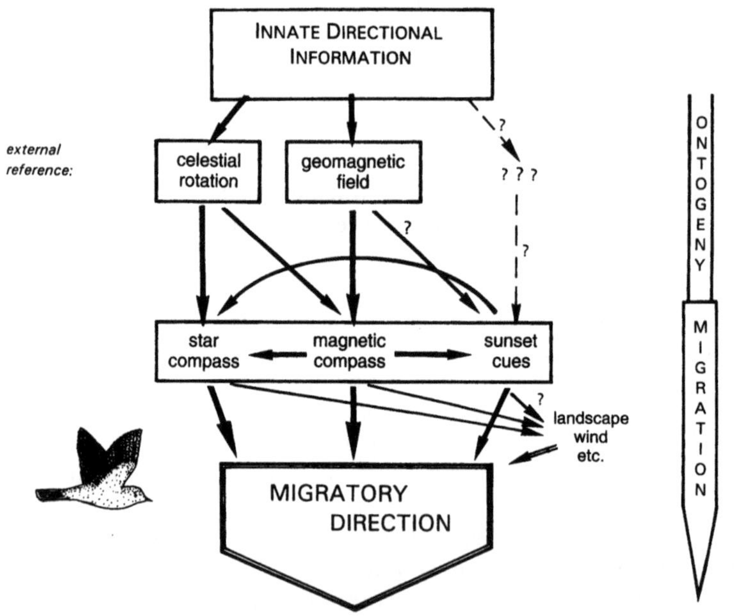

Fig. 5.8. Model summarizing the interactions of various cues in the orientation of night-migrating birds. The *question marks* indicate relations that have not yet been clearly demonstrated; *dashed arrows* indicate a relation that was considered only theoretically. For further explanations, see text. (After W. Wiltschko and Wiltschko 1988)

cerned. Thus, when birds start their migratory flight, celestial rotation provides the most reliable cue for indicating directions. However, as the birds move on, the familiar stars lose altitude and disappear below the horizon, while new ones appear which need to be calibrated. The magnetic field provides a most useful reference for these calibrating processes, as it represents a reliable source of directional information at lower latitudes. – Details of this change in the relative importance of magnetic and celestial cues are not yet available. In nature, experience associated with natural migration, like overcast nights, changes in the appearance of the starry sky, etc. may stimulate this process (see W. WILTSCHKO et al. 1989).

Because of their manifold interactions, magnetic compass, star compass, and sunset cues should be considered as components of an integrated system of migratory orientation. Such an integration of mechanisms stabilizes the system and may help birds to overcome situations where one of the mechanisms is impaired. For example, magnetic cues are not impaired by adverse weather conditions. Celestial cues, on the other hand, may help long distance migrants to master the situation at the magnetic equator where the magnetic compass becomes ambiguous. Sunset cues and stars may be used to distinguish between both ends of the magnetic axis, until the magnetic compass can be used again, now with a 'poleward' course (cf. Fig. 4.6). Planetarium experiments showed indeed that Bobolinks continued to be oriented in southerly directions when the inclination of the magnetic field was gradually changed from +67 ° to – 67 °. Under stars, the birds maintained a constant southerly heading in the horizontal field, whereas they were disoriented in diffuse light (BEASON 1992).

5.2.2.4 Day Migrants

Only very few day migrants have been studied in view of their orientation mechanisms. The Australian Yellow-faced Honeyeater, *Lichenostomus chrysops*, is the only species so far whose orientation has been systematically tested for a magnetic compass, a sun compass and the use of polarized light. A magnetic compass was demonstrated; the honeyeaters also responded to changes in the magnetic field when they could see the natural sky (MUNRO and WILTSCHKO 1993b). Clock-shifting affected the behavior, but the observed deflections did not correspond to the ones expected if a sun compass was used (MUNRO and WILTSCHKO 1993a). Changing the axis of polarization had no effect when magnetic information was available; in a compensated magnetic field, the honeyeaters were oriented as long as either the sun or natural polarized skylight was present (MUNRO and WILTSCHKO 1995). Together, these data suggest that all three cues are involved, with the magnetic compass being the primary cue.

Aside from the honeyeater, few data are available in Starlings, *Sturnus vulgaris*, and Meadow Pipits, *Anthus pratensis*. Both species were well oriented under overcast, and in both species, clock-shift experiments in autumn resulted in disorientation rather than in the predicted deviation (R. WILTSCHKO 1980, 1981; ORTH and WILTSCHKO 1981). In spring, starlings changed their directional tendencies when the sun was deflected with mirrors (KRAMER 1950), and responded to clock-shifting with the deflection indicating the use of the sun compass (R. WILTSCHKO 1980). Thus, data on the relative role of the sun compass and the magnetic compass are not entirely consistent; at least in autumn, the sun does not seem to be of major importance.

Theoretical considerations show that the changes in geographic latitude in the course of migration might pose a severe problem for sun compass orientation. The dramatic changes in the sun's arc would require continuous readjustments of the compensating mechanisms; changes in longitude would require readjustments of the internal clock. Because of this, the sun would not make a good reference for the innate information on the migratory direction, and thus cannot be expected to be a major orientation cue during migration. The few data available are in accordance with this, indicating that the magnetic compass is also a most important mechanism for day migrants.

5.2.2.5 Changes in Migratory Direction

Many migrants do not reach their winter quarters following straight routes, but detour major ecological barriers, which results in curved or bent routes. For example, European migrants may bypass the Alps and the Mediterranean Sea, flying via Iberia or Asia Minor (ZINK 1977; Fig. 5.9, left). Similar phenomena occur in other parts of the world as well. In Australia, Yellow-faced Honeyeaters from Victoria migrate along the Great Dividing Range first on northeasterly, later on northwesterly courses into central Queensland (MUNRO et al. 1993; Fig. 5.9, right). This means that the birds have to change their migratory course after they have reached a certain region. In laboratory tests, the corresponding changes occurred in some species spontaneously at about the time when the free flying birds alter their direction. This indicates innate components and control by the endogenous time program (GWINNER and WILTSCHKO 1978; HELBIG et al. 1989; MUNRO et al. 1993). In nature, additional external triggers might also be involved, as a time program alone might not be accurate enough (see Sect. 6.3.1).

It is assumed that when routes are not straight, innate directional information is available for a sequence of courses which are followed one after the other. The interactions between celestial rotation and the magnetic field in establishing the first course have already been described (cf. 5.2.2.1); little is known on how the later courses are encoded. Two species that change their migratory direction, the night-migrating Garden Warblers, and the day-migrating Yellow-faced Honeyeater, do so in closed rooms when the magnetic field provides the only directing cue (Fig. 5.9). This shows that the magnetic field also serves as a reference for the second direction (GWINNER and WILTSCHKO 1978; MUNRO et al. 1993). Whether the second course is likewise encoded with respect to celestial rotation is unknown.

Similar questions arise with respect to the spring migratory direction. Theoretically, birds could just reverse their autumn course, but ringing recoveries

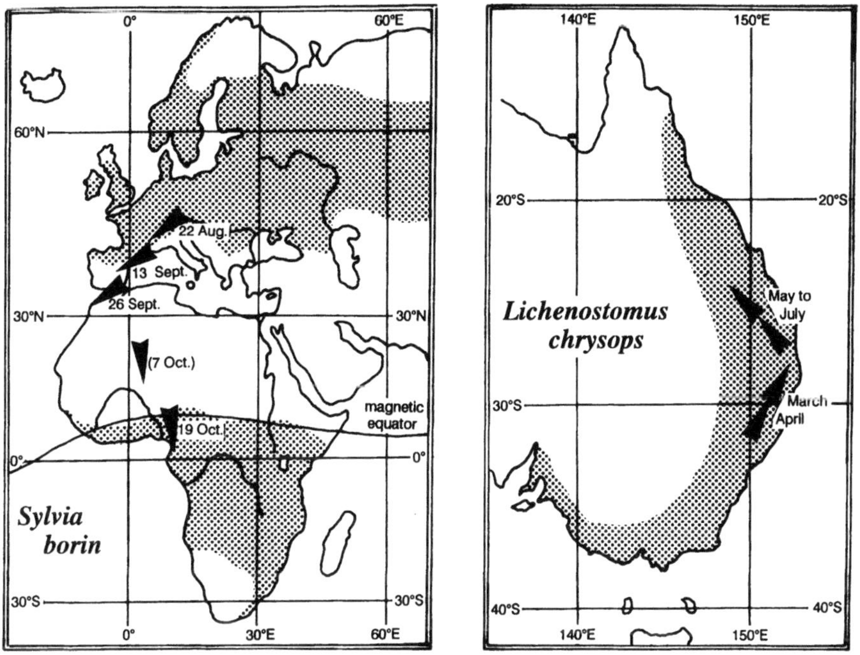

Fig. 5.9. Changes in the migratory direction of Garden Warblers, *Sylvia borin,* migrating from Central Europe to Africa south of the Sahara, and of Yellow-faced Honeyeaters, *Lichenostomus chrysops,* migrating from southeastern Australia northward along the Great Dividing Range. *Left diagram* The *dates* mark the median time of passage of Garden Warblers at various banding stations (the date in the Sahara was extrapolated). *Arrows* indicate the mean directions of hand-raised birds tested in cages in the local magnetic field in southern Germany; these mean directions are based on headings recorded from 10 days before to 10 days after the respective date. (After GWINNER and WILTSCHKO 1978). *Right diagram* Yellow-faced Honeyeaters are assumed to pass the Brisbane area end of April/beginning of May; *arrows* indicate the mean directions recorded in cages in the local geomagnetic field in northern New South Wales from March to April and from May to the beginning of July. (After MUNRO et al. 1993)

suggest that many species take a different route in spring. Hence, the spring direction might be independently encoded. The magnetic field has been shown to serve as reference also in spring (e.g. W. WILTSCHKO 1968; GWINNER and WILTSCHKO 1980). Experiments by PETTERSON et al. (1991) and SANDBERG et al. (1991b), on the other hand, suggested that celestial cues are necessary and sufficient for normal orientation in autumn, but not in spring. This indicates the increased importance of the magnetic compass for spring migration.

Spring migration, however, is always a return to a familiar region; even young birds select their future breeding site already in autumn before they migrate (e.g. LÖHRL 1959; SOKOLOV et al. 1984). The same is true when experienced migrants return to their former wintering area to spend a second or third winter. Thus, spring migration and later autumn migrations include homing, at least in the last part of the route after the birds have reached their home region or wintering region, respectively. Here, migrants are assumed to give up flying their migratory courses and switch to the mechanisms described for homing pigeons.

5.2.3 Maintaining Directions – Staying on Course

When birds have decided upon a direction and started their flight, they have to maintain the given course over a certain period of time. Some authors (e.g. EMLEN 1975: MIHELSONS and VILKS 1975) tended to take this as a third, separate step in orientation. They argued that cues used to locate and to maintain a direction need not necessarily be the same, pointing out that much simpler cues might be sufficient to stay on course once it was found.

The magnetic compass seems to have some weaknesses when maintaining directions is concerned. Numerous reports claim that orientation becomes more scattered when visual cues are absent and the magnetic field is the only available cue. This is true for orientation during the day as well as during the night, and for free flying birds as well as for caged migrants in laboratory tests.

5.2.3.1 Free-Flying Birds

When pigeons were first released under overcast skies, their bearings were so scattered that the sun was taken to be an essential factor for orientation (e.g. MATTHEWS 1953; WALLRAFF 1966b). At this time, it was customary to release pigeons for training flights only when the weather was fair. KEETON (1969), in contrast to his colleagues, trained his birds under overcast conditions and obtained well oriented behavior also when the sun was not visible (cf. Fig. 5.1). This suggests that pigeons normally use the sun also for maintaining directions; KEETON's birds had probably learned to substitute other cues like landmarks, etc. to help them stay on course. Data on very young inexperienced pigeons (KEETON and GOBERT 1970) also indicate a first important role of the sun in maintaining directions.

Similarly, radar data and direct observations revealed oriented bird migration under overcast (e.g. BELLROSE and GRABER 1963; DRURY and NISBET 1964; HELBIG et al. 1987). Often, however, the tracks were more scattered and less linear than those recorded under clear sky, especially when landmarks were likewise unavailable, e.g. on dark nights or between opaque cloud layers (see STEIDINGER 1968; GRIFFIN 1973).

The behavior of migrants crossing a strong magnetic anomaly, however, indicates that birds also realize at least marked changes in the magnetic field when other cues are available. At a Swedish anomaly, with intensities up to 60% above normal and irregular changes in declination and inclination, migrants usually maintained their course; yet observers noted frequent changes in altitude, most often descents with a loss of altitude of 100 m or more, and a tempo-

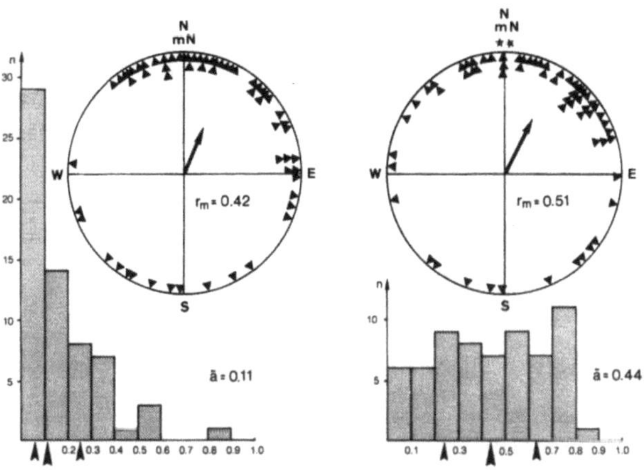

Fig. 5.10. Effect of the presence of stars on the accuracy of locating directions and staying on course: orientation of European Robins, *Erithacus rubecula*, tested in the local geomagnetic field indoors without celestial cues *(left)* and outdoors under the clear natural sky *(right)*. *Symbols in the circular diagrams* as in Fig 5.4. The vector length r_m represents the accuracy of locating directions, its numerical value is given *inside the circle*. The values *a*, calculated from the distribution of activity of each test night, express how well the selected direction is maintained. Their frequency distribution is given *below the circular diagrams*, with the median value and the 1st and 3rd quartile marked by *larger and smaller arrows*, respectively. The median value *a* is also given numerically. For further explanations, see text. (Data from R. WILTSCHKO and WILTSCHKO 1978a)

rary breaking-up of flock formations during these descents. On an overcast night with poor visibility, some birds flew around the area for more than an hour, suggesting orientational difficulties when no visual cues were available (ALERSTAM 1987).

5.2.3.2 Cage Experiments

When the experimental analysis of migratory orientation began, many authors hesitated to accept 'non-visual' orientation because of large scatter in the absence of visual cues (cf. WALLRAFF 1972; cf. Sect. 4.1.1). However, they considered only the distribution of acitivity within single test periods. A comparison of the behavior of birds tested in the local geomagnetic field with and without visual cues clearly shows that the distribution of the birds' headings is hardly affected, but the activity is considerably more concentrated when visual cues are present (Fig. 5.10).

Table 5.2 summarizes an analysis based on cage data from European Robins obtained under various magnetic conditions, in the absence of stars and under natural and artificial stars. It is obvious that the distribution of headings, represented by the vector length, *r*, depends on whether or not meaningful magnetic information is available, whereas the concentration of activity within the test cage, represented by *a*, depends on the presence or absence of stars.[5] Obviously, stars do not lead to better accuracy in locating the migratory direction, but they facilitate the maintenance of that direction, which is expressed by a more pronounced concentration of the bird's activity in the preferred cage sector. Artificial 'stars' which per se were without significance had the same concentrating effect. This was true even when, in the absence of meaningful magnetic orientation, the birds' headings were not oriented (R. WILTSCHKO and WILTSCHKO 1978a).

Table 5.2. Effect of magnetic field and stars on the accuracy of directional selections and the concentration of activity of European Robins, *Erithacus rubecula*

	Vector length, r			Concentration, a		
	Natural stars	Artificial stars	Stars absent	Natural stars	Artificial stars	Stars absent
Natural geomagnetic field	0.54	–	0.45	0.42	–	0.08
Experimental magnetic field	0.51	0.57	0.45	0.37	0.41	0.07
Compensated magnetic field	–	0.07	0.10	–	0.34	0.08

Experimental magnetic fields were in the range between 43 000 nT and 54 000 nT, providing meaningful directional information; the compensated fields were too weak to be used for orientation. The artifical stars did not contain any significant information. (Data from W. WILTSCHKO 1968, 1978; W. WILTSCHKO and WILTSCHKO 1972, 1975b, 1976)

Similar findings were obtained with day migrants: the vector lengths *r* of starlings under overcast and under sun were 0.54 and 0.53, respectively, while the concentration increased from 0.24 under overcast to 0.41 under sun (R. WILTSCHKO 1980). PERDECK (1957) reported that visible landmarks increased the concentration of a starling's activity in a similar way.

5.2.3.3 Visual Help to Stay on Course

Under natural conditions, many more cues are available that might help birds to maintain their course. Transfer processes similar to the ones described above between compass systems might be used to give directional significance to other suitable environmental factors. Wind (BELLROSE 1967) and landscape features (e.g. VLEUGEL 1955) have been discussed in this respect. Birds might calibrate any factors with directional characteristics and use them as secondary compass cues; the ones included in Fig. 5.8 stand symbolically for all of them. Thus, the apparent weakness of the magnetic compass in maintaining a straight

[5] Formally, both, r and a are mean vectors and can take values between 0 and 1. With respect to the vector length r, 0 means that the headings are equally distributed around the circle, whereas 1 means that all headings coincide (cf. Sect. 2.3). With respect to the concentration a, 0 means that the activity was equally distributed on the eight perches of the cage, whereas 1 means that all activity was concentrated on one perch only.

course does not pose a real problem in the wild. In nature, birds seldom have to maintain their flight direction with the magnetic compass alone – there are almost always visual cues available. Under overcast, landmarks and landscape features may facilitate staying on course as celestial cues do on clear days (W. WILTSCHKO and WILTSCHKO 1991a).

5.3 Migrations and Homing of Fishes and Amphibians

Besides birds, fish and amphibians are the two vertebrate groups whose orientation has received most attention. Their spatial behavior is fairly well known, which is an important prerequisite for understanding their orientation tasks and the strategies and mechanisms they might apply. Many fish are stationary, establishing territories to which they return when displaced. Others migrate over distances ranging from a few kilometers to several thousands of kilometers, reaching dimensions comparable to the migrations of birds (cf. HARDEN-JONES 1968; BAKER 1978; TESCH 1980). These may be regular migrations, as part of an annual cycle, or they may be singular migrations at the beginning and end of life. The spawning migrations of salmons and eels between rivers and the open ocean provide well-known examples for the latter type. The movements of amphibians, in contrast, mostly take place within a few kilometers. As adults, most species are terrestrial; in spring, they migrate regularly in order to reenter ponds and streams where they mate and spawn.

Due to the ecological situation, the range of cues normally accessible to amphibians and fish differs to some extent from those available to birds. Amphibians move on the ground, often among dense vegetation; hence, they have to move around obstacles, and celestial cues are less available. Fish move in a three-dimensional open space like birds, but the specific environment of rivers, lakes, and oceans provides them with orientation cues which are unavailable to terrestrial animals, e.g. currents, temperature gradients, salinity gradients, direction-resolvable accelerations of ocean swell and tidal cycles. Also, odorous substances may be transported over great distances, and their distribution is more regular and predictable in water than in air. Visual cues, on the other hand, rapidly fade as the light level decreases with increasing depth. This is true for landmarks in the water as well as for celestial cues. Nevertheless, a sun compass was demonstrated in several fish species, but these studies mainly involved fishes living close to the surface (cf. SMITH 1985).

The magnetic field is available to water-dwelling animals and terrestrial animals alike, and it has been frequently discussed as a factor of great importance in the orientation of migrating fish and amphibians. Yet in none of these groups has its role been as thoroughly analyzed as in birds. Especially cue-conflict experiments are rare, so that little is known about the relative importance of the magnetic compass compared to other cues. – This chapter describes the orientation of eels and salmons in the various phases of their life cycles, and the homing of toads as examples for orientation based on a multitude of cues. Potential uses of the magnetic compass will be pointed out and possible interactions with other factors will be discussed insofar as they are in accordance with the available data.

5.3.1 Orientation of Atlantic Eels (Gen. *Anguilla*)

Species of the genus *Anguilla* are catadromous, i.e. they hatch in the sea and migrate into fresh or brackish water, where they spend several years growing. When they mature, they return to the ocean to spawn. The two Atlantic species, the European Eel, *Anguilla anguilla*, and the American Eel, *Anguilla rostrata*, have the most extended migrations (BAKER 1978). Considerable research efforts were devoted to elucidate their life history. Because they migrate at great depth, where hardly any other cues are available, it is assumed that they rely mainly on magnetic cues.

5.3.1.1 The Life Cycle of Eels

Both Atlantic species hatch in the Sargasso Sea (20 °-30 °N, 50 °-80 °W); their *leptocephalus* larvae migrate to the American and the European coasts, which they reach after 1 and 3 years, respectively. From the temporal and spatial distribution of larvae in catches, it is obvious that young eels follow the currents of the Golf Stream, but this passive transport appears to be supported by active swimming and active control of direction (TESCH 1980). After reaching the continental slope, the larvae change their body shape and become elvers or glass eels. Later, they move into brackish water and develop into yellow eels.

When entering an estuary, eels become positive rheotactic and swim actively against the current. Some individuals migrate considerable distances upstream. A period spent in fresh or brackish water follows, which may last between 7 and 19 years. During this time, yellow eels are largely stationary. When displaced, they return to the place of capture, some of them even after displacement of over 200 km across the open sea (TESCH 1967; DEELDER and TESCH 1970). In laboratory experiments, however, yellow eels seldom showed any pronounced directional tendency. Tank experiments attempting to determine what orientation cue they use often suffered from the lack of oriented behavior as a baseline for manipulation (cf. Sect. 4.2.1.1). After reaching a critical size, eels develop into silver eels, the mature sea-faring form that returns to the Sargasso Sea to spawn. During the first part of their migration downstream, they follow the water current; passive drifting is supported by some negative rheotactic responses.

Directionally oriented migrations seem to begin only after the silver eels have reached saline water. Silver eels equipped with transmitters were found to take rather straight routes. As long as they were on the continental shelf, Euro-

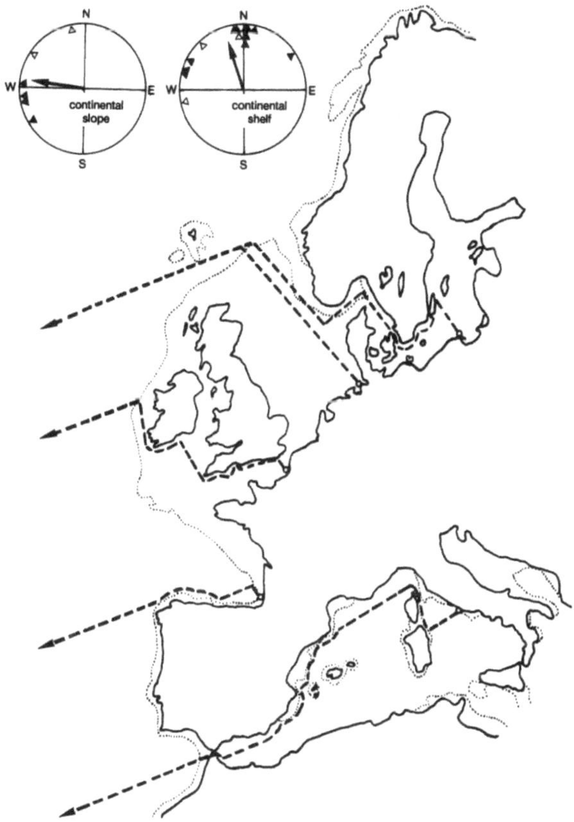

Fig. 5.11. Hypothesis of the spawning migration of European Eels, *Anguilla anguilla*, in different areas of the east Atlantic shelf and offshelf areas. On the shelf, the eels are assumed to move with NNW compass courses; when coastlines are encountered, they follow the shoreline until they can move with NNW headings again. Offshelf, the eels prefer WSW courses which possibly shift to more southerly headings later. *Circular diagrams at the upper left* indicate the headings of eels tracked on the continental shelf and above the continental slope. *Open symbols* mark headings of individuals with part of their movements over the respective other ground. (After Tesch 1980; map adjusted so that North is at the top)

pean silver eels preferred northerly or northwesterly directions. On the continental slope, however, they changed to more westerly courses which were close to the direction leading to the Sargasso Sea (Fig. 5.11; Tesch 1974b, 1978, 1980). They appeared to continue their migration at depths below 600 m (Tesch 1989), possibly following a deep-sea countercurrent of the Golf Stream to save energy. Very little is known about this last part of migration. Three silver eels, caught in Europe and treated with hormones in order to accelerate their maturation, preferred southwesterly and southerly directions when released in the region north of the Sargasso Sea (Tesch 1989). Thus, increasingly southerly tendencies as time progresses, or even some mechanisms of homing, might be involved in reaching the spawning region.

5.3.1.2 Preprogrammed Migration

The migration of eels in fresh water, their ascent and descent in the rivers are obviously controlled by rheotactic behavior. The factors controlling the vast oceanic migrations, however, are still largely unknown and have been subject to frequent speculations. For example, for a long time, it has been discussed whether the larval migration represented mere passive displacement or whether active orientation was involved (cf. TUCKER 1959). Yet the observed distribution of larvae of the European Eel cannot be explained by passive transport alone when recent findings on Atlantic currents are considered. This suggests active control of direction. TESCH (1980) discussed innate compass tendencies as they are found in migratory birds. Even if a more or less uniform east-northeasterly tendency were assumed, the large size of the spawning area, together with some distributing functions of the Golf Stream, would explain the geographic extension of the range of eels.

The return journey to the Sargasso Sea likewise seems to represent a sequence of preprogrammed courses. The northerly and northwesterly tendencies observed in the German Bight suggest that North Sea eels reach the Atlantic north of Scotland (cf. Fig. 5.11). TESCH (1978) interpreted this finding in view of the geological age of the northern opening of the North Sea, which is older than the English channel. This would imply innate courses. The shift to southwesterly courses when reaching the continental slope is discussed in connection with greater depth. During tracking experiments (STASKO and ROMMEL 1974; TESCH 1978) frequent diving excursions were observed, which TESCH (1978) attributed to orientation activities. He suggested that increasing pressure encountered during diving excursions may act as a trigger for the change in course after leaving the continental shelf.

This leads to the question as to which cue the respective courses are established. The great depth at which most migrations take place, as well as the great longitudinal and latitudinal displacement, would seem to exclude celestial compass mechanisms. Indeed, orientation of American silver eels was found to be independent of celestial cues (e.g. MILES 1968). This leaves the magnetic field for providing a reference for innate directional information and for controlling migration. A magnetic compass was frequently discussed to play a major role in the orientation of eels, during larval migration as well as for the return to the Sargasso Sea (e.g. TESCH 1978, 1980). Direct evidence for its use during oceanic migrations is still lacking, however, as *lepthocephali* or silver eels in the marine migratory phase have never been tested in altered magnetic fields (cf. Sect. 4.2.1.1).

5.3.2 Orientation of Salmons (Salmonidae)

In contrast to eels, salmons are anadromous, hatching in rivers, creeks, and streams, and migrating to the ocean where they spend one or more years until they mature. Species that spawn in the lower reaches of river systems begin their seaward migration shortly after hatching, while others first spend a certain period of time in fresh water before they move downstream to the sea. The duration of the oceanic phase varies, depending on species and growth rate. When salmons mature, they return to fresh water to spawn. Normally, they enter their natal river system and spawn in the same tributary where they themselves hatched. This may involve an upstream migration of several hundred kilometers. Tagging suggests very high homing rates in the individuals that do return to spawn (cf. QUINN 1990). Specimens of the Pacific genus *Oncorhynchus* die after reproduction, whereas a certain number of Atlantic Salmons, *Salmo salar*, survive, return to the sea and, after another oceanic phase, reenter the rivers for another spawning period (see BAKER 1978; MILLER and BRANNON 1982).

Massive commercial interest advanced extensive research programs on migration of salmons; as a result, their spatial behavior is better known than that of many other animals. The best-studied group are Pacific salmons, with considerable efforts having been devoted to their orientation abilities. Most data are available from the Sockeye Salmon, *Oncorhynchus nerka*, a species with a most interesting life cycle: it spawns in western Canada and Alaska, far inland in the upper tributaries of complex river systems adjacent to lakes, and the young spend 1 or 2 years in a freshwater lake before they finally begin their seaward migration.

5.3.2.1 Migration to the Nursery Lake

After hatching and emerging from gravel nests, young Sockeye fry must find a suitable nursery lake. Different populations migrate at different times of the day, and while many populations move downstream, others move upstream. The geographic directions of migration are different for each population, depending on local geography (see Fig. 5.12). Hence, the orientation behavior must be finely tuned to the population's specific demands.

Rheotactic Behavior and Magnetic Responses. The control of this first migration is partly known. Water current is a most important cue. BRANNON (1967) observed a spontaneous tendency to move upstream in hatchery-reared Sockeye fry from populations spawning in a lake outlet, while fry from populations spawning in a lake inlet showed corresponding tendencies to move downstream (Fig. 5.12). BRANNON (1972) described a case where such responses were modified by water velocity: fry from a certain creek were found to move downstream at higher and upstream at lower water velocities[6]. This behavior reflected the specific situation in the natural environment of the population, where fry first had to swim downstream in a fast-running tributary until they reached the lake outlet, then upstream through the slow-flowing outlet to enter the nursery lake.

[6] The water velocities were always well below the swimming speed of salmon fry.

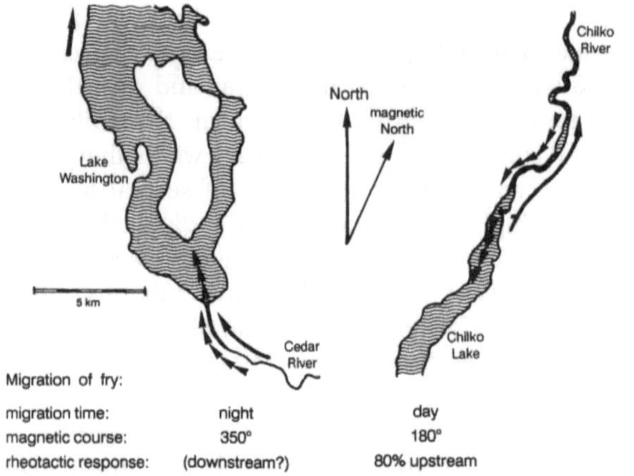

Fig. 5.12. Migration of Sockeye Salmon fry, *Oncorhynchus nerka*, in two North American river systems. *Left* Cedar River to Lake Washington, Washington, USA; *right* Chilko River to Chilko Lake, British Columbia, Canada. *Arrows* mark the direction of water flow, *series of arrowheads* indicate the migration of salmon fry. (After QUINN 1980). The spontaneous directional preferences relative to the magnetic field indicated below were determined by QUINN (1980), spontaneous rheotactic responses of Chilko River fry by BRANNON (1972)

In addition to rheotactic responses, Sockeye fry show specific directional preferences. Tests with fry of different origin (e.g BRANNON 1972; SIMPSON 1979) revealed compass tendencies in population-specific directions, which were mostly in good agreement with the general courses leading from the place of origin to the respective nursery lakes and, within the lakes, to the nursery areas (cf. Table 4.5 and Fig. 5.12). These tendencies are based on the magnetic field, since they could be altered by changing magnetic North (QUINN 1980, 1982a). They are often discussed as genetically determined tendencies (e.g. BRANNON et al. 1981), although orientation in the expected direction was not always found in hatchery-reared fry of known origin (see DODSON 1988 for discussion).

The specific nature of possible interactions between current and the magnetic field is still open. Both could serve as cues for innate behavior, or the response to current could help to establish a course with respect to the magnetic field. In complex routes involving directional changes, it seems possible that different water velocities might act as triggers, eliciting different preprogrammed responses not only to current, but also to the magnetic field.

The Use of Visual Cues. A comparison of the directional behavior with and without a view to the sky indicates that celestial cues are also involved (QUINN 1980, 1982a). Some population-specific differences became evident: fry of a night-migrating population responded to a change in magnetic North regardless of whether or not the sky was visible, but the vectors were longer when the test tank was uncovered (cf. Table 4.5). Fry from day-migrating populations, however, seemed to ignore a change in magnetic North. They responded to such changes only under overcast or when the test tank was covered, which shows that in this case, mainly visual cues controlled the direction of migration. QUINN (1982a) discussed their orientation as a parallel to bird orientation, suggesting that day-migrating fry might have learned to use a sun compass in the period before they were caught and tested.

The orientation of young Sockeye fry thus includes innate and learned responses to cues of different nature (see Table 5.3). In nature, chemical cues from rivers and nursery lakes may likewise play a role (BRANNON et al. 1981), and temperature may affect timing, so that the migration to the nursery lake is controlled by a multifactorial system in a rather complex way (for summary, see GROOT 1982; SMITH 1985).

5.3.2.2 Seaward Migration

After developing into smolts, Sockeye Salmons begin their migration down the rivers to the sea. Schools of smolts have been observed to approach the lakes' outlets on rather direct routes. Tests with smolts of this stage in tanks revealed directional tendencies which corresponded to the migratory directions in the wild. These tendencies changed in the course of time. The sequence of directions proved to be different even in various populations from the same lake system, but in each case, the observed tendencies led to more or less direct routes through the extended lake system to the outlet. These directions are assumed to be innate. Attempts to analyze the orientation mechanisms revealed a sun compass and a response to the pattern of polarized light. However, the smolts oriented in appropriate directions also under total overcast, indicating non-celestial mechanisms (GROOT 1965). When Sockeye smolts were tested in artificial magnetic fields with North turned 90 ° counterclockwise, they continued their normal orientation under clear and overcast skies. A significant directional change could only be observed when the tank was covered with black plastic, i.e. when all outside visual cues were excluded (QUINN and BRANNON 1982). This suggests that some outside factor had been calibrated for orientation use.

Once the smolts have reached the river, they can rely on currents for orientation by simply following the waters downstream. If they have to cross more lakes, however, they face the same problems as before. Altogether, the orientation of smolts on their seaward migration seems to be similar to the one described for fry, with innate directional tendencies and magnetic compass, rheotactic behavior, celestial cues, and maybe landmarks working together (see Table 5.3).

5.3.2.3 The Time at Sea and the Return to the Home Estuary

The duration of the stay at sea and the distribution of the various *Oncorhynchus* species in the northern Pacific Ocean are summarized by ROYCE et al. (1968). Some species stay near the home coast; others, among them Sockeye Salmons, seem to follow the currents of the Alaska stream and adjacent currents into the northern Pacific Ocean. It is unclear whether spontaneous directional tendencies are additionally involved and lead to an oriented dispersion. Oceanic features like salinity and temperature gradients (BLACKBOURN 1987), as well as availability of food, also seem to affect their distribution. Populations of different origin mix in the open sea (NEAVE 1964). When salmons are ready to spawn, they return to the estuary of their natal river system. The strategies that they might use have been controversially discussed. A random model requiring only a very low degree of orientation (SAILA and SHAPPY 1963) was questioned because some of the initial assumptions had been incorrect (QUINN and GROOT 1984); in particular, the percentage of returning adults that reached their own home estuary had been grossly underestimated (QUINN 1990)[7]. Also, the model was not in accordance with available tagging data (HIRAMATSU and ISHIDA 1989). Findings on the European *Salmo* also indicate that the fish do not leave their feeding area in random directions, but head towards their home coast (HANSEN et al. 1993). The orientation mechanisms are largely unknown. A 'map and compass' model analogous to the one of homing birds has been proposed by QUINN (1982b, 1984; see Sect. 6.1.2); it would allow salmons to determine their position and derive the compass course leading to the home estuary. BLACKBOURN (1987), in contrast, suggested that the return of salmons is based on a fixed migratory direction, which leads them back to their home coast (see also JAMON 1990).

For directional orientation of the oceanic stage of return migration, the magnetic compass has been suggested. In theory, celestial cues may also be used, yet the normal weather situation in the northern Pacific Ocean with abundant overcast makes a non-celestial compass indispensable (BRANNON 1982; QUINN 1982b). Data from ultrasonic tracking of returning Atlantic Salmons indicate the involvement of magnetic cues: fish that had been equipped with coil systems altering the magnetic field at 2-h intervals changed their swimming speed and their direction after magnetic changes significantly more often than during periods of steady fields (WESTERBERG 1982). Olfaction does not appear to be involved in this stage of migration (HANSEN et al. 1987).

Data on Atlantic Salmon suggest that the return to the home coast and finding the home river system may involve different mechanisms. After reaching the coast, salmons seem to know their location relative to the home estuary and head for it. However, only salmons that had migrated as smolts down the river are able to do so; fish that had been displaced to the oceanic feeding ground returned to the home coast, but not to their home estuary. This indicates that important orientational information for the latter step is obtained during smolt migration (HANSEN et al. 1993). These findings, together with the assumption that salmons reach their home coast by heading in fixed directions

[7] The percentage of adult salmons returning to their own home stream has been estimated in two ways, namely,(1) on the basis of all tags, and (2) on the basis of only those tags that were recovered. Both calculations lead to very different return rates (see Sect. 5.3.2.5). The first method ignores mortality between tagging and recovery; the second includes only individuals that survived and reached a coast, but it may be biased by irregular sampling efforts.

(BLACKBOURN 1987), suggest a migration strategy similar to that discussed for birds returning in spring, where flying innate courses is supplemented by navigational processes (cf. Sect. 5.2.2.5).

5.3.2.4 Return to the Home Stream

When salmons have reached their home estuary, olfactory cues become most important. Because of local differences in soil, vegetation, etc., each tributary has its own distinctive odor. The hypothesis that anadromous fish use these odor bouquets to recognize their own home stream was first suggested by HASLER and WISBY (1951). Meanwhile, numerous experiments support the idea that young salmons become imprinted on their home odors, and that adult salmons use this information to distinguish their natal tributary from other streams flowing in the same river and lake system (e.g. HASLER and SCHOLZ 1983; for summary, see SMITH 1985). NORDENG (1977) proposed that odors from juvenile members of their own population might serve a similar function, but although such odors are attractive for migrating salmons (GROOT et al. 1986), they do not seem to be used for recognizing the natal river system (QUINN 1990). In the home river, chemical information, together with a positive rheotactic response, guide mature salmons to their spawning grounds: the fish swim upstream in the presence of home stream odors, and turn back and search when the home odors are absent (see HARDEN-JONES 1968; HASLER and SCHOLZ 1983). Details of such a strategy are described by a model of JOHNSTON (1982). This behavior will eventually lead the fish to the source of the home odors and thus to their spawning site.

Theoretically, reaching the spawning ground could be facilitated by directional preferences based either on spontaneous tendencies as they are described for smolts migrating downstream, or on the individual experiences made during the fish's own downstream migration (cf. HANSEN et al. 1993). Such compass tendencies have been discussed (GROOT 1982; Table 5.3), but as the orientation of mature salmons in the fresh water phase is poorly studied, it is still open whether they are involved.

5.3.2.5 Population-Specific Migration Programs

Possible cues involved in the various migration steps of the life cycle of Pacific salmons have been summarized by GROOT (1982; see Table 5.3). The migration to the nursery lake, into the sea and back are controlled by a multifactorial system, many aspects of which have not yet been properly analyzed. The magnetic compass was shown to be involved in the first migrations by juvenile salmons; hence, its use in return migration has been discussed, without experimental data being available.

Altogether, the homing mechanisms of salmons prove highly efficient: while the overall return rate of tagged fish is not very high, the distribution of the tags suggests that almost all fish that do return to fresh water return to their own home stream. The percentages vary between species and populations, but are well above 90% for the *Oncorhynchus* species and for *Salmo salar* (QUINN 1990); for Sockeye Salmons, estimates are as high as 99% (QUINN et al. 1987). As a result of such high rates of homing, there is very little gene flow between the various populations and subpopulations of a salmon species, even within the same river system.

Considerable differences between populations of Sockeye Salmon, involving size and developmental rates (MEAD and WOODALL 1968), spawning dates and temperature (BRANNON 1984) have indeed been reported; differences in rheotactic responses, time of migration, compass courses and use of celestial cues in fry and smolts are described above. All these differences indicate that the species consists of many isolated stocks. MILLER and BRANNON (1982) mentioned at least 45 separate populations in the Fraser river system; they pointed out that while spawning time may vary between populations by as much as 5 months,

Table 5.3. Orientation cues which Sockeye Salmons, *Oncorhynchus nerka*, might potentially use during the various migrations of their life cycle

Migration	Life stage	Cues used	Mode
From gravel bed to nursery lake	Fry	Current	I
		Odors	I
		Landmarks?	I
		Geomagnetic field	II
		Celestial cues	II
From nursery lake downstream to the sea	Smolt	Current	I
		Landmarks?	I
		Geomagnetic field	II
		Celestial cues	II
Oceanic migration	Immature	Oceanic factors?	I
		Geomagnetic field?	II
		Celestial cues?	II
To home coast and home estuary	Maturing adult	Grid 'map'?	III
		Geomagnetic field?	II
		Celestial cues?	II
		Odor bouquets	I
Within home river system upstream to spawning ground	Mature adult	Current	I
		Odor bouquets	I
		Geomagnetic field?	II
		Celestial cues?	II

? indicates that the respective factor is discussed without experimental evidence supporting its use being available yet. 'mode' indicates how the respective factor is used, following the classification by Griffin (1952): I = direct cues, II = compass orientation, III = true navigation. (After Groot 1982, modifications by the authors indicated by italics)

the peak of fry emergence occurs within a few weeks, with the developmental program of each population finely tuned to the specific local situation around its own spawning site. This requires genetic isolation between populations. The same is true for population-specific migration programs, including appropriate rheotactic responses and sequences of magnetic courses adapted to the shape of lakes. In breeding and cross-breeding experiments, the return rates of salmons with both parents from the local stock, from mixed stock, and fishes from alien stock were 86, 46 and 22%, respectively (Bams 1976; Hershberger, cited by Miller and Brannon 1982). This documents an important role of genetic factors in homing.

Thus, in salmon, detailed endogenous migration programs ensure high rates of homing to the natal stream. This, in turn, by isolating the populations from one another, ensures that these detailed programs are maintained. The programs themselves consist of innate as well as imprinted responses to a variety of environmental factors summarized in Table 5.3, of which the magnetic field is but one (Groot 1982).

5.3.3 Homing of Toads (Gen. *Bufo*)

A typical lifetime track of most amphibians also includes migrations. Almost all amphibians spend an aquatic larval phase in ponds, lakes, creeks, or rivers. After metamorphosis, most of them leave the waters in order to begin a terrestrial life. In the majority of species, the adult animals stay terrestrial and return to ponds or rivers only to spawn. During these migrations in spring every year, distances of up to 3 km, occasionally more, may be covered (e.g. Heusser 1960). Many amphibians show a high degree of site fidelity, especially with respect to their spawning sites. Additionally, some amphibians migrate regularly from summer homes to hibernation sites (for summary, see Sinsch 1990b). Other species leave the aquatic habitat only for the juvenile period and return later to a pond or river, becoming more or less aquatic for the rest of their life (e.g. Phillips 1987a).

Orientation has been studied in three behavioral situations, namely, orientation on an axis perpendicular to the home shore (so-called *Y-axis orientation*, Ferguson 1967), migration to and from breeding sites, and homing after displacement. Orientation on the axis land-water was found to involve stable compass courses which could be adapted to a new home if necessary. The newt *Notophthalmus* can find the shoreward direction using a magnetic inclination compass (Phillips 1986a,b; cf. Fig. 4.11); the influence of other cues was not studied in this species. Data on orientation mechanisms used during migration are also scarce.

The best analysis of the factors involved in the orientation of amphibians relates to homing experiments performed with several species of anurans, mostly toads of the genus *Bufo*. This approach made use of the strong tendency of males to return to their breeding ponds after displacement. The distances normally did not exceed 300 m, since the initial orientation was found to deteriorate rapidly with increasing distance. European Toads, *B. bufo*, however, still showed a certain tendency towards the home pond when displaced over 3 km (Sinsch 1987).

5.3.3.1 Cues Used in Homing to the Breeding Pond

Frogs and toads were tested in their natural environment. They were either released in the center of a round, fenced enclosure (e.g. OLDHAM 1965), or their movements were recorded by marking their track with a thin thread that was wound from a sewing machine bobbin (e.g. SINSCH 1987).

In a large-scale study, SINSCH (1987, 1990a,b, 1992) tested European Toads and Natterjack Toads, *B. calamita*, in view of potential orientation factors such as magnetic cues, visual cues, olfactory cues, and acoustic cues. The results are summarized in Table 5.4. Magnetic information was found to play an important role, as both species were disoriented by bar magnets glued to their heads (cf. Fig. 4.12). Especially the strong effect on Natterjack Toads is rather surprising, since the animals could hear the chorus[8] from their home pond during these experiments. Cloud cover did not affect orientation, but blindfolding significantly decreased the straightness of tracks in both species, indicating that visual cues, which need not be celestial, play an important role in staying on course. However, while blindfolding had a marked effect on the orientation of Natterjack Toads, it hardly affected the directions of European Toads (Table 5.4). Olfactory cues are also involved (SINSCH 1987, 1992). The percentage of toads that returned was hardly affected by any of the above-mentioned manipulations, however; they only prolonged the homing times considerably (HEUSSER 1960; SINSCH 1990a, 1992).

Table 5.4. Factors affecting the homeward orientation of displaced male toads of the genus *Bufo*

Control and	European Toad, *Bufo bufo*						Natterjack Toad, *Bufo calamita*					
treatment	N	α_h	r_h	ΔC	st	ΔC	N	α_h	r_h	ΔC	st	ΔC
Brass bars	14	+8°	0.73***		0.77		6	−7°	0.75***		0.62	
Magnets	14	−53°	0.08 n.s.	***	0.74	n.s.	6	−46°	0.35 n.s.	**	0.53	n.s.
Clear sky	13	+11°	0.74***		0.87							
Overcast sky	29	−6°	0.67***	n.s.	0.85	n.s.						
Control	7	+11°	0.76**		0.84		6	−7°	0.76***		0.60	
Blindfolded	7	−2°	0.71**	n.s.	0.61	**	6	−99°	0.25 n.s.	***	0.39	**
Untreated	7	−16°	0.72***		0.85		7	+5°	0.88***		0.62	
Anosmic	8	−135°	0.08 n.s.	***	0.80	n.s.	7	−46°	0.39 n.s.	**	0.51	n.s.
Chorus at home site							5	0°	0.79**		0.63	
No chorus							7	−17°	0.77***	n.s.	0.60	n.s.

Second-order statistic based on the vectors of various releases: N = number of releases, α_h, r_h = direction and length of the mean vectors relative to the home direction, asterisks at r_h indicate a significant common tendency of the vectors (HOTELLING test); st = mean value for straightness of tracks. Δ C indicates differences to the controls by the HOTELLING two sample test and the MANN WHITNEY test, respectively. n.s. = not significant, * = $p < 0.05$, ** = $p < 0.01$, and *** = $p < 0.001$. (Data from Sinsch 1987, 1992)

[8] In contrast to most *Bufo* species, male Natterjack Toads *B. calamita* produce fairly loud calls which can be heard over distances of 500 m and more.

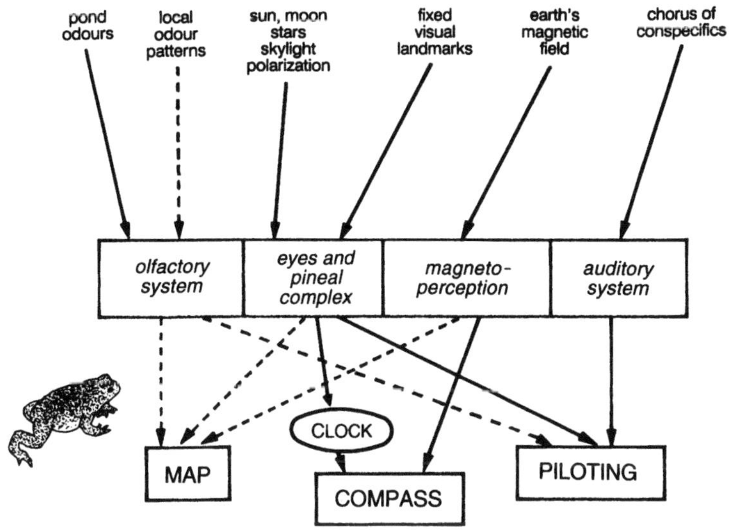

Fig. 5.13. Model summarizing the interactions of various cues in the orientation system of anurans. The nature of the factors, the perception systems involved and the possible role of the cues in the 'map and compass' concept (KRAMER 1961) are indicated. *Dashed lines* represent assumed relationships. (SINSCH 1990b)

SINSCH (1990b) summarized the nature of the orientation cues, the sensory system processing them and their assumed role in the orientation system in a model presented in Fig. 5.13. He discussed anuran orientation as an analogue to KRAMER'S (1957, 1961) 'map and compass' model and suggested a multifactorial system where various cues complement, but also to some extent replace each other. The similarity of some of these considerations with the ideas on homing in birds (cf. Fig. 5.3) is striking. The magnetic field is taken to provide directional information for a compass, even if a possible role as a 'map' factor cannot be excluded. Odors and the distribution of visual landmarks are assumed to form the 'map', and calls from conspecifics, pond odors, and some visual cues may guide the animals directly to the breeding ponds ('piloting' in Fig. 5.13).

5.3.3.2 Differences Between Species

Although most amphibians may have the basic elements of their orientation system in common, a considerable variation in the weighting and ranking of cues is indicated even in the closely related species of the genus *Bufo*. The different responses to blindfolding of European Toads and Natterjack Toads were already mentioned (cf. Table 5.4); similar findings have been reported e.g. with respect to olfactory deprivation, which affected the behavior of the two species mentioned above, but had little effect on the American Toad, *B. americanus* (OLDHAM 1965). Whether the role of magnetic cues also differs between species, is still open. In the two species used here as an example, magnets resulted in disorientation, but in the Peruvian species *B. spinulosus*, they caused a deflection of headings (for summary, see SINSCH 1990a).

SINSCH (1990a) discussed the observed differences between species concerning the relative importance of cues as being related to these species' general biology. Sex differences in orientation responses were also found; e.g. female Natterjack Toads show much less site fidelity than males, but a stronger response to a conspecific chorus. In addition, animals may choose different strategies depending on the distance of displacement. Green Frogs, *Rana clamitans*, displaced more than 1.5 km, showed a pronounced tendency to move downhill and/or towards conspecific calls, whereas they headed towards their own home pond, apparently ignoring other choruses and features of the terrain, when they were released closer to the pond (OLDHAM 1967).

Thus, the orientation of amphibians, despite their limited spatial range, also involves several possible strategies and various cues of different nature. The magnetic field provides mechanisms for direction finding which are used in shoreward orientation, homing, and probably also in migration. The system appears to be highly flexible, i.e. capable of adaptation to ecological demands.

5.4 Orientation at the Borderline Land-Sea

The borderline land-sea is characterized by specific ecological conditions which form narrow bands along the beach, parallel to the waterline. These ecological niches, which are characterized by specific degrees of humidity and other abiotic and biotic factors, are almost linear: they may extend several hundred kilometers along the coast, at the same time being of only a few decimeters or meters in width. Their position is not fixed, but varies with the tides. As seashores are dynamic structures, they may be shifted together when wave action, e.g. in the course of storms, changes the shape and the course of beaches (see A.C. BROWN and MCLACHLAN 1990). Hence, animals living at the borderline land-sea face orientation tasks which differ from the ones discussed so far: they must find a zone with favorable conditions rather than a specific site in an extremely variable environment.

Sandy beaches along the seashore are inhabited by several species of arthropods. Many of them are sandhoppers, i.e. small crustaceans of the order Amphipoda; isopods, decapods and even members of other arthropod groups are also found (for summary, see HERRNKIND 1983; A.C. BROWN and MCLACHLAN 1990). The activity of these animals centers around home zones which offer favorable conditions for their specific needs. Here, they rest during the daytime, mostly remaining burrowed in damp sand near or above the high tide line, under debris, etc. From these zones they venture out on their mainly nocturnal foraging excursions leading them either inland or downshore, depending on the species. The movements of these beach-dwelling amphipods are mainly confined to an axis perpendicular to the shoreline. For example, individuals of the Beachhopper *Talitrus saltator* may move up to 50 m inland on foraging trips. Moving seaward will then bring them back to a zone favorable for resting, even if they do not return to the precise spot from where they started. Landward and seaward movements along this axis also occur when the location where the animals momentary stay becomes too moist or too dry for comfort.

5.4.1 Orientation Factors Involved

This simple system of spatial movements stimulated the analysis of the orientation mechanisms involved. The test design took advantage of the tendency of sandhoppers to move seaward on dry and to move landward on wet substrates. *Zonal orientation*, as it was called by JANDER (1975), was thoroughly studied by the Italian research group of Leo PARDI; today, it represents one of the best understood orientation systems (for summary, see PARDI and ERCOLINI 1986). In general, two types of orientation cues may be distinguished: (1) compass mechanisms indicating the direction of the axis land-sea with respect to an external reference, also designated as 'global' cues because they are available everywhere, and (2) local cues indicating the direction towards land and sea directly.

5.4.1.1 Compass Orientation

Specimens of *Talitrus saltator* from the Tyrrhenian coast were found to prefer the seaward direction of their home beach even when tested inland or at the Adriatic coast. This pointed out that the behavior was based on some type of compass orientation. Manipulations such as reflecting the sun with mirrors or shifting the internal clock altered this directional tendency in a predictable way, indicating that beachhoppers use a sun compass for direction finding during daytime (for summary, see PARDI and SCAPINI 1987). Tests at night likewise revealed good orientation as long as the moon was visible (PAPI 1960)[9]. As beachhoppers seemed to be disoriented under overcast (e.g. ERCOLINI and SCAPINI 1972), celestial mechanisms were believed to be the only means of direction finding in amphipods.

HOWEVER, VAN DEN BERCKEN et al. (1967), testing *Talitrus saltator* of French, Dutch and Welsh origin, reported orientation in the absence of visual

[9] This led to the assumption of a moon compass (PAPI 1960; ENRIGHT 1972). Such a compass would require much more complex compensation mechanisms than the sun compass, because aside from the daily movements, monthly movements would have to be compensated. It is unclear whether the moon is a truly independent compass mechanism. The possibility that directional information from other systems might have been transferred to the moon and that the magnetic field may be a possible compass mechanism were not considered in the respective studies.

cues, and ARENDSE (1978) demonstrated magnetic compass orientation in Dutch specimens (cf. Sect. 4.3.2.2). In Italian *Talitrus saltator*, tests for a magnetic compass continued to produce inconclusive results (SCAPINI and QUOCHI 1992; UGOLINI 1994). Yet a tropic amphipod species from the Kenyan coast, *Talorchestia martensii* (PARDI et al. 1988; UGOLINI and PARDI 1992; cf. Fig. 4.16) was found to use a magnetic compass for orientation along the axis land-sea. Interestingly, the absence of visual cues enhance existing bimodal tendencies.

5.4.1.2 The Origin of Landward and Seaward Tendencies

When the sun was visible, laboratory-born specimens of *Talitrus saltator* without previous sun experience preferred a direction which corresponded to 'seawards' on the home beach of their parents. Cross-breeding experiments with *Talitrus saltator* from various Italian beaches showed that the preferred courses have an innate basis, being genetically transferred from one generation to the next (PARDI and SCAPINI 1987). Orientation performance is improved by experience, however: adult sandhoppers generally show less scatter than young ones (UGOLINI et al. 1988). Also, the innate courses can be subject to modifications by experience. For example, when beachhoppers from the seashore are washed into a back-dune lagoon, they adjust their axis of orientation to the new shoreline. Adults from such a lagoon were found to orient along the new axis, whereas their offspring bred in the laboratory showed a preference for the course that would have been correct at the nearby seashore (e.g. SCAPINI et al. 1988; UGOLONI and MACCHI 1988; UGOLINI et al. 1991). In summary, innate information and experience are integrated to determine the course for orientation along the axis land-sea (for summary, see PARDI and ERCOLINI 1986; UGOLINI et al. 1988).

The external reference for innate information may be the sun compass, as demonstrated in experiments with laboratory-born specimens of *Talorchestia deshayesi* and *Talitrus saltator* (PARDI and ERCOLINI 1986; SCAPINI and QUOCHI 1992), or the magnetic field, as indicated by experiments with inexperienced specimens of *Talorchestia martensii* (PARDI et al. 1988).

5.4.1.3 A Multifactorial System

Local cues associated with the specific situation of the border land-sea were also found to be involved in the orientation process. Three main types of factors have been identified: landscape feature, slope of the terrain, and the spectral distribution of sky light. Sandhoppers on a dry substrate tend to orient away from elevated parts of landscape and towards flat terrain. Likewise, such animals tended to move downhill when the inclination of the slope exceeded 5 °. The different compositions of sky light also seemed to play a role as accessory factors. In general, *Talitrus saltator* tested under the natural sky at their home beach were better oriented than those tested inland or at a beach with a

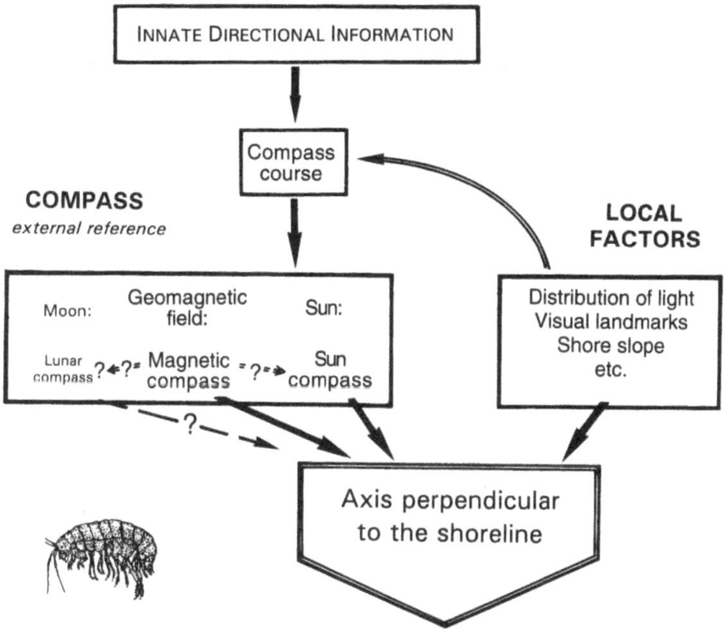

Fig. 5.14. Model summarizing the interactions of various cues in the orientation of littoral amphipods. *Open arrows* indicate learning processes and the modification of the set direction by experience; *solid arrows* represent the flow of information during the actual orientation process. Relationships not yet clearly demonstrated bear *question marks*. For further explanations, see text. (Draft by UGOLINI, modified)

reversed relationship of land and sea (for summary, see PARDI and ERCOLINI 1986; UGOLINI et al. 1988).

The various factors and their contribution to zonal orientation are summarized in Fig. 5.14. Innate information provides some basic direction, which is subsequently modified by experience and adapted to the sandhopper's specific home situation. The magnetic field, the sun (and possibly also the moon) can be used to locate the respective course. Additionally, local cues indicating the border land-sea directly may be involved. Since these last-mentioned factors always directly indicate the shortest route towards the sea, they can replace the celestial compasses on overcast days. Additionally, they may help to modify the orientation axis, adapting it to a new situation when sandhoppers are displaced and/or the beach changes as a result of wave action (UGOLINI 1989).

5.4.2 Relative Significance of Various Factors

Attempts to analyze the interaction of the various factors soon revealed that their relative importance varied greatly between species and even between populations of the same species. This is true for compass cues, or 'global' cues as compared to local cues, as well as for the relationship between sun compass and magnetic compass.

For example, landscape features were found to overrule compass directions in *Orchestoidea* tested at the Californian coast (HARTWICK 1976). Landscape features also play an important role for *Talitrus saltator* at the British coasts, where they could counteract the information from the sun, leading to disorientation when the cues were conflicting (WILLIAMSON 1954; UGOLONI and CANNICCI 1991). On Italian beaches, however, landscape features seem to be of only minor importance to *Talitrus*; here sun compass orientation dominates over all other cues, and conflicting information from local cues caused only a moderate increase in scatter. In general, the relative importance of the various factors seems to reflect the local situation: local factors are used when obvious and easily available, but on the beaches as in Italy, compass orientation seems to be preferred.

The relative significance of the sun compass and magnetic compass also shows great variability. For Italian *Talitrus saltator*, the sun compass was found to be of utmost importance, while magnetic orientation, first observed in northern European populations, could be demonstrated only with difficulties (cf. Sect. 4.3.2.2). In *Talorchestia martensii* living on tropical beaches, on the other hand, a magnetic compass was easily demonstrated. Deflecting magnetic North affected the orientation not only in the laboratory in the absence of celestial cues (cf. Fig. 4.16), but also in outdoor tests under sun (PARDI and ERCOLINI 1986; UGOLINI and PARDI 1992). At the same time, shifting the internal clock did not induce the clear shift as observed in Italy (Fig. 5.15; upper diagrams). In both situations where the sun compass and the magnetic compass gave conflicting information, a marked increase in scatter was observed; the distributions of data suggest that some individuals used their sun compass, whereas others preferred the magnetic compass. A crucial role of the magnetic compass in the tropics is also indicated by the observation that compensating the magnetic field affected the orientation under sun by increasing scatter, which was most pronounced around noon, when the sun was close to zenith (Fig. 5.15, lower diagrams; UGOLINI and PARDI 1992).

The different responses to shifting the internal clock – in Italy, all beachhoppers responded, whereas only a certain portion did so at the Kenyan coast – were observed also when laboratory-born specimens were tested. At the same time, the laboratory-born *Talorchestia martensii* were well oriented in the absence of visual cues (PARDI et al. 1988). This indicates that the increased importance of the magnetic compass also refers to its function as a reference system for innate information.

In the tropics, the sun's arc undergoes large seasonal variations, with culmination changing between north and south. PARDI et al. (1988) pointed out that this might pose a severe problem to animals using a sun compass. They saw the increased importance of magnetic information in the tropics as a necessary adaptation, suggesting that the magnetic compass might be used to recalibrate the sun compass and to adjust the compensation mechanisms. However, the large increase in scatter observed in compensated magnetic fields (UGOLINI and

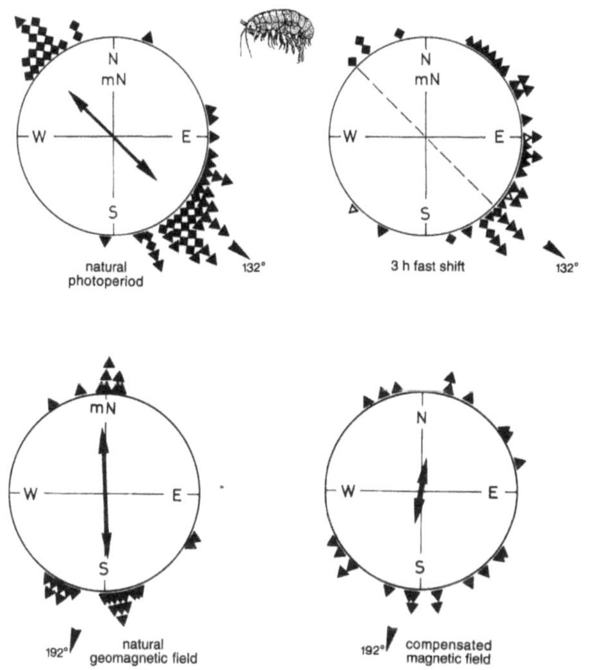

Fig. 5.15. Experiments indicating the relative importance of the magnetic compass and the sun compass in the equatorial sandhopper *Talorchestia martensii*; outdoor tests under sunny sky. *Upper diagrams* Orientation in the local geomagnetic field; *left* sandhoppers living in the natural photoperiod; *right* sandhoppers that had been subjected to a 3-h fast shift of their internal clock. *Symbols* as in Fig. 4.16. (After PARDI et al. 1988) *Lower diagrams*, *left* Orientation of sandhoppers in the local geomagnetic field; *right* in a compensated magnetic field. *Symbols* as in Fig. 4.16. (After UGOLINI and PARDI 1992)

PARDI 1992; UGOLINI 1994) indicates that the sun compass alone is hardly sufficient to guarantee orientation during the entire day. Possibly, the compensation mechanisms are unable to cope with the extreme situation in the tropics, in particular around noon, when the sun azimuth changes very rapidly.

Altogether, the findings discussed here show that zonal orientation is highly flexible, with mechanisms adapted to cope with a variety of problems arising from the specific conditions. The magnetic compass which was long overlooked, represents one component of a multifactorial system. Its relative importance, like that of all the other factors, varies according to the situation at the sandhoppers' home beach.

5.5 The Role of the Magnetic Compass Compared to That of Other Cues

From the case histories described in this chapter, one common feature becomes evident: (1) orientation behavior is controlled by a multitude of factors, and (2) there is no fixed hierarchy of cues that is generally valid. Very often, different types of cues provide the same type of information so that the available information is at least in part redundant. At the same time, the flexibility of the systems is remarkable when rating and ranking of the various cues are considered. The specific role of a given factor seems to be in optimal correspondence with the situation in the respective environment. These adaptations may take place on a genetic basis reflecting long-term action of natural selection, or on a more short-time basis when the preferences for certain factors are modified by experience.

5.5.1 Control of the Direction of Movement

While serving as compass to establish the actual direction of movement, the magnetic field competes with other factors also providing directional information, such as the sun, the stars, and local cues. In the first case, different types of compass information compete with each other, whereas local cues provide information of a different nature, as they indicate the direction to a goal directly.

5.5.1.1 Compass Orientation Versus Local Cues

Because of their usually limited range, local cues play a most important role in orientation for animals that move over short distances, e.g. amphibians and amphipods. In situations where such cues allow the immediate establishment of the direction towards the goal, following them appears to be the simplest strategy. Hence, one might expect that animals preferentially respond to local cues whenever they are available. Indeed, direct factors often prove to be dominant, but in some situations, compass orientation is nevertheless preferred. The behavior of the various species of littoral amphipods exemplifies how different conditions of the environment, in particular the easy availability of local cues, may affect their importance. At the Californian and British coasts with their cliffs, landscape features proved dominant (HARTWICK 1976) or could at least counteract information from the compass mechanisms (UGOLINI and CANNICCI 1991). On flat Italian beaches, in contrast, landscape and shore slope seldom provide well-defined cues; here, sandhoppers tend to prefer compass orientation which is normally very reliable on such beaches.

The behavior of Natterjack Toads, on the other hand, provides a good example for an adaptation of the preferred orientation mechanisms to the species' general biology. Natterjack Toads are able to use temporary ponds for breeding, and both sexes seem to follow different strategies (Sinsch 1992). When displaced from a breeding pond, males show a strong tendency to home to the site of capture – a behavior which promotes the optimal utilization of suitable habitats, thus minimizing intraspecific competition. For male Natterjacks, the magnetic compass plays an important role as indicated by the marked effect of magnets, even when direct cues like calls from the home pond were available. Females, in contrast, are faced with the task to find a suitable breeding pond while expending only as much energy as necessary. This can be done by migrating to a remembered breeding site on the most direct route. However, if other suitable ponds have formed in the vicinity, it would be more advantageous to use those; moving towards a conspecific chorus is the best way to find such pools. Hence, in female Natterjack Toads, the preference for local cues like calls seems adaptive. In Green Frogs, the distance of displacement determined the strategy used, as the frogs oriented towards their home pond when displaced over short distances, but followed local cues when farther away (Oldham 1967). Two interpretations of this difference in behavior are possible: it might indicate problems in establishing the home course at greater distances, or it might reflect a change in motivation. If it requires a considerable expenditure of time and energy to reach the own home pond, settling at any pond in the vicinity indicated by local cues may be advantageous.

5.5.1.2 Magnetic Compass Versus Celestial Compass Mechanisms

In most cases, however, information from the magnetic compass competes with alternative directional information provided by the sun compass and, for night-migrating birds, by the star compass. The relative importance of the magnetic compass and the celestial compass mechanisms was also found to vary considerably. Normally, the sun compass seems to be preferentially used. This is demonstrated by the clear effect of shifting the internal clock in pigeons and other birds (cf. Fig. 5.1, right), in *Talitrus saltator* and in many other animals, and the observation that populations of day-migrating salmons do not follow a shift in magnetic North as long as the sky is visible. Yet in situations where using the sun compass or interpreting the sun's arc meets with difficulties, e.g. in regions with frequent periods of overcast sky or when large latitudinal shifts are involved, the magnetic field provides the most important directional cues.

This applies also to the special situation found in the tropics. The changes in the sun's azimuth at various times of day differ greatly, and the seasonal changes in culmination between north and south would require a very sophisticated mechanism of compensation for the sun's movement. Apparently, these problems are so severe that *Talorchestia martensii* can no longer rely on the sun compass alone. In this tropical sandhopper, the magnetic compass was found to play a much more important role than, for example, in related species at the Italian coast (UGOLINI and PARDI 1992; UGOLINI 1994). Unfortunately, the behavior of *Talorchestia* is the only case where cue-conflict experiments have been performed with animals living close to the equator; hence, it must remain open whether a dominant role of the magnetic compass is a general feature of directional orientation in the tropics. A possible parallel is suggested in the migratory orientation of birds: the magnetic compass appears to gain importance when the birds migrate towards lower latitudes (W. WILTSCHKO and WILTSCHKO 1991b). Also, long distance migrants responded faster to a shift in magnetic North than European Robins, a species that winters in the Mediterranean (W. WILTSCHKO and WILTSCHKO 1975a,b). However, when celestial cues are used in bird migration, the shift in latitude itself might constitute an even greater problem than the situation at the equator.

In view of the clear results of clock-shift tests, many authors emphasized the dominant role of the sun and used terms like 'hierarchy' of cues or mechanisms to characterize the relationship between sun compass and magnetic compass (e.g. EMLEN 1975; PARDI and ERCOLINI 1986; ABLE 1991a). This term, however, appears to disregard the interactions between the compass mechanisms which cannot be described in terms of a fixed hierarchy: in pigeons, the sun compass normally dominates over the magnetic compass when the actual flight direction is located, but during ontogeny, the sun compass is established with the help of the magnetic field. An even more complex relationship is indicated between magnetic field and stars in the orientation of night-migrating birds. Data suggest that in both cases, the celestial mechanisms are partly independent and allow the birds to locate their courses independently from the magnetic compass. However, in the long run, they are controlled by the magnetic field. This clearly suggests that the various factors involved do not represent independent sources of information which are simply added up; instead, the information is integrated, with the directional significance of the celestial factors being adjusted to that of the magnetic field to bring the entire system into harmony. Little is known about possible ontogenetic interactions between the various compass mechanisms in other animals. A similar relationship between factors as described for migratory birds, however, has also been discussed for salmons and amphipods (e.g. QUINN 1982a; PARDI et al. 1988).

5.5.1.3 Maintaining a Course

The prominent role of celestial compass mechanisms leads to the question why animals that are endowed with a magnetic compass show such a strong tendency to rely on directional information provided by other, mainly visual cues. The magnetic compass has some obvious advantages over celestial mechanisms, as it is a much simpler mechanism not requiring complex compensation and learning processes, and it is independent of weather conditions. If such a compass is available, why do animals undergo the effort to a establish a sun compass, a star compass or even transfer directional information to landmarks where they have to allow for parallax?

This question has not yet found a definite answer. So far, we can only state that it seems to be a common trait of animals to use the magnetic compass *together* with visual cues. One of the possible reasons might lie in the obvious difficulties to stay on course when visual information is not available. The magnetic compass, well suited to locate courses, seems to have some serious weaknesses when the task is simply to stay on course. This is documented by the more pronounced orientation usually observed in the presence of visual cues, which need not be celestial. The behavior of migrating birds in cage tests (cf. Fig. 5.10 and Table 5.2), and increased deviations from straight routes in the movements of blindfolded toads (cf. Table 5.4) are not the only examples. BRANNON et al. (1981) and QUINN (1982a) observed more pronounced directional tendencies in salmon fry when visual cues were available (cf. Table 4.5); BAKER and MATHER (1982) and BAKER (1987b) found less scatter in their tests with noctuids in the presence of celestial cues. The same is true for the experiments with *Talorchestia martensii*; the data suggest that straighter routes are taken when the sun is visible (PARDI et al. 1988). The immense scatter frequently found in tests under 'non-visual' conditions reflects this general phenomenon, which handicapped, and in some cases still handicaps, the demonstration of magnetic orientation.

The reasons for the difficulties in staying on course by magnetic information alone may lie in some aspects of magnetoreception that are still unknown (see Chap. 9). However, it must be emphasized that such resulting problems are typical only for laboratory tests. In nature, they hardly exist, because there are almost always some visual cues available that can be used to stay on course: celestial cues when the sky is clear and terrestial cues, e.g. landmarks, under overcast. Magnetic and visual cues of various nature are integrated components of a complex system of directional orientation (W. WILTSCHKO and WILTSCHKO 1991a).

5.5.2 The Magnetic Field as a Directional Reference

The magnetic field serves as a directional reference for innate directional information and for acquired courses of various origin. In many species, however, it is not the only such reference. The sun compass and factors like celestial rotation may provide alternative reference systems that can be used in a similar way.

5.5.2.1 Genetically Encoded Courses

The best-studied case of competing reference systems is the information on the innate migratory direction in night-migrating birds. In several species, it was found to be encoded with respect to magnetic North as well as with respect to celestial rotation, i.e. the same information is represented twice in the same species (cf. Fig. 5.4). Under natural conditions, these two sets of information will coincide and produce a common response. In experiments where they gave different information, a change in their relative importance when the young birds mature and start migration became evident. In view of the large distances involved in migration, this change in control seems to reflect the usefulness and reliability of stellar and magnetic cues at various geographic latitudes.

The same might apply to solar and magnetic cues in the tropics, as indicated by the directional reference for innate information on the axis land-sea in littoral amphipods. In Italy, the sun compass provides the most important reference, whereas at the Kenyan coast, the magnetic compass serves this function. Whether the innate course is generally represented with respect to both mechanisms in Kenyan specimen of *Talorchestia martensii*, is unclear, because experiments with laboratory-born amphipods in compensated magnetic fields are still lacking. Clock-shift experiments with such amphipods, however, indicate that only a certain portion of young *Talorchestia* prefers to rely on the sun compass (Pardi et al. 1988).

In salmon fry, possible spontaneous directional tendencies relative to the magnetic field complement spontaneous rheotactic tendencies. In nature, these two tendencies would work synergistically; cue-conflict experiments have not been performed. Innate compass responses in other animals are largely unknown, as the ontogeny of orientation behavior has received little attention so far.

5.5.2.2 Acquired Courses and Learning Processes

With respect to acquired compass courses, our knowledge is likewise limited. The origin of acquired courses has been examined in marine turtle hatchlings and in Italian *Talitrus saltator*; the magnetic compass and the sun compass, respectively, provide the directional reference. Cue-conflict experiments regarding these questions have not been performed. Learning processes establishing mechanisms like the navigational 'map' in pigeons and other birds have also not been analyzed in detail. The 'map' is assumed to be a *directionally* oriented mental image of the distribution of navigational factors (see Fig. 6.1); hence, it must be calibrated by a compass. The 'map' then indicates home courses which can be located with the sun compass or with the magnetic compass. This implies that both compass mechanisms are somehow linked with the 'map'. However, since the magnetic compass is involved in the learning processes establishing the sun compass, and possibly also in those establishing the 'map', the magnetic compass may serve as a basic reference which ties these mechanisms together in a meaningful way.

5.6 Summary

The role of the magnetic compass in multifactorial orientation systems is described, using homing and migration of birds, migration of eels and salmons, homing of toads, and the orientation of littoral amphipods at the axis land-sea as examples.

In homing pigeons, the role of the magnetic compass varies with increasing age and experience. In young birds, the magnetic compass is a most important mechanism: it is the first compass available, and it is used for determining the home course by a navigational strategy called route reversal, where the net direction of the outward journey is recorded and reversed. It also provides a reference for the learning processes establishing the sun compass and possibly also for those establishing the navigational 'map'. In adult birds, the sun compass seems to dominate over the magnetic compass; however, the magnetic compass still controls the behavior to some extent and probably continues to serve as reference for adapting the sun compass to seasonal changes in the sun's arc and for updating the 'map'. In night-migrating birds, the magnetic compass serves as a directional reference for the innate migratory direction. The migratory direction is also encoded with respect to celestial rotation and, during the premigratory period, this information dominates that from the magnetic field. During migration, however, the magnetic compass becomes the most important mechanism. On clear nights, birds use a star compass the directional significance of which is under the long-term control of the magnetic compass. The magnetic field also serves as reference for preprogrammed directional changes in the course of migration and for the spring direction.

The magnetic compass is discussed as directional reference for possible innate directional tendencies in Atlantic eels. In young Sockeye Salmon, spontaneous rheotactic tendencies are complemented by magnetic tendencies to guide fry to their nursery lakes. Later, the magnetic compass may be used when the smolts migrate downstream to the sea. Whether it is also involved in oceanic movements or the return migration to the spawning ground in freshwater creeks is unclear. In amphibians, the magnetic compass appears to play an important role in homing of displaced toads of the genus *Bufo*, where it interacts with olfactory cues, auditory cues and landmarks. Species- and sex-specific differences in the role of the various orientation cues are indicated.

In the orientation of littoral amphipods, the sun compass and the magnetic compass are used to orient on an axis perpendicular to the home shore. In Mediterranean populations, the sun compass was found to be the crucial orientation mechanism, whereas the magnetic compass appears to be more prominent in populations from northern Europe, where the use of a sun compass is difficult because of frequent overcast, and in the tropics because of the dramatic changes in azimuth. Local factors like landscape features and shore slope are likewise involved; their role also varies between species and between populations.

The examples described show that the orientation is usually controlled by a multitude of factors in a highly complex way. There is no fixed hierarchy of cues; the relative importance of the various factors varies according to the environmental situation. When controlling the actual direction of movement, information from the magnetic compass competes with information from the celestial compasses and local cues indicating the goal direction directly. Frequently, the magnetic compass is dominated by the sun compass, but in situations where the sun compass meets with difficulties, the magnetic compass plays the more important role. – For maintaining directions, visual cues are of great importance. Generally, orientation is found to be more pronounced in the presence of visual cues, irrespective of whether these are the sun, the stars or terrestrial landmarks. This suggests that the magnetic compass alone is not well suited for maintaining directions. In nature, this problem is of little importance, since visual cues of some kind are usually available.

The magnetic field provides a directional reference for innate and acquired directional orientation; celestial rotation and the sun compass can also serve this function. In migratory birds, the migratory direction is represented with respect to the magnetic field and to celestial rotation. In littoral amphipods, both the sun compass and the magnetic compass can serve as a reference for innate information on the axis land-sea; the relative importance of the two systems varies between populations and between species, reflecting the conditions in the home region. In most other animals, orientation behavior has not been analyzed under this point of view.

Part 3: Non-Compass Responses

Chapter 6

Non-Compass Orientation by the Magnetic Field

A number of orientational responses to the magnetic field described over the last decades are basically different from compass orientation, as they are not based on the vector quality of the magnetic field. Instead, they seem to depend on the *spatial distribution* of magnetic parameters such as total intensity and inclination. In most regions, the changes in these values with distance are rather small (cf. Sect. 1.1.2.1); as differences must exceed a certain threshold to be detected, the use of this type of magnetic information appears to be meaningful only when large distances are involved. Hence, it seems to be restricted to highly mobile animals that have extended migrations or home from distant sites.

Orientation by the spatial distribution of magnetic parameters is much more difficult to demonstrate than compass orientation, because the manifestation of any effect can hardly be predicted. It is almost impossible to supply animals with meaningful 'false' information in this respect; hence, direct evidence in the sense that a defined change in magnetic information results in a specific change in orientation is not available. Data suggesting the magnetic nature of the effects in question are mainly correlations with natural changes in magnetic parameters, observed breakdowns of the behavior when the magnetic field is disturbed or other kinds of indirect evidence. In general, our knowledge on non-compass orientation with the help of the magnetic field is rather limited. Although data clearly indicate that magnetic information is involved, details on the precise relationship with specific magnetic parameters are only vaguely known. Nevertheless, the most prominent hypotheses describing orientation by the spatial pattern of the magnetic field are presented in this chapter, together with data supporting them.

6.1 Magnetic Components in Navigational 'Maps'

The idea that animals make use of the spatial distribution of magnetic parameters to determine their *position* dates back to the last century, when Viguier (1882) first pointed out such a possibility. Being aware of the mysterious ability of homing pigeons to return after displacement, scientists began to speculate about possible orientation mechanisms. The familiar navigational charts with their graduation served as a model for the concept of a 'map' based on a grid of

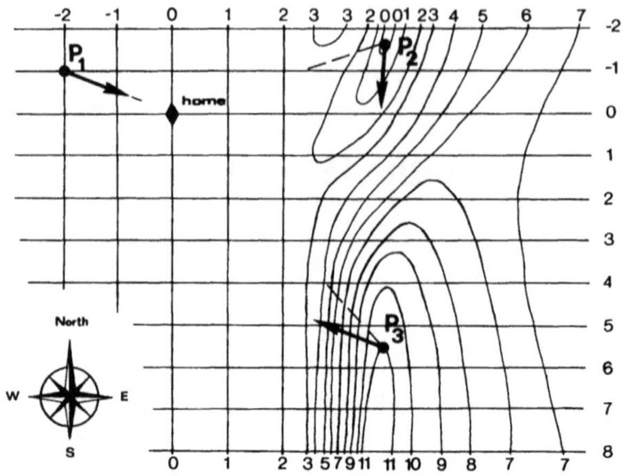

Fig. 6.1. Model of the hypothesis assuming environmental gradients as components of the navigational 'map'. The isolines of two gradients are given in relative units, with home values set to 0. The birds 'know' that the E-W gradient decreases towards west; hence values lower than the home values indicate that home lies east, etc. *Left side* At site P_1, where the gradients show a regular course, the birds determine their home course correctly; *right side* irregularities in the course of gradients lead to initial errors, which are later corrected. (After W. WILTSCHKO and WILTSCHKO 1982)

gradients. Since several magnetic parameters show fairly regular worldwide gradients (cf. Sect. 1.1.2), they became obvious candidates for at least one coordinate of such a 'map'.

Most considerations on navigational maps refer to homing in pigeons, because this is by far the best-studied phenomenon. The 'map' is taken to be a *directionally oriented mental image* of the spatial distribution of specific environmental factors. It is established by experience (cf. Sect. 5.2.1.2); pigeons acquire the respective knowledge during the first months of their life. After displacement, they determine their home direction as a compass course by *comparing* values of the navigational factors at the release site with remembered home values; i.e., in the example given in Fig. 6.1, the birds are aware that factor A decreases towards the west; hence, when encountering values of A which are lower than the home value, they know that their loft lies east. As pigeons appear to be able to use 'map' information also at distant, unfamiliar sites, it is generally assumed that the map factors are gradients that can be extrapolated beyond the familiar range. Unpredicted irregularities may lead to deviations from the true home course (see Fig. 6.1, right). A more detailed description of the 'map' concept and its implications is given by WALLRAFF (1974) and W. WILTSCHKO and WILTSCHKO (1978, 1987).

A number of factors have been discussed as potential components of the navigational map, among them magnetic parameters. Today, the pigeons' 'map' is assumed to involve a multitude of cues which interact in various ways (cf. Fig. 5.3, left). A general discussion of all evidence related to the 'map' is beyond the scope of this book; the following section will focus on findings which suggest an involvement of magnetic parameters in the 'map' of homing pigeons.

6.1.1 Magnetic Parameters in the 'Map' of Homing Pigeons

First hypotheses assuming magnetic 'map' factors proposed a combination of total intensity and inclination (VIGUIER 1882) and, more than half a century later, vertical intensity and the Coriolis force, a force associated with the earth's rotation (YEAGLEY 1947). Early attempts to verify a role of magnetic information in homing of displaced birds (e.g. CASAMAJOR 1927; GRIFFIN 1940; YEAGLEY 1947, 1951) were inconclusive, however. Hence, the possibility of magnetic map factors was largely disregarded until a number of magnetic effects in pigeon homing were described in the 1970s. In the 1980s, it became rather popular to speculate on magnetic maps (e.g. GOULD 1980, 1982; B. MOORE 1980; QUINN 1984; PHILLIPS 1986b).

Summaries on magnetic map effects have been published by C. WALCOTT (1980a, 1982, 1991), LEDNOR (1982), WAGNER (1983), WALLRAFF (1983), GOULD (1985), W. WILTSCHKO et al. (1986) and W. WILTSCHKO and WILTSCHKO (1988). The respective data were obtained under sunny sky, i.e. when the sun compass was available; hence, they are generally assumed to represent an interference with the map rather than with the magnetic compass. According to experimental design, the available evidence can be grouped into three categories, namely, (1) effects of local anomalies, (2) effects of temporary changes and (3) effects of experimental manipulations.

6.1.1.1 Behavior at Magnetic Anomalies

If pigeons used magnetic map factors, their orientation should be affected when local magnetic values deviate markedly from the general course of the magnetic field in the region. This is indeed the case. Effects of magnetic anomalies on pigeon races have been repeatedly reported (e.g. CASAMAJOR 1926; TALKINGTON 1967). First systematic studies were performed by WAGNER (1976) and FREI and WAGNER (1976) who released pigeons at weak anomalies in Switzerland where the field changed in a fairly regular way, with total intensity deviating between 70 and 100 nT from the expected values and gradients of ca. 12 nT/km. The birds seemed to depart in the direction of the maximum magnetic gradient that approached their home values. Frequently, considerable deviations from the home course (Fig. 6.2) and occasionally even flying in the opposite direction were observed.

These findings seemed to suggest that the pigeons followed local magnetic gradients, a strategy that differs fundamentally from the postulated use of magnetic coordinates in a 'map'. However, the fact that after 20 s the birds already headed in the direction in which they finally vanished argues against following

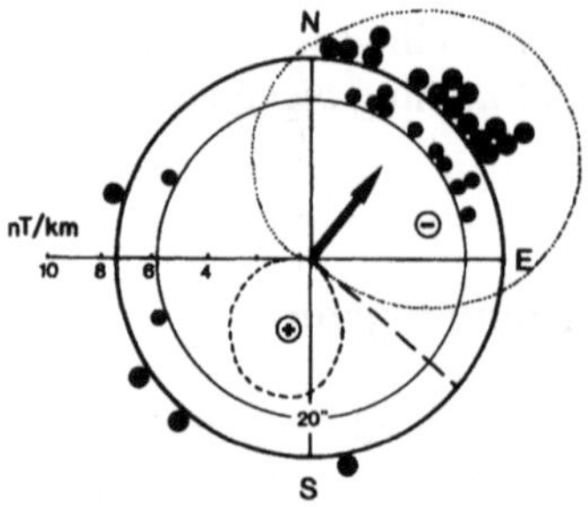

Fig. 6.2. Behavior of pigeons released at a weak magnetic anomaly in Switzerland. The total intensity at the release site was 33 nT higher than at the loft; the *dashed* (increasing) and *dotted* (decreasing) lines indicate the local magnetic gradients at the release site, with the scale of the left. The home direction is marked by a *dashed radius*. The *symbols at the periphery of the outer circle* indicate the vanishing bearings of individual pigeons, the respective *symbols at the inner circle* indicate the bearings 20 s after release. The *arrow* represents the mean vector of the vanishing bearings proportional to the radius of the circle. (After WAGNER 1976)

gradients, because 20 s would be too short to detect gradient directions by flying around. Later releases at another Swiss anomaly with gradients up to 50 nT/km also produced considerable deviations from home, yet without a pronounced trend to follow maximum gradients (FREI 1982). Clock-shift experiments at a weak anomaly in Germany resulted in the typical deflection indicating the use of the sun compass (KIEPENHEUER 1986), which also speaks against the idea that the observed orientation was forced upon the birds by magnetic topography. Rather, it seemed to be caused by a false estimate of the home course.

C. WALCOTT (1978) and KIEPENHEUER (1982) released pigeons at stronger, magnetically 'rugged' anomalies, with irregular intensity changes and local gradients up to 3500 and 8000 nT/km, respectively. The vanishing bearings of these birds showed a marked increase in scatter up to disorientation. Both authors found a negative correlation between the rate of change in magnetic intensity at the respective sites and the vector lengths (Fig. 6.3). GRAUE (1965) had already reported a similar observation, yet without giving data. Thus, strong anomalies, in contrast to weaker ones, seem to make orientation impossible. Magnets attached to the pigeons' back did not have any effect, i.e. the birds remained disoriented (C. WALCOTT 1980b).

The return rate, however, was hardly affected, which means that most pigeons were able to overcome the initial problems and reorient themselves (C. WALCOTT 1978; KIEPENHEUER 1982). When released within the anomaly a second time, the birds were no longer disoriented (KIEPENHEUER 1982). This was true even if they were not released at exactly the same site, as long as the new site was within the same anomaly. Experience with other anomalies did not improve orientation, however (LEDNOR and WALCOTT 1988), suggesting that local aspects are crucial. Two interpretations of these findings might be considered: either practice enables birds to cope with initially contradicting informa-

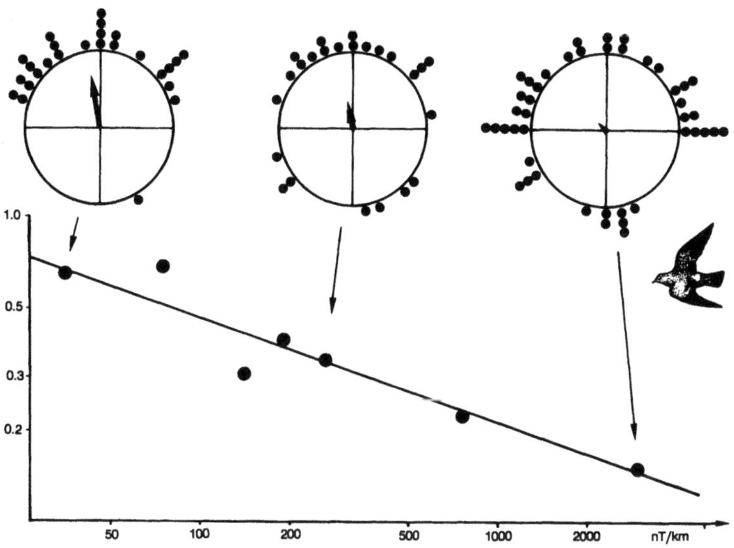

Fig. 6.3. Orientation at various sites at and near the strong magnetic anomaly at Iron Mine Hill near the Massachusetts/Rhode Island border, USA, where total intensity is about 6% above the normal regional value (see Fig. 6.4, right, for a magnetic survey map of the area). The vector lengths (*ordinate*, logarithmic scale) are correlated with the maximum difference in intensity within 1 km in the direction toward home (*abscissa*, logarithmic scale). With a coefficient of correlation of 0.96, the correlation is significant. The data of three releases are given as examples; the home direction is *upward*; otherwise *symbols* as in Fig. 6.2. (Data from C. WALCOTT 1978)

tion, or the birds simply remember the magnetic 'signature' of that specific anomaly and associate it with the home course of the previous successful homing flight. In the latter case, the local values of the anomaly would no longer constitute a coordinate of the 'map', but would represent some magnetic 'landmark' whose position with respect to the loft is memorized.

Interestingly, strong anomalies have a disorienting effect only when pigeons start their homing flight. Tracks of birds that crossed the anomaly at Iron Mine, Rhode Island, with intensity highs up to 6% of the background field were unaffected (Fig. 6.4; C. WALCOTT 1978). The same is true of the vanishing bearings of birds that had been transported through an anomaly before they were released in magnetically normal terrain (KIEPENHEUER 1986).

As a whole, the data obtained at magnetic anomalies are largely in accordance with the view that anomalies lead to mistakes in determining the home direction, because they represent a local distortion of the 'map'. The disorientation observed at strong anomalies is usually attributed to local magnetic values deviating so much from normal that they cannot be spontaneously interpreted. In view of this, it is not surprising that magnets do not improve orientation.

An analysis of releases in northern Germany (DORNFELDT 1991), however, seemed to suggest that the pigeons were better homeward oriented in areas with

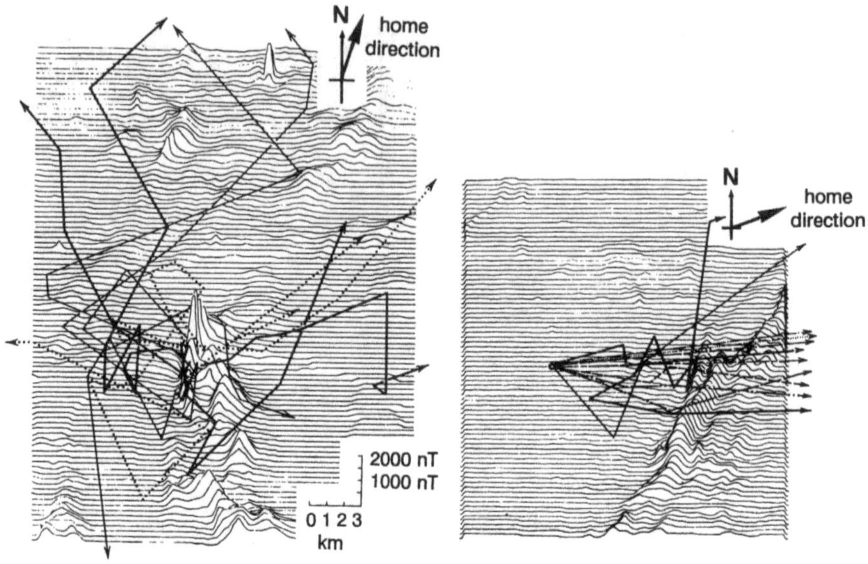

Fig. 6.4. Flight paths of pigeons at strong magnetic anomalies. *Left* Birds released at the Iron Mine Hill anomaly (cf. Fig. 6.3); *right* birds that had to cross a strong magnetic anomaly on their homing flight. The regional intensity values show a steady increase of about 4 nT/km to the NNW; the magnetic relief in the figure is based on deviations from this regular pattern. (After GOULD 1982, based on data from C. WALCOTT 1978 and unpubl.)

irregularly varying magnetic conditions than in those with more regular magnetic fields. The reasons are unclear; possibly, birds can become familiar with the magnetic topography of their home region and use specific features to recognize certain sites.

The effects at magnetic anomalies described so far appear to suggest that magnetic parameters are an essential part of the avian navigational system. This is not generally true for all pigeons, however. Studies by C. WALCOTT (1986, 1992) in the northeastern USA indicate that the role of magnetic map factors is determined by the loft location, i.e. by the (magnetic) conditions under which the pigeons grew up. Pigeons from a loft in Massachusetts had been disoriented at strong magnetic anomalies in the New England region, whereas birds from Ithaca, New York, were unaffected by the same anomalies. Further experiments revealed that the behavior of pigeons having grown up in lofts only 2.5 km apart differed at the strong magnetic anomaly at Iron Mine, Rhode Island (Fig. 6.5). Interestingly, birds from the loft at Codman Farm in magnetically 'smoother' territory (88 nT/km) were oriented, while the birds from the loft at Fox Ridge Farm with a gradient of 450 nT/km were not. Since both groups of birds were from identical stock and had been trained together, these findings indicate that the situation at the loft itself is of crucial importance. C. WALCOTT (1992) also reported that birds having grown up at Fox Ridge Farm and released in the strong anomaly at Worchester, Massachusetts, were dis-

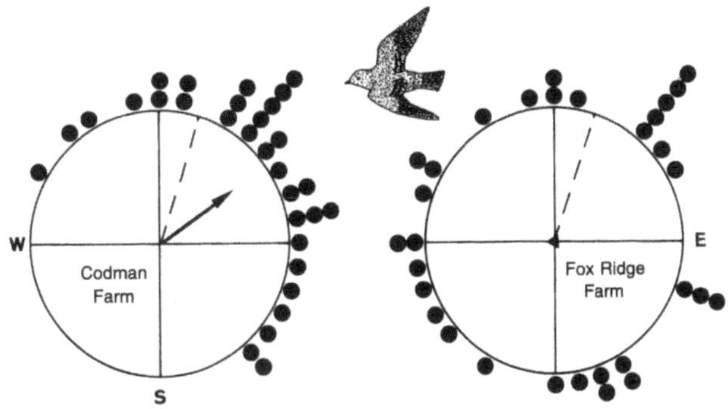

Fig. 6.5. Orientation of pigeons released at the strong magnetic anomaly at Iron Mine Hill (cf. Figs. 6.3 and 6.4, left). The test birds were from identical genetic stock and had been trained together, but were housed at two different sites in Massachusetts, 2.5 km apart. *Symbols* as in Fig. 6.2. (After C. WALCOTT 1992)

oriented in the period from 1976 to 1981, while they showed good homeward orientation in 1991.

An approach by LEDNOR (1982) likewise did not indicate the use of magnetic map factors by pigeons from Ithaca, New York. Comparing the pattern of *release site biases*[1] around the Ithaca loft (WINDSOR 1975) with predictions derived from distributions of potential navigational factors, he found that important aspects of the pattern could be explained by assuming that the loft lay on a local 'high'. However, the regional data on total intensity available from charts revealed a very different picture, providing no explanation for the pattern of biases observed in the Ithaca region (LEDNOR 1982). GOULD (1985), after examining the patterns of magnetic variables around Ithaca, pointed out that the specific Ithaca situation might make inclination a more plausible factor.

6.1.1.2 Effects Associated with Temporal Changes in the Magnetic Field

Animals using magnetic map information would have to consider the daily variations of the magnetic field, because – at least over the distances which most animals move – the spatial variations and the temporal fluctuations fall into a similar range (see Sect. 1.1). Hence, temporal variations of the geomagnetic field should be expected to affect orientation, in particular when they differ from the usual course. Because of the enormous amount of variance involved, large data sets are required to establish effects of temporal fluctuations.

Magnetic Storms and K-Variation. Many reports on the effect of increased magnetic fluctuations relate to homing performance in pigeon races. YEAGLEY (1951) was the first to call attention to a correlation between homing speed and magnetic activity as reflected by the relative sunspot number on the days prior

[1] Release site biases (KEETON 1973) are deviations from the homeward course which are regularly observed in untreated pigeons at the sites in question.

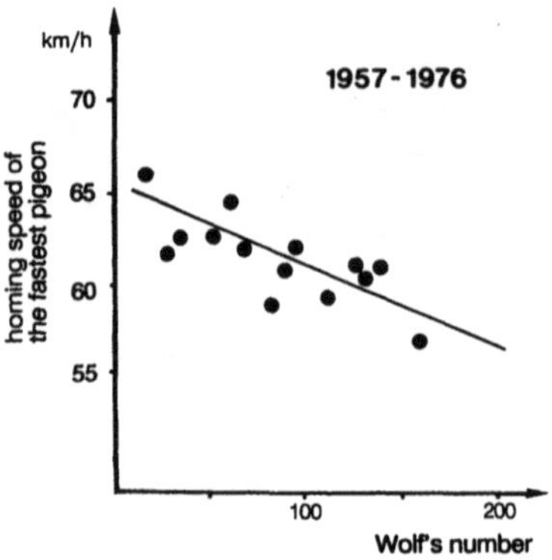

Fig. 6.6. Results of the annual pigeon races from Paolo to Parma, Italy, in the years 1957 to 1976. Correlation between homing speed of the fastest pigeon and Wolf's number, an index for sun activity reflecting the number of sun spots, which, in turn, is correlated with magnetic disturbances. With a coefficient of correlation of – 0.778, the correlation is significant. (After SCHREIBER and ROSSI 1978)

to the races. Analyses of Italian racing data by SCHREIBER and ROSSI (1976, 1978) revealed a similar relationship (Fig. 6.6). DORNFELDT (1977), analyzing German data, claimed, in contrast, that speed increased with increasing disturbances, yet without giving the range of variation of the two variables. SCHIETECAT (1988), analyzing Belgian data in view of meteorological and magnetic conditions, found some correlation between homing performance and the geomagnetic index A_k which summarizes magnetic fluctuations in the course of the day. He stated that in good weather, magnetic fluctuations had little effect, while under less favorable conditions they may lead to slower homing and higher losses. A.I. BROWN et al. (1986) obtained inconsistent results when analyzing several sets of American racing data: in some cases, increasing K-values led to faster, in others to slower homing, or there was no correlation at all.[2]

Experiments indicate that magnetic storms also affect initial orientation. The vanishing bearings of groups of pigeons repeatedly released from the same sites were found to be correlated with the K-values which characterize fluctuations of the geomagnetic field (see Sect. 1.1.3). At two sites in upstate New York, USA, with different home directions, increasing K-values induced *counterclockwise shifts* of the mean direction (see Fig. 6.7, left). The rate varied at the two sites, being 0.8 ° and 3.3 ° per unit of K_{12} (KEETON et al. 1974)[3]. The fact that these correlations disappeared when the pigeons were released with magnets indicates that it was indeed a magnetic effect (Fig. 6.7, left; T. LARKIN and KEETON 1976).

A similar study in Frankfurt a.M., Germany, showed a K-correlation at only one of two sites simultaneously studied. It resulted in a *clockwise* shift of the mean direction by 0.9 ° per unit of increasing K_{12}, i.e., a shift in the opposite direction from the one observed in upstate New York. Increasing K_{12} also led to

[2] CARR et al. (1982) discussed poor homing performance in American pigeon races on 17 June 1972, in connection with a severe magnetic storm occurring on that day. The authors interpreted the data as suggesting that birds might base their judgement of latitude on magnetic intensity. It is unclear, however, to what extent the magnetic storm really was the cause of the slow homing. It is striking that pigeons in the western United States that were still en route during the onset of the storm were, in general, less affected than pigeons in the east that could have been home already.

[3] K_{12} is the sum of the K-values of four 3-h periods prior to the beginning of the release.

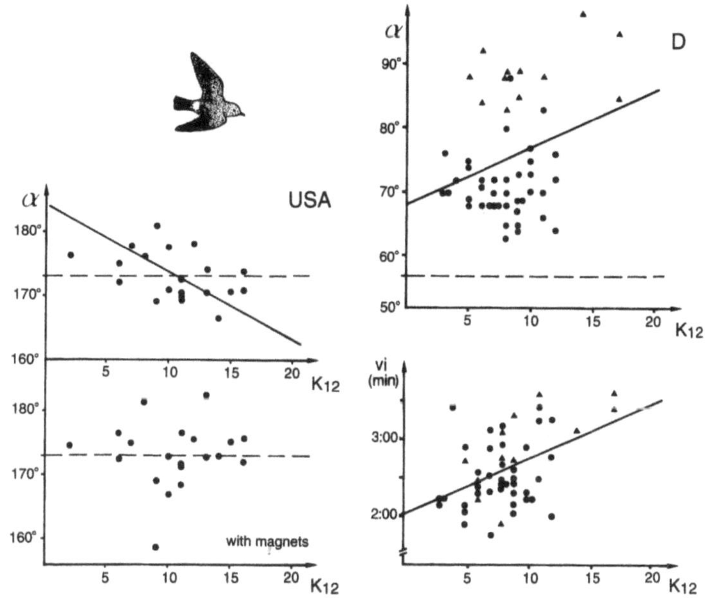

Fig. 6.7. Data on initial orientation of groups of pigeons repeatedly released from the same site, plotted as a function of the amount of magnetic fluctuations within the 12 h prior to the release, as characterized by K_{12}, the sum of the *K*-values of the preceding four 3-h periods. *Left diagrams* Vanishing bearings (α) of pigeons released at a site 75 km N of their home loft at Ithaca, N.Y., USA; the home direction 173 ° is indicated by a *dashed line*. The birds were alternately equipped with brass bars (*above*) and with magnets (*below*). With brass bars, their bearings show a significant correlation with K_{12}, with a coefficient of correlation of – 0.403; with magnets, this correlation was no longer significant with a coefficient of correlation of + 0.038. (After T. LARKIN and KEETON 1976). *Right diagrams* Data of pigeons released 57 km SE of their home loft in Frankfurt a.M., Germany; the home direction 57 ° is indicated by a *dashed line*. Their vanishing bearings (α, *above*), with a coefficient of correlation of +0.299 , and the vanishing intervals (*vi*, *below*), with a coefficient of correlation of +0.457 , are both correlated with K_{12}. *Triangles* indicate the data obtained during the first part of the test series; *dots* the data obtained later. (After KOWALSKI et al. 1988)

shorter vector lengths of the group and to longer median vanishing intervals; these relationships were even more pronounced than the ones with the mean direction (Fig. 6.7, right). An analysis of the behavior of individual birds revealed that only 10 out of 32 pigeons showed a significant correlation between vanishing bearings and K-variation; 17 of the non-significant correlations had a positive, 5 a negative sign (KOWALSKI et al. 1988).

An analysis of the orientation behavior at various sites around Tübingen, Germany, also emphasized the site-specific nature of these phenomena, as the mean vector length at 3 out of 25 sites was correlated with the amount of magnetic fluctuations, whereas no such relationship was observed at the other sites (SCHMIDT-KOENIG and GANZHORN 1991).

Taken together, these findings suggest that K-correlations, although widespread, cannot be observed at every site, and that not all birds are equally affected. Interestingly, in all of the above-mentioned cases the best correlation between the mean of the vanishing bearings and the K-values did not emerge with the K-values at the time of release, but with K_{12}, which summarizes the magnetic activity in the 12 h before release. The same is true for some of the correlations with homing speed in pigeons races. This suggests that not the actual magnetic data alone are important, but that the variation during a period prior to release can affect the way pigeons process magnetic information during release.

Regular Daily Variations of the Geomagnetic Field. W. WILTSCHKO et al. (1986) reported two additional effects that could be suppressed by magnets. These effects were discussed in connection with temporal variations of the geomagnetic field; they seem to suggest that regular daily variations are somehow incorporated into the navigational system.

The first case relates to repeated releases of a group of pigeons at the same site twice or three times per day. In several such series, the vanishing bearings at local noon were significantly further clockwise than those at about 6:00 local time (see Fig. 6.8, left); when the birds were again released at about 18:00 local time, their vanishing bearings were very similar to the ones observed at 6:00. Magnets made these differences disappear by shifting the bearing in the morning and in the evening 25 ° clockwise; the noon bearings were shifted by a much smaller amount (Table 6.1).

Table 6.1. Change in vanishing bearings of pigeons in the course of the day and the effect of magnets

		Times of release				
	N	6:00	Δα	12:00	Δα	18:00
15 pigeons showing daily variations of their bearings						
Without magnet	14	143 °	+ 19 °	162 °	– 17 °	145 °
With magnet	14	168 °	+ 1 °	169 °	+ 1 °	170 °
Difference induced by magnet		+ 25 °		+ 7 °		+ 25 °
18 pigeons not showing daily variations of their bearings						
Without magnet	14	147 °	+ 1 °	148 °	– 1 °	147 °
With magnet	14	158 °	+ 1 °	159 °	– 1 °	158 °
Difference induced by magnet		+ 11 °		+ 11 °		+ 11 °

N = number of test days with three releases; Δ α = differences in mean direction. (Data from FÜLLER 1986; cf. W. WILTSCHKO et al. 1986)

During the years 1981, 1983, and 1984, such differences in bearings between morning and noon in the range between 8 ° and 16 ° were observed at three different sites; the noon data lay always *clockwise* from the morning data, regardless of their relation to the home direction. When the respective tests were continued in the years 1985, 1986, 1989, and 1990, however, such variations were no longer observed (M. BECKER et al. 1991). Figure 6.8 compares data recorded at the same site 30 km north of the loft in 1984 and 1989. Note that the bearings in 1989, when no difference was observed, were close to the direction where the noon bearings in 1984 had been.

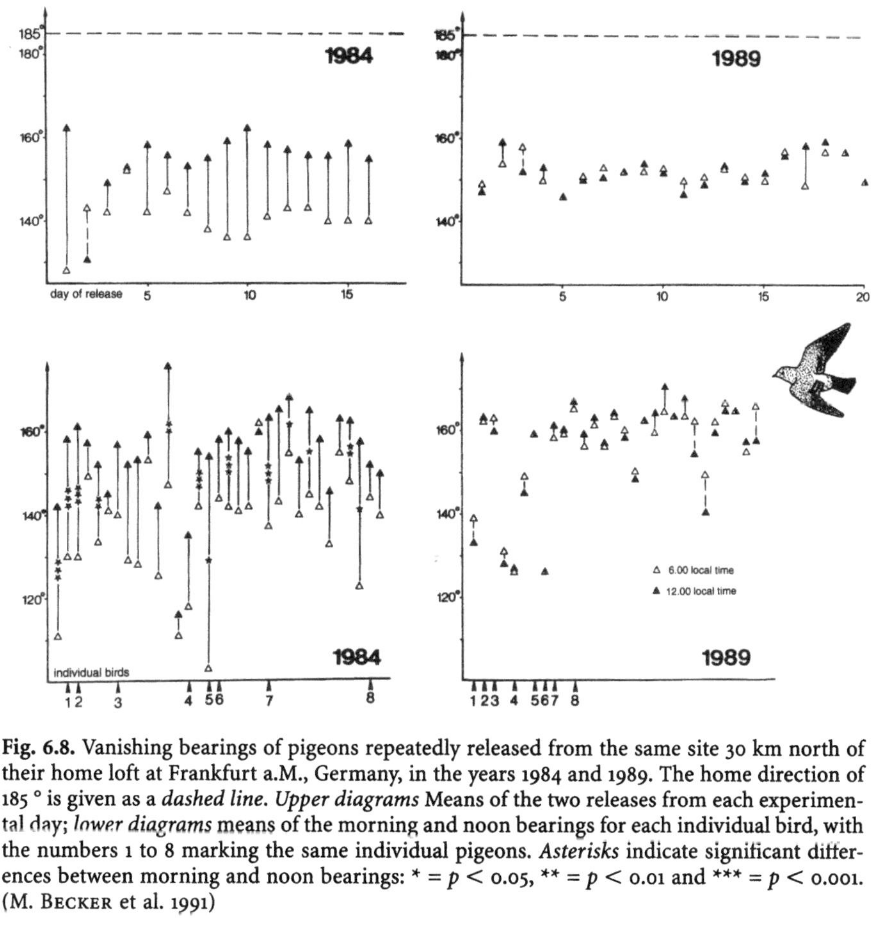

Fig. 6.8. Vanishing bearings of pigeons repeatedly released from the same site 30 km north of their home loft at Frankfurt a.M., Germany, in the years 1984 and 1989. The home direction of 185 ° is given as a *dashed line*. *Upper diagrams* Means of the two releases from each experimental day; *lower diagrams* means of the morning and noon bearings for each individual bird, with the numbers 1 to 8 marking the same individual pigeons. *Asterisks* indicate significant differences between morning and noon bearings: * = $p < 0.05$, ** = $p < 0.01$ and *** = $p < 0.001$. (M. Becker et al. 1991)

In these studies, remarkable individual differences between pigeons were observed. Some birds showed differences of 30 ° and more, while others showed hardly any difference at all (Fig. 6.8, lower left part; cf. W. Wiltschko et al. 1986). In 1983 and 1984, largely the same birds were tested at three sites; pigeons that showed pronounced variations at one site did not necessarily show them also at the others; on the contrary, such birds were a small minority. Yet, on the other hand, only 3 out of 22 pigeons did not show significant variations at any of the three sites (Füller 1986). In 1989, however, some of the same individuals that had shown pronounced variations in 1984 did not do so any longer (Fig. 6.8, lower part; M. Becker et al. 1991).

The second type of effect concerns pigeons that were clock-shifted 6 h slow[4] and were kept under these 'permanently' shifted conditions for more than 2

[4] A '6 h slow shift' means that the internal clock of the experimental birds was 6 h behind natural time, i.e. their subjective day began 6 h after sunrise and ended 6 h after sunset.

months. During that period, they were released with their subjective day out of phase with the natural day (W. WILTSCHKO et al. 1984). A number of homing flights during the first days served to adopt the sun compass to the experimental situation; after that, any difference between the shifted birds and the controls had to be interpreted as indicating the use of 'map' factors which undergo changes in the course of the day. The shifted birds continued to deviate by about 20 ° – 30 ° clockwise from their controls when released at sites where they had not been before during the permanent shift period (Fig. 6.9, left). These deviations disappeared when the birds were released from the same site a second time. After readjusting their internal clock to the natural day, the experimental pigeons showed an approximately 30 ° counterclockwise shift at sites where they had been during the permanent shift period (Fig. 6.9, right). This deviation was the mirror image of the first and, again, disappeared on the second flight from the same site (W. WILTSCHKO et al. 1984). Magnets could suppress both types of shift (Fig. 6.9), indicating that the effect was of a magnetic nature (W. WILTSCHKO et al. 1986; W. WILTSCHKO and WILTSCHKO 1988).

It should be noted that in these tests experimental pigeons and controls were released alternately, so that they experienced identical magnetic conditions. The fact that their responses differed suggests that birds are aware of the normal temporal course of magnetic parameters and usually take it into account. The findings can be best explained by assuming that the experimental birds, while living under shifted conditions, first misinterpreted site-specific magnetic factors because of their mistake in time. On a subsequent flight from

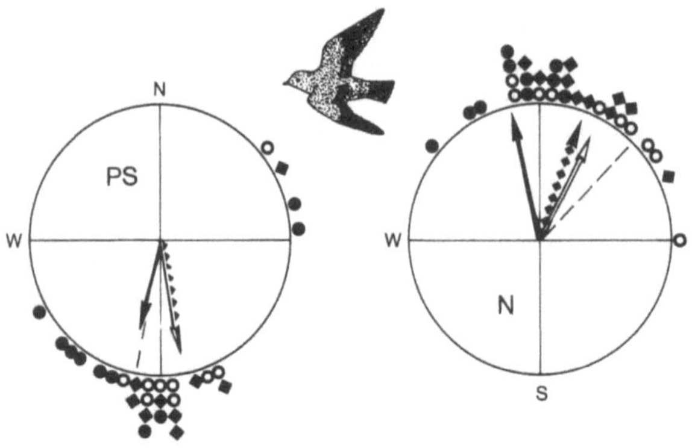

Fig. 6.9. Deflections observed in the vanishing bearings of homing pigeons associated with the regular daily variations of the magnetic field. The home direction is indicated by a *dashed radius*. *Left diagram* Deflections induced by a lasting 6 h slow clock-shift (*PS*). At a site unfamiliar under these conditions, the clock-shifted birds (*round solid symbols*) show an about 30 ° clockwise deviation from the mean of the controls (*open symbols*). Shifted birds wearing magnets (*diamond-shaped solid symbols*) do not show these deviations. *Right diagram* After shifting their internal clock back to the natural photoperiod (*N*), the former shifted birds (*round solid symbols*) show a corresponding counterclockwise deviation from the mean of the controls (*open symbols*) when released at sites that have become familiar to them under the condition of the lasting clock-shift. This deviation, too, is suppressed by magnets (*diamond-shaped solid symbols*). *Symbols* as in Fig. 6.2. (Data from W. WILTSCHKO et al. 1986)

the same site, they had adapted their way of interpretation at that specific location to the experimental situation, which, after normalization, caused the reverse deviations (W. WILTSCHKO et al. 1986; W. WILTSCHKO and WILTSCHKO 1988).

Considerations on Temporal Variations. Altogether, the effects associated with temporal variations are difficult to interpret and in part seem to contradict each other. The fact that the K-effect depends on K_{12}, the sum of variations prior to the release rather than on the actual K-value, and the deviations produced by living under the conditions of a 'permanent' shift seem to suggest that pigeons expect a certain temporal course of the magnetic field during the day. However, whereas the K-effect indicates that experiencing magnetic variations during the last 12 h may affect these expectations, the behavior of the 'permanently' shifted pigeons speaks against such an assumption. The differences between bearings in the morning and those at noon observed over several years, on the other hand, seem to suggest that the pigeons do not compensate for the regular daily variations.

The manifestation of above-mentioned effects, if they occur, appears to be independent from the home direction. For example, in upstate New York, the K-effect produced counterclockwise deviations at two sites with home directions 100 ° apart. Similarly, the noon bearings were clockwise from the morning bearings at three sites in Germany with different home directions. The sense of rotation induced by the 'permanent' shift was likewise independent of the home direction in New York as well as in Germany: always clockwise during the shifted period and always counterclockwise after normalization. A plausible interpretation for these findings has not been forwarded yet.

6.1.1.3 Experimental Manipulations

A number of experiments were aimed at demonstrating or disproving a role of magnetic parameters by depriving the birds of meaningful magnetic information using various kinds of manipulation. Under sun, the resulting effects were mostly rather small; either an increase in scatter and/or deviations from the means of controls were observed.

Magnetic Manipulations During Release. The most traditional way of interfering with magnetic information, already suggested by the French physicist MAURAIN in 1926, is releasing birds with magnets. Because of its simplicity, it is still used today. Table 6.2 summarizes the effects observed in pigeons and species of wild birds. Several of these experiments, especially the older ones, are inconclusive because of insufficient sample size. Nevertheless, a clear pattern emerges: homing performance is largely unaffected. Only YEAGLEY (1947) claimed slower homing in his experimental pigeons; however, this difference is marginal.

Table 6.2. The effect of magnets on displaced birds

Bird species	Place of magnet(s)	Effect on		Reference
		in.or.	homing	
Columbiformes, Columbidae:				
Pigeon, *Columba livia*	head?	-	- ?	Casamajor (1927)
» » »	wings	+?	+?	Yeagley (1947)
» » »	wings	- ?	- ?	Yeagley (1951)
» » »	wings	?	-	Gordon (1948)
» » »	neck; wings	-	-	Matthews (1951)
» » »	wings	?	-	van Riper and Kalmbach (1952)
» » »	back	+	-	Keeton (1971, 1972)[(*)][a]
» » »	right wing	-?	-?	Lamotte (1974)
» » »	back	+	?	T. Larkin and Keeton (1976)[b]
» » »	head	+	-	Visalberghi and Alleva (1979)
» » »	back	+	- ?	R. Wiltschko et al. (1981)*
» » »	head, neck, wings	- ?	?	Wallraff and Foa (1982)
» » »	head, neck, wings	+	- ?	Ioalè (1984)*[a]
» » »	head; back	+?	?	Ranvaud et al. (1986, 1991)
» » »	back	+	?	W. Wiltschko et al. (1986)[c]
» » »	back	+	?	W. Wiltschko et al. (1987a)*
» » »	back	-	-	B. Moore (1988)[a]
Ciconiformes, Ciconiidae:				
White Stork, *Ciconia ciconia*	head	?	?	Wodzicki et al. (1939)
Charadriiformes, Laridae:				
Black-backed Gull, *Larus fuscus*	legs	?	-	Matthews (1952)
Ring-billed Gull, *Larus delewarensis*	head	+	?	Southern (1972a)
Passeriformes, Hirundinidae:				
Barn Swallow, *Hirundo rustica*	head	-	-	Bochenski et al. (1960)
Procellariiformes, Procellariidae:				
Shearwater *Calonectris diomeda*	head, neck, wings	-	- ?	Massa et al. (1991)

in.or. = initial orientation; ?, not recorded; ? at + or – indicates unclear results. * indicates significant effects on initial orientation, which according to the circumstances of release (under overcast) or the test design (no properly developed sun compass) are usually considered to be compass effects; (*) indicates that this applies to part of the data.
[a] See Table 4.3; [b] see Fig. 6.7; [c] see Table 6.1.

A number of effects on initial orientation were observed. Some of these effects were recorded under circumstances where the sun compass could not be used, which suggested a compass effect rather than a 'map' effect; they are indicated in Table 6.2. Part of the data of KEETON (1971, 1972), the data of VISALBERGHI and ALLEVA (1979), T. LARKIN and KEETON (1976) and W. WILTSCHKO et al. (1986), however, cannot be interpreted in view of a compass effect, as they involved behavior under sun. The data sets of T. LARKIN and KEETON (1976) and W. WILTSCHKO et al. (1986) refer to the repeated releases from the same sites, already described in Section 6.1.1.2. The findings of RANVAUD et al. (1986, 1991) are difficult to classify, as the experiments were performed at the magnetic equator where the birds' magnetic compass is bimodal, at times when the sun crossed zenith. The effect of magnets varied between years: in 1983, magnets seemed to cause bimodality during morning and afternoon, which was no longer observed in 1984 and 1986; in 1984, on the other hand, the magnets tended to have a disorienting effect at noon which was not observed in 1983 and 1986 (RANVAUD et al. 1991).

Fields produced by coils around the pigeons' head (cf. Fig. 4.7) affected initial orientation fairly regularly also under sun, even if the effect was mostly rather small. C. WALCOTT (1977) found that the same treatment which had reversed the pigeons' orientation under overcast (C. WALCOTT and GREEN 1974) caused only small changes under sun; i.e. superimposed fields of 10 000 to 30 000 nT increased the scatter slightly; fields of 60 000 nT sometimes seemed to induce small deviations. VISALBERGHI and ALLEVA (1979) reported similar results. This type of experiment was resumed by LEDNOR and WALCOTT (1983), who equipped pigeons with coils producing a magnetic field of varying intensities. The mean bearings of these birds and those of controls showed differences in the range of 8 ° and 40 °; the sense of rotation seemed to be uncorrelated with the polarity of the field produced by the coils.

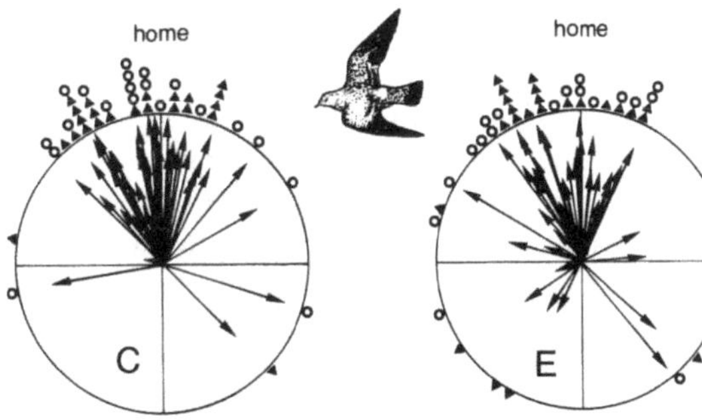

Fig. 6.10. Initial orientation of pigeons subjected to magnetic treatments when released under sun; second order analysis of published data. *Left* Controls; *right* experimental birds. *Solid triangles* Data from tests in which experimental birds carried small bar magnets (data from MATTHEWS 1951; KEETON 1971, 1972; VISALBERGHI and ALLEVA 1979; WALLRAFF and FOA 1982. For details on the placement of magnets, see Table 6.2); *open circles* data from tests in which experimental birds carried small battery-operated coils around their head, cf. Fig. 4.7 (data from C. WALCOTT 1977; VISALBERGHI and ALLEVA 1979). (After WALLRAFF 1983, modified)

In his 1983 review, WALLRAFF subjected all published data on magnetic manipulations of otherwise untreated pigeons released under sun to a second-order analysis (see Fig. 6.10). He found a significant difference in variance. The average vector length of experimental pigeons was about 25% shorter than that of controls. The mean directions of experimental birds also tended to be slightly more scattered.

PAPI and IOALÈ (1986) subjected pigeons to alternating fields of 0.14 Hz, with sinusoidal and square-shaped pulses (see Fig. 6.13). The treatments began 2 h before the birds were released and ended when they returned to the loft. Both had a disorientating effect on initial orientation, which was more pronounced when a rectangular pulse was applied. Homing performance was largely unaffected.

Aftereffects of Treatments Prior to Release. A number of studies performed in Italy subjected pigeons to various magnetic treatments before release, ending when the bird was set free. Initially, the magnetic field was manipulated during transportation to the release site, but later it became evident that the effect was not associated with transportation under experimental conditions (see Fig. 6.11); i.e. the effects were not based on the disruption of information on the outward journey. The changes in magnetic conditions were much more dramatic than those used in the experiments discussed in Section 5.2.1.2, involving strong reductions in total intensity and alternating fields of various frequencies and impulse shapes.

Displacing pigeons in iron containers which reduced the magnetic field to about 0.5% had a dramatic effect on initial orientation, inducing large deflections from the heading of controls and increasing scatter considerably (PAPI et al. 1978; BALDACCINI et al. 1979; BENVENUTI et al. 1982). Subjecting pigeons to alternating fields of completely irregular shape produced by three pairs of coils perpendicular to each other, each running on a different frequency, also resulted in considerable deviations from the mean of the controls and/or in marked increases of scatter. It was irrelevant whether the irregular fields were applied at the home loft, during transportation, or at the release site; however, the amount of disturbance was roughly correlated with the duration of the treatment and seemed to decrease as time elapsed after its termination (Fig. 6.11; PAPI et al. 1983). A later study, however, seemed to suggest that the effect persisted at a similar level at least for the 5-h duration of an average release. Repeated releases at the same site produced very similar deviations from the home course, suggesting that the effect was site-specific to some extent (TEYSSÈDRE 1986). These effects of irregularly alternating fields seemed to depend on the availability of atmospheric odors (WALLRAFF et al. 1986a; PAPI and IOALÈ 1988).

At other lofts, however, alternating fields of irregular shape hardly affected initial orientation: there was no significant effect on German or American pigeons (WALLRAFF 1980; WALLRAFF et al. 1986b; BENVENUTI 1986). This surprising finding that the identical treatments have different effects in different countries led to the question about possible reasons. Experiments with pigeons of German stock raised in Italy, together with young pigeons of Italian stock, indi-

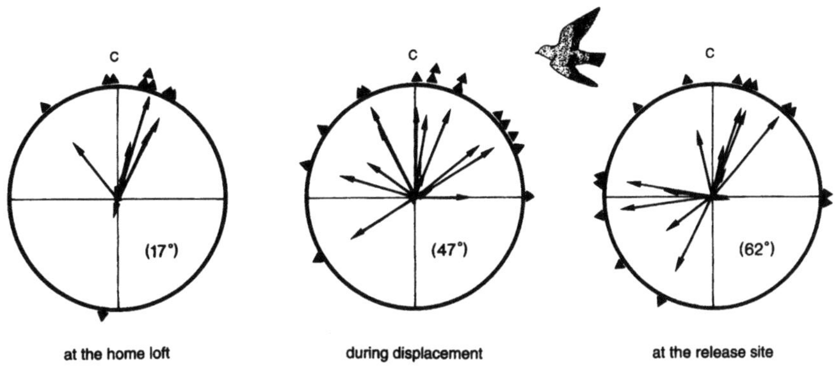

Fig. 6.11. Effect of treating Italian pigeons with irregularly alternating magnetic fields during different phases before release. The *arrows* represent the mean vectors of the experimental birds with respect to the mean of the untreated controls, *C*, which is set upward. The *numbers in parentheses* indicate the median deviation induced by the treatment; treatments at the home loft (*left*), during transportation (*center*), and at the release site (*right*) also imply increasing intervals of time passed after the end of treatment. (Data from Papi et al. 1983)

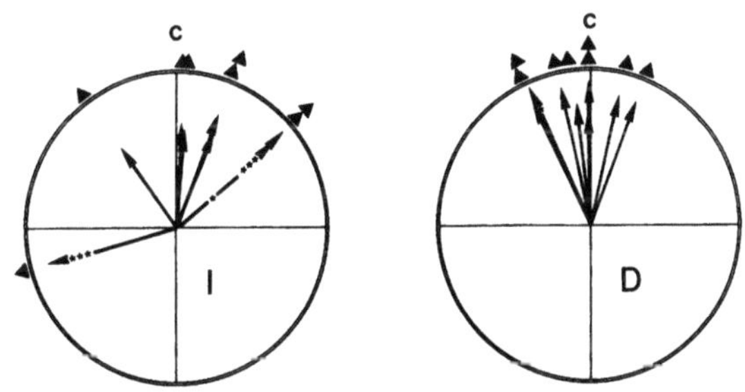

Fig. 6.12. Effect of a 3-h treatment with irregularly alternating magnetic fields before release on pigeons of Italian stock (*I, left diagram*) and of German stock (*D, right diagram*) raised together in Italy. *Arrows with asterisks* indicate that the respective vector was significantly different from that of the controls. *Symbols* as in Fig. 6.11. (Data from Benvenuti and Ioalè 1988)

cated that genetic dispositions are involved (Fig. 6.12; Benvenuti and Ioalè 1988).

In contrast to the irregular fields used in the experiments already described, Ioalè and Guidarini (1985) treated their pigeons with well-defined magnetic fields which oscillated along one axis only. They reported an interesting dependence on frequency and impulse shape: when frequencies of 0.14 Hz were used, a sinusoidal or triangular wave did not affect initial orientation, whereas a rectangular wave (i.e. an impulse shape able to transfer compass information, cf. Fig. 4.5) induced a considerable counterclockwise deviation (Fig. 6.13). Further experiments revealed that from 0.42 Hz upward, sine waves also caused deflections, the size of which increased with increasing frequency in a quasi-logarithmic way (Fig. 6.13). The direction of the induced deflection was clock

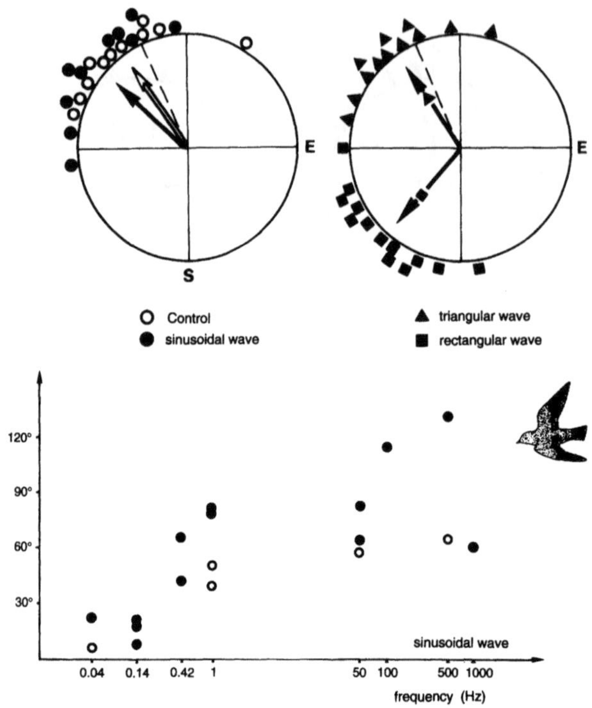

Fig. 6.13. Effect of pulse shape and frequency of alternating fields applied horizontally on the vanishing directions of pigeons in Italy. *Upper diagrams* Fields of 0.14 Hz with various pulse shapes. The *symbols at the periphery* indicate the vanishing bearings of individual pigeons; the *arrows* represent the respective mean vectors (After IOALÈ and GUIDARINI 1985). *Lower diagram* Correlation between the frequency of sinusoidal alternating magnetic fields and the respective deflection of the vanishing bearings. *Open symbols* Release site in the north, *solid symbols* release site in the south of the loft; in both cases, the deflection is towards west. (IOALÈ and TEYSSÈDRE 1989)

wise to the north and counterclockwise to the south of the home loft[5] (IOALÈ and TEYSSÈDRE 1989).

Extreme Fields and Treatments Designed to Affect Magnetite Particles. Another type of magnetic treatment was specifically aimed at interfering with magnetoreceptors in order to prevent access to meaningful magnetic information. Such a treatment was first used on Leach Petrels, *Oceanodroma leucorhoa* (Aves, Procellariiformes: Hylobatidae), whose heads were held for 30 s in a strong field of about 0.05 T. Homing performance was not affected, however (GRIFFIN 1940).

[5] Because this leads to westerly directions in both cases, IOALÈ and TEYSSÈDRE (1989) attributed the deflection to westerly 'preferred compass directions', a spontaneous tendency of their birds to fly west. It is doubtful whether this is a correct description of the behavior, because at high frequencies, the mean directions at the southern release site were shifted counterclockwise beyond the home direction of the northern site. This clearly speaks against a simple preference of western directions. Treatment with irregularly alternating fields (PAPI et al. 1983) did not result in such westerly tendencies in birds from the same loft.

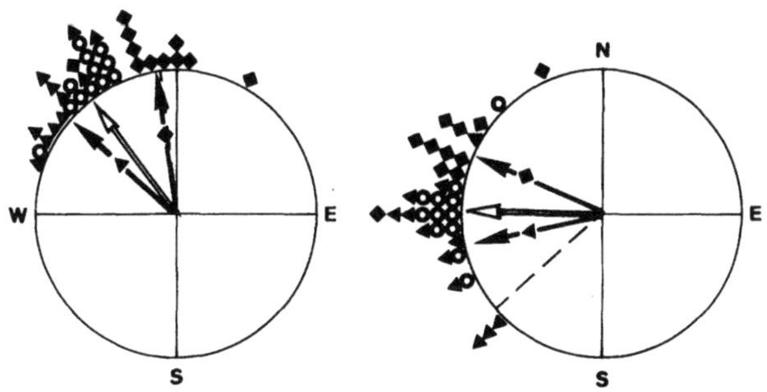

Fig. 6.14. Orientation of pigeons from the loft in Frankfurt a.M., Germany, treated with a short, strong magnetic pulse of 0.5 T and 5 ms duration, at a site 108 km south of the home loft (home direction 353 °, *left*) and at a site 129 km northeast of the home loft (home direction 229 °, *right*). *Open symbols* Untreated controls; *solid triangles* pigeons treated with the head forward; *solid diamonds* pigeons treated with the head turned 90 ° to the right. The distributions of the two treatments are significantly different from each other; in the *left diagram*, both are also significantly different from the one of the untreated controls; in the *right diagram*, the pigeons magnetized with the head turned 90 ° are different from the controls. *Symbols* as in the upper part of Fig. 6.13. (Data from W. WILTSCHKO and BEASON 1990)

A possibly long-lasting aftereffect of a similar treatment was observed after subjecting French pigeons for 1 min to an ultrahigh magnetic field of 10 T (about 200 000 times the strength of the geomagnetic field). Released the same day at different sites, the experimental birds deviated between 30 ° and 145 ° clockwise from untreated controls. Released again 4 weeks after treatment, the experimental birds showed a weak effect in only one of four releases (KIEPENHEUER et al. 1986). It is unclear whether or not this was a true aftereffect of the treatment. This type of experiment was continued by C. WALCOTT et al. (1988) in the USA. The authors attempted to affect the magnetization of ferromagnetic material which had been found in the pigeons' head (C. WALCOTT et al. 1979). The treatment was designed to alter the magnetic moment of single-domain particles by applying strong fields above 0.1 T and fields with steep gradients of 8 T/cm (see Sect. 9.3.3.2), but when the pigeons were released, their orientation seemed to be unaffected.

W. WILTSCHKO and BEASON (1990) treated German pigeons with a very short, strong magnetic pulse – too short to move the magnetite particles, but strong enough to reverse their polarity. At distances beyond 80 km, the vanishing bearings of birds treated with this pulse deviated from those of untreated controls in the range of ca. 20 °. The effect was highly variable, however, and appeared to be site-specific to some extent; at some release sites, it was negligible, while it was comparatively large at others, reaching up to 50 °. The direction of the deflection was found to depend on the direction of magnetization (Fig. 6.14). Some effect was still observed over the following days, but then it seemed to disappear. North American Bank Swallows, *Riparia riparia* (Passeriformes: Hirundinidae) were disoriented after the same treatment (BEASON 1994; see Sect. 9.3.3.2).

Interpreting Effects of Magnetic Manipulations. The small effects observed in experiments involving magnets or fields produced by coils during flight are of a similar size as the effects associated with temporal variations of the magnetic field and might be interpreted in view of magnetic 'map' factors. The aftereffects of treatments with magnetic fields having non-physiological qualities, however, remain mysterious. One obvious possibility is that they might reflect some unspecific trauma caused by the sheer strength of the treatment which was overcome only after a certain time had passed (see PAPI et al. 1992). This interpretation is plausible when disorientation or an increase in scatter is observed; however, it seems hardly applicable in the case of deflections, which involve oriented behavior by all members of the group in directions different from those of untreated controls. The site specificity of such effects (IOALÈ and TEYSSÈDRE 1989; W. WILTSCHKO and BEASON 1990) and the correlation observed by IOALÈ and TEYSSÈDRE (1989; cf. Fig. 6.13, lower part) also seem to argue against non-specific effects.

Any interpretation of the aftereffects of treatments with non-physiological magnetic fields is necessarily handicapped by the fact that the basis of their effectiveness is still unclear. Several hypotheses have been proposed, and the effects mentioned in Section 6.1.1.3 will be discussed again in Sections 9.3.3 and 9.3.4 in view of their possible significance for specific types of receptor mechanisms.

6.1.2 An Assessment of the Role of Magnetic Factors in the Avian 'Map'

When VIGUIER (1882) first suggested a magnetic 'map', he proposed what he thought to be a general explanation for avian navigation. Today, we are aware that orientation cannot be explained by one factor alone, and the questions are now whether magnetic parameters are normally involved and what their nature and their specific role might be. The interpretation of experimental findings is complicated by the multifactorial nature of the 'map', which does not allow an easy assessment of the role of any one factor. With respect to magnetic information, two aspects are especially striking, namely, the great variability of magnetic effects and the almost negligible effects on homing. Also, the findings on magnetic map factors do not present a consistent picture as yet.

6.1.2.1 Variability

One general characteristic of magnetic map effects is the immense variability observed at all levels. For instance, the response to weak anomalies differed between regions (WAGNER 1976; FREI 1982; DORNFELDT 1991), not all pigeons were disoriented at strong anomalies, and the effect seemed to vary between years (C. WALCOTT 1992). K-correlations could not be observed at all sites (KOWALSKI et al. 1988; SCHMIDT-KOENIG and GANZHORN 1991), and differences between morning and noon bearings could not be observed in all years (M. BECKER et al. 1991). Moreover, even when these effects were demonstrated, not all individuals within a group showed a K-effect or a variation between morning and noon (KOWALSKI et al. 1988; M. BECKER et al. 1991). The same appears to apply to the effect of magnetic treatments. RANVAUD et al. (1991) observed great differences in the effect of magnets between years. When, as part of the K-effect study, pigeons were released alternately with magnets 'North up' (i.e. the north pole of the magnet towards the head) and magnets 'South up', the difference caused by the two treatments was less than 3 ° in the majority of birds; in some individuals, however, it was up to 30 ° (T. LARKIN and KEETON 1976). Alternating magnetic fields with irregular shapes had a strong disorienting effect on Italian pigeons, whereas the orientation of German and American pigeons was not significantly affected.

This does not mean, however, that these effects are usually small and fade into the general noise. On the contrary, when they occur, they are highly significant and quite spectacular; yet, they do not occur everywhere, everytime. Magnetic map effects share this peculiarity with other findings associated with map information. A prominent example is the effect of olfactory deprivation, which varies between regions, and, within a region, may differ between sites and even between individual birds, depending on the way the pigeons are raised (see KEETON 1980; W. WILTSCHKO et al. 1987c; R. WILTSCHKO and WILTSCHKO 1989). Apparently, these variabilities reflect the multifactorial nature of the 'map', which allows birds several degrees of freedom in the choice of their navigational factors.

Little is known as to why birds prefer one type of cue over others in a given situation. The various responses to magnetic anomalies were found to be related to the local conditions at the sites where the birds grew up; small differences within 2.5 km led to marked differences in behavior (C. WALCOTT 1992). The varying responses to irregularly alternating fields could be attributed to genetic predisposition (BENVENUTI and IOALÈ 1988). Yet, in most cases, the reasons for the variability remain unknown. Especially if members of the same group of pigeons from the same stock and the same loft having comparable experiences during the first months of life are involved, differences in behavior are hard to explain. In olfactory deprivation, birds housed only 50 m apart and trained together showed different responses; in this case, differences in the wind exposition of the loft proved crucial (R. WILTSCHKO et al. 1989). Thus, it cannot be excluded that seemingly insignificant differences in experience, for example during spontaneous free flights, might have a lasting effect on the relative significance of navigational factors by establishing an individual ranking, e.g. inducing preferences for certain types of cues and hesitation to rely on others.

6.1.2.2 Effects on Homing

Another characteristic of magnetic phenomena is that homing performance is hardly ever affected. This is not surprising when the initial orientation is only slightly altered, but it is noteworthy in cases where magnetic treatments cause marked deviations from the controls which should lead to an initial detour. In general, pigeons seem to be able to cope with magnetic manipulations fairly easily. Radio tracking of pigeons deviating about 60 ° counterclockwise from the home course after being treated with irregularly alternating magnetic fields revealed that most birds turned towards home within 4 km (TEYSSÈDRE 1986). Unfortunately, the pigeons were released at a site very close to the sea so that it is unclear whether they overcame the effect of the treatment, whether they turned to non-magnetic orientation cues, or whether they simply followed the coastline.

Interestingly, homing performance is also hardly affected in cases where the effects of magnets are usually attributed to compass orientation (cf. Table 6.2). Magnets, in particular, mask the geomagnetic information by adding a constant value so that the total field depends on the direction of flight. Here, the birds might quickly learn to overcome this difficulty and to regain access to magnetic information. KEETON (1972) reported an interesting observation about young pigeons that had been equipped with magnets and released for training flights. When these birds were later released in critical tests under overcast without magnets, they were as well oriented as the controls that had worn brass bars. Yet they took significantly longer to vanish from sight, indicating that they needed time to readjust to normal magnetic information. However, it would seem much easier to extract compass information than to read the tiny difference in local magnetic values from the artificially distorted field.

6.1.2.3 Open Questions About 'Map' Effects

The experiments described above indicate that magnetic factors are involved in avian navigation, yet a consistent picture does not emerge. Some findings appear to be at variance and even contradict each other. The effect of magnets provides a good example for the difficulties one faces when magnetic map effects are to be interpreted. Under sun, magnets mostly cause slight deviations, a certain increase in scatter, or both, but they do not markedly change initial orientation (see WALLRAFF 1983). The magnets normally used are so strong (e.g. pole strength in the range of 0.025 T, KEETON 1971; 0.045 T, VISALBERGHI and ALLEVA 1979) that they should prevent the birds from obtaining meaningful magnetic information; yet pigeons carrying magnets are not disoriented under sun, in contrast to many pigeons released at strong anomalies. This seems paradoxical, since even strong anomalies represent a much weaker distortion of the magnetic field than a magnet. It might be argued that pigeons had already determined their home course *before* they were equipped with the magnets at the release site, and that they were unable to do so at strong anomalies. In this case, however, it is unclear why magnets should have any effect at all.

Magnets, on the other hand, cancelled effects associated with temporal variations of the magnetic field; in fact, these effects are considered to have a magnetic nature *because* they are suppressed by magnets. In these cases, pigeons released with magnets were well oriented, although they must be assumed to be deprived of magnetic map information. When the birds were repeatedly released from only one site, one may argue that because of their familiarity with the release site, they had no problem relying on other means. Yet when pigeons lived under the condition of a 'permanent' 6 h slow shift, magnets suppressed the deviation from the controls at slightly unfamiliar sites (cf. Fig. 6.9). In this case, pigeons presumably deprived of meaningful magnetic information were oriented like the birds that had all natural cues available. These findings are puzzling, because they imply that magnetic information is normally used when available, but at the same time they show that magnetic cues can be given up without loss. Non-magnetic factors proved to be sufficient to ensure oriented behavior comparable to that of the controls. This is also indicated by the almost negligible effect on homing performance (cf. Sect. 6.1.2.2). In view of this, however, it is difficult to understand why pigeons should be disoriented at strong magnetic anomalies without assuming that non-magnetic cues are likewise distorted.

A phenomenon associated with temporal variations of the magnetic field poses another problem: the effects of K-variation and 'permanent' shift manifest themselves in deflections rather than in a constant directional tendency (cf. Sect. 6.1.1.2). For example, birds living under the conditions of a 6-h slow shift, when released at noon (which was their subjective morning), encountered total intensities lower than expected. Lower intensity values are normally found to the south and hence might be expected to induce northerly flight directions. Instead, we find clockwise *deflections*, i.e. a northerly shift to the east, but a southerly shift to the west (W. Wiltschko et al. 1986). Kiepenheuer et al. (1986) reported that treatment with an ultrastrong magnetic field also induced a clockwise shift at release sites in various directions from home.

These findings are puzzling and cannot be interpreted in view of the a gradient 'map' as described in the introductory section of 6.1 (cf. Wallraff 1974). The classic concept of a gradient map would predict that a value altered in a constant way induces a directional tendency, not a rotation, as it obviously does in some magnetic phenomena. The effects described above thus might suggest yet another involvement of magnetic information in the map step of homing, the nature of which is totally unknown. Interpretations other than the traditional ones have to be sought.

6.1.2.4 *Significance of Magnetic 'Map' Effects*

It is not easy to assess the general role magnetic factors might play in the pigeons' navigational 'map'. It is clear that they represent just one type of information among others in a multifactorial system. The birds obviously have alternatives. The effects on initial orientation seem to suggest that magnetic information is normally involved in the orientation process, while the negligible effect on homing indicates that magnetic cues are not essential.

Several authors reviewed the evidence on magnetic map factors and discussed their possible role in avian navigation. WALLRAFF (1983, p. 655), after analyzing the effect of magnetic treatments (cf. Fig. 6.10), considered their contribution to the whole 'map' system to be "*marginal at best*". GOULD (1985) considered magnetic map factors to be a realistic possibility, yet points out problems arising from the nature of the potential magnetic map factors. Other authors (e.g. C. WALCOTT 1991) hesitate to accept an important role of magnetic cues in the 'map' because the available evidence shows the described amount of unexplained variability and seems to disagree in several aspects.

At the moment, a final statement on the role of magnetic parameters in the pigeons' 'map' is indeed not possible. Taken together, the positive evidence can hardly be dismissed, but a convincing hypothesis explaining the known details of magnetic phenomena in a consistent way has not yet been forwarded. We are still far from understanding how magnetic map factors might work.[6]

6.1.3 Magnetic 'Map' Factors in the Orientation of Other Vertebrates

Position finding based on a comparison of local magnetic values with home values requires that the animals can detect the differences. Therefore, magnetic map factors appear to be usable only when the distances covered are large enough to involve a certain amount of change in the respective parameters. The usefulness should improve with increasing distance. So it is not surprising that a magnetic map has been repeatedly discussed for animals that show large-scale migration movements in the open ocean. The navigational task of salmons returning to their natal river system after spending a long period at sea has been described in Section 5.3.2; QUINN (1984) suggested that salmons may use a map based on such parameters as magnetic inclination and declination to determine their home course. He even designed a critical experiment that would allow one to distinguish between true navigation and other possible mechanisms. Tests have not been carried out, however. LOHMANN (1991) and SALMON and WYNEKEN (1994) considered magnetic parameters as possible components of a map that would allow marine turtles to determine their position with respect to their nesting beach.

[6] Additional theoretical problems arising from the nature of the available magnetic variables also apply to the strategy of following magnetic contours and will be discussed in Section 6.4.

Magnetic map factors have also been discussed for animals whose movements involve shorter distances. When young alligators, *Alligator mississippiensis* (Crocodylia), were displaced up to 35 km, animals in their second year of life showed homeward-directed orientation which proved to be independent of the outward journey, suggesting that they made use of some kind of navigational 'map' (RODDA 1984). Analyzing his data in view of magnetic parameters, RODDA found a significant correlation of the directions chosen by the alligators in an arena with minute changes in the angle of inclination in the range of 10^{-3} degrees. Such a correlation was not observed during a period of unusual magnetic storm activity. RODDA (1984) claimed that these effects are associated with a short-term assessment of geomagnetic information, because the best correlations were found with the magnetic fluctuations at the time of release and not with those in the period of 1 h prior to release. Correlations with the variation within the preceding 12 h, as in pigeons, were not considered.

The use of magnetic factors has been discussed for a few other vertebrate species that normally move over short distances only. One is the newt *Notophthalmus viridescens* (Urodela: Salamandridae); after a displacement over 20 to 30 km, the newts preferred the direction towards their home pond in an arena (PHILLIPS 1986a,b). PHILLIPS and BORLAND (1994), observing that shoreward orientation and homing are affected differently by treatments such as inversion of the vertical component (cf. Sect. 4.2.2.1) and light of specific wavelengths (see Sect. 9.2.3.2), interpeted their findings in view of magnetic map factors. SINSCH (1990b) discussed the use of a magnetic map as a theoretical possibility in the orientation of toads of the genus *Bufo* (Anura), pointing out the problems resulting from small gradients and the limited range of the toads' movements. BOVET and colleagues (1988) released wood mice, *Apodemus sylvaticus* and *A. flavicollis* (Rodentia: Muridae), at sites within the same magnetic anomaly where FREI and WAGNER (1976) observed large deviations from home in their pigeon study. In general, the results provided only limited support to the use of magnetic map factors; yet fewer mice released in the immediate surroundings of the magnetic trough of the anomaly homed.

6.2 Orientation Along the Magnetic Relief

A different navigational strategy, namely, migration along magnetic contours on routes minimizing changes in total intensity, has been suggested for some marine animals. At sea, the magnetic field is generally less irregular than on land. Systems of roughly linear magnetic ridges and valleys run parallel to the mid-oceanic ridges where new crust is formed (VACQUIER 1972; cf. Sect. 1.1.2). With some dislocations, they may extend over thousands of kilometers and thus may serve as guidelines for large-scale north-south movements (KIRSCHVINK et al. 1986). However, even on smaller scales, the magnetic relief might structure an otherwise featureless environment and thus provide orientational cues.

6.2.1 The Migration of Cetaceans

Many cetaceans show seasonal movements between higher and lower latitudes. During summer, the polar and subpolar seas are areas of high productivity and offer abundant food, whereas in winter, ice coverage and the equartorward movements of prey species force the cetaceans to leave. Also, the females give birth to their calves in the warmer waters at temperate and subtropical latitudes. Regular annual migrations are reported for the majority of mysticetes. The movements of odontocetes are less pronounced, especially in species with a largely tropical distribution, but a considerable number of odontocetes also show latitudinal shifts of distribution in the course of the year (see BAKER 1978 for more details).

6.2.1.1 Evidence from Strandings and Sightings

First indications that magnetic parameters might be involved in the orientation of cetaceans emerged from stranding records. When KLINOWSKA (1985a, 1986) thoroughly analyzed all reports on strandings along the British coastlines, she found some marked differences between *passive strandings* of dead bodies and *active strandings* of live animals: most dead and decomposed bodies belonged to species living in coastal waters; the stranding sites were widely distributed,

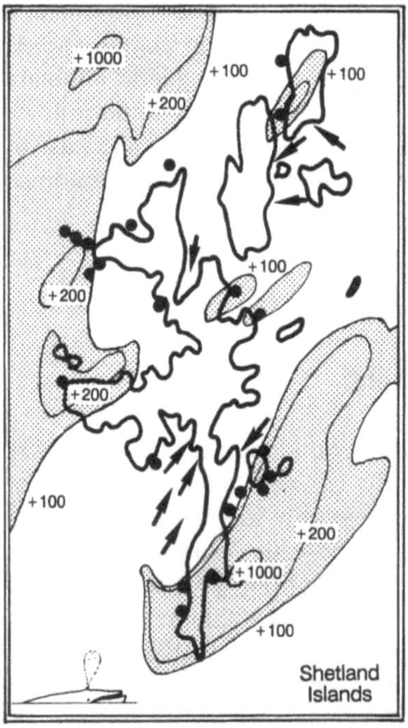

Fig. 6.15. Cetacean stranding sites at the Shetland Islands in the northeast of Scotland. *Arrows* mark live strandings, *circles* indicate sites where decomposed bodies were found. Areas of locally anomalous high intensity are *hatched*, contours are marked in nT. (After KLINOWSKA 1985a)

Fig. 6.16. Profiles of the magnetic field around the 16 stranding positions of the Atlantic White-sided Dolphin, *Lagenorhynchus acutus* (Odontoceti: Delphinidae). The stranding events are arranged so that their geographic and magnetic positions coincide at the center of the coordinate system. For this analysis, the coastal area was divided in small segments of roughly 4 × 4 km; away from the center of the coordinates, magnetic intensity changes are plotted as a function of the coastline distance within each segment, each line ending at the end of the respective segment. (KIRSCHVINK et al. 1986)

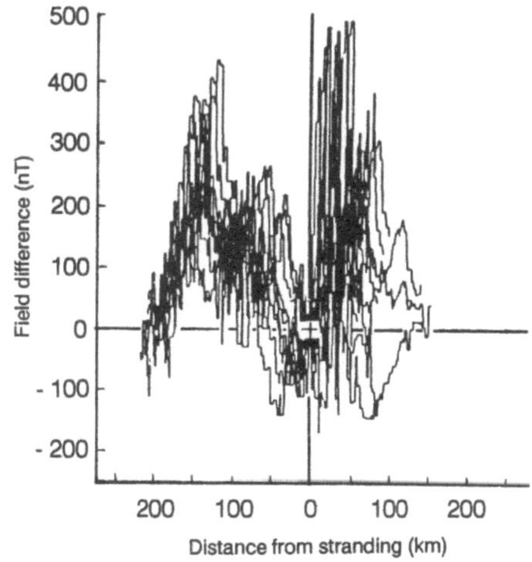

with their locations reflecting tides and currents, numbers of observers (e.g. coast guard establishments), and fishing activity. Active strandings, in contrast, were characterized by an overrepresentation of species normally living in the open ocean; they are grouped at certain locations, some of which have witnessed such events more than once. These sites are independent from tides, currents, etc. and have no common local hydrography or geography. For example, the 136 live strandings reported in Britain in the 20th century were on steep rocky shores as well as on gently sloping beaches or sandy and muddy estuaries. However, the sites share a common magnetic feature: live strandings mainly occurred at locations where 'valleys' of relatively low magnetic intensity cross the coastline or are blocked by islands, often in the vicinity of an area with local high intensity (Fig. 6.15). Locations with 'drive' fishery tradition, i.e. beaches on the Faroe, the Shetland and the Orkney Islands as well as sites on the Newfoundland coast where herds of pilot whales are driven ashore by local fishermen in boats, also share this characteristic (KLINOWSKA 1989). An analysis of temporal geomagnetic records suggested a relationship between magnetic disturbances and active strandings, which often occur within 1 or 2 days after major magnetic storms. The exact relationship varies for different regions, being 1 day in the north (Orkney Islands, Scotland, etc.) and 2 days at the British south coast (KLINOWSKA 1985b).

A similar analysis of stranding data from the east coast of North America between Maine and Texas (KIRSCHVINK et al. 1986; KIRSCHVINK 1990) was in agreement with the British findings: live strandings were found mainly near locations where local magnetic minima intersected the coastline. Such a tendency was also significant for 10 of the 14 species or species groups analyzed separately (Fig. 6.16; KIRSCHVINK 1990). A temporal relationship between live strandings and magnetic storms did not emerge in the American data (KIRSCHVINK, pers. comm.). BRABYN and MCLEAN (1992), in contrast, found that mass strandings at the New Zealand coast were associated with certain features of the

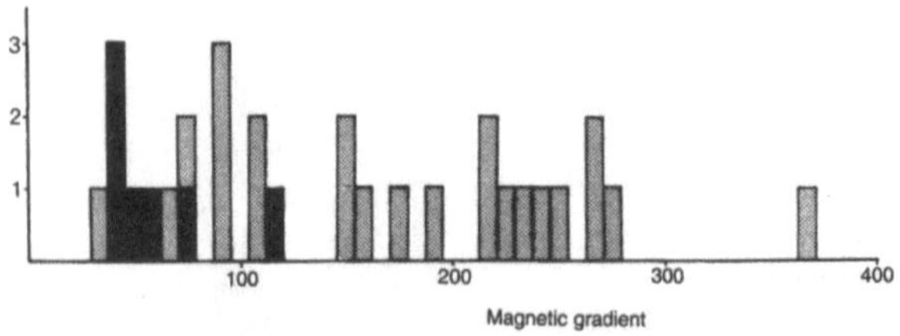

Fig. 6.17. Frequency distribution of segments (of 2 × 2 km) with the respective magnetic gradients, inspected by airplane in autumn. The *dark bars* indicate the segments where non-feeding Fin Whales, *Balaenoptera physalus* (Mystacoceti), were observed. (After WALKER et al. 1992)

coastal topography, whereas a relationship to geomagnetic 'valleys' could not be established. It should be noted, however, that the marine magnetic features around New Zealand are radically different from those in the North Atlantic, lacking the regular north-south running 'ridges' and 'valleys'.

The next step was to look for information on the behavior of live whales and correlate sighting data with magnetic topography, namely, local intensity and local gradients. WALKER et al. (1992) analyzed 94 sightings of Fin Whales, *Balaenoptera physalus* (Mysteceti), recorded by the American Cetacean and Turtle Assessment Program between October 1978 and January 1982 along the east coast of North America. The study area ranged from Cape Hatteras, North Carolina, to the Gulf of Maine, and included continental shelf waters between the shore and 9.3 km seaward of the 1829 m isoline. An overall correlation was not found. However, when the data were separated according to season, significantly more winter sightings than expected by chance were observed at locations with low magnetic intensity. Also, sightings in winter, spring and fall were associated with low magnetic gradients (Fig. 6.17). Excluding the data from animals that were observed to be engaged in feeding activities led to more pronounced relationships. A correlation with the bathymetric variables tested for comparison was not found for any of the data sets (WALKER et al. 1992). When tracks of two odontocetean species recorded by EVANS (1974) were analyzed in view of magnetic topography, they were also found to follow magnetic contours (KLINOWSKA 1986).

6.2.1.2 Following the Relief as Navigational Strategy

The findings that cetaceans which strand alive are usually outside their normal oceanic habitat and the relation of stranding events to magnetic topography suggested that live strandings are caused by navigational errors of an orientation system relying on magnetic parameters. This, together with the data on the distribution of whales at sea, led to the hypothesis that pelagic cetaceans orient their oceanwide migrations with the help of the marine magnetic relief. In the open ocean, with hardly any other cues available, the regular course of magnetic 'ridges' and 'valleys' parallel to the oceanic ridges would provide important guidelines for a roughly north-south migration. KLINOWSKA (1986) assumed that whales follow magnetic isolines along the slopes of valleys and avoid steep positive anomalies. KIRSCHVINK et al. (1986; KIRSCHVINK 1990), in contrast, suggested traveling along magnetic minima as a most useful strategy. Minima are relatively continuous, while local maxima show a more irregular pattern. Navigational errors, in part caused by irregularities of daily variations, might cause errors leading whales towards the coastlines. From magnetic charts of the sea areas around Britain and the time relationship between magnetic storms and strandings, KLINOWSKA (1985b, 1988) determined a specific location 300 km from Cornwall and one north of Shetland where such errors can easily occur because magnetic offshore formations branch off towards the coast. This is also true for the 'Blake Spur' anomaly discussed by KIRSCHVINK et al. (1986), which has a fork near Florida (KIRSCHVINK, pers. comm.).

Details of this navigational hypothesis are not yet clear, however. KLINOWSKA (1985b, 1986, 1987, 1989) suggested that magnetic topography is used as a 'map', and that animals monitor their position and progression on this map. Yet the strategies of following isolines and of making use of a map are so fundamentally different that they would seem to exclude each other; i.e. the magnetic relief would guide animals directly, whereas the use of a map means that magnetic data are interpreted to determine courses, which requires foreknowledge on their spatial distribution (cf. Sect. 6.1). It is unclear how the cetaceans would acquire the necessary knowledge on magnetic topography, or, if that knowledge were available, why they should follow magnetic isolines instead of taking direct routes from one location to the next.

6.2.2 Orientation of Sharks Near Seamounts

Studies on the spatial behavior of Scalloped Hammerhead Sharks, *Sphyrna lewini* (Elasmobranchii, Carcharhiniformes), in the Gulf of California revealed that many individuals stay together as a group around a seamount during most of the day. Late in the afternoon, single individuals and/or small groups venture out into the surrounding pelagic environment, apparently on nocturnal foraging trips, from which they return early in the morning (KLIMLEY and NELSON 1984; KLIMLEY 1993).

Some sharks equipped with transmitters were found to move along fairly straight routes. It became evident that various individuals frequently followed, at least in part, similar routes, suggesting something like 'preferred roads' for movements in the vicinity of the seamount. Yet the tracks showed no relationship to features like current or bottom topography. A comparison with the local pattern of magnetic intensity revealed that more tracks than expected by chance were associated with marked changes in the steepness of the magnetic gradients. The actual tracks were often found to pass near maxima and minima (KLIMLEY 1993). From these findings, KLIMLEY (1993) concluded that the hammerhead sharks show a tendency to follow features of the magnetic relief. He termed the respective behavior 'geomagnetic topotaxis' and pointed out similarities to human pioneers who followed local topography when venturing across an unfamiliar area.

It is unclear whether other shark species show similar behavior. An analysis of several tracks of Blue Sharks, *Prionace glauca* (Elasmobranchii, Carcharhiniformes: Carcharhinidae), recorded along the Atlantic Coast of North America did not reveal a relationship with magnetic contours (CAREY and SCHAROLD 1990).

6.2.3 The Relief Hypothesis: Problems and Open Questions

The hypothesis that marine animals make use of the magnetic relief as a guide appears attractive, because such a strategy seems to be meaningful in an oceanic environment where other potential cues are scarce and/or not readily accessible. Orientation with the help of features of the magnetic relief has also been considered for young marine turtles during their pelagic phase (SALMON and WYNEKEN 1994). However, this strategy means that the routes taken are determined by magnetic topography. Animals following the magnetic relief cannot move randomly between arbitrary locations, and they cannot head directly for any specific goal. Their movements would be restricted to the routes leading along the given 'ridges' and 'valleys'[7]. This may not be so important for cetaceans on their large-scale migrations between higher and lower latitudes; such limitations seem to be rather disadvantageous, however, for animals on foraging trips.

The critical problem of the relief hypothesis, however, lies in the question of how animals moving along magnetic 'ridges' and 'valleys' are able to identify the future course of the magnetic relief. In contrast to the topographical structures frequently quoted as an analogy, the course of magnetic features cannot be perceived over distances. An animal cannot anticipate in which direction a magnetic ridge or valley will continue. Even large whales would be too small to detect gradients between their head and the end of their body. So any direction of movement has to be chosen on a trial-and-error basis. It might be promising to proceed with the previous heading, assuming that the contours are roughly straight. When intensities change, however, it is not clear on which side the lower intensity or the smallest change is to be found. The animals would have to search, wandering to and fro. The severity of this problem is difficult to assess, as the minimum intensity difference detectable by marine animals is unknown. None of the authors promoting the relief hypothesis have suggested

[7] This orientation strategy, with no specific goal defined, does not fit into either of the two categories depicted in Fig. 2.3. In whale migration, routes immediately depending on the magnetic relief show affinities to direct orientation; the seasonal pattern, on the other hand, is similar to bird migration with its innate tendencies to follow specific compass headings. The orientation proposed for sharks would be analogous to the use of linear landmarks, which serve as auxiliary mechanisms in many orientation processes.

a convincing solution to this problem so far. Thus, the strategy of following magnetic valleys, although it may appear simple at first glance, per se is not without difficulties.

Another question concerns auxiliary mechanisms. For example, a magnetic compass could provide a means to maintain a straight heading and thus may facilitate finding the continuation of a given 'valley'. KLIMLEY (1993, p. 19) points out that "*it is essential to distinguish* (magnetic) *topotaxis from a compass sense...*", yet without considering that the two mechanisms complement each other and might be used together. KLINOWSKA (1987, p. 47), on the other hand, with regard to cetaceans, explicitly states: "*They are not using the directional information of the Earth's field, as we do with our compasses,...*", and (p. 48) "*No compass is involved*". The available evidence, however, being of correlative nature, is insufficient to exclude this possibility. From a theoretical point of view, a magnetic compass would be most helpful for migrating whales: magnetic valleys are axial formations, and any animals following them must be aware of the direction in which they are heading. Hence, animals that migrate seasonally between higher and lower latitudes must generally be expected to possess a compass. It is true that such a mechanism has not yet been described for cetaceans, but it has been demonstrated in other marine animals (cf. Sect. 4.2). In view of this, it would be rather surprising if the cetaceans did not make use of one.

6.3 Magnetic Parameters Controlling the Routes of Extended Migrations

On worldwide migrations, animals inevitably encounter changing values of total intensity and inclination. For example, migrating birds moving equatorward in fall meet decreasing values of both parameters, which increase again beyond the magnetic equator (cf. Sect. 1.1.1). Hence, it seems possible that this kind of information is incorporated in migratory programs which control the migration route. For example, by indicating the position at critical points, specific magnetic conditions might ensure that animals do not proceed beyond certain locations, thus preventing them from entering unsuitable habitats.

6.3.1 Magnetic Conditions Acting as Triggers During Bird Migration

In migrating birds, two situations have been described where specific magnetic conditions modify the course. The first case concerns migrants that bypass extended areas having unfavorable conditions (cf. Sect. 5.2.2.5); these birds must alter their course after reaching a certain region. In several species, laboratory

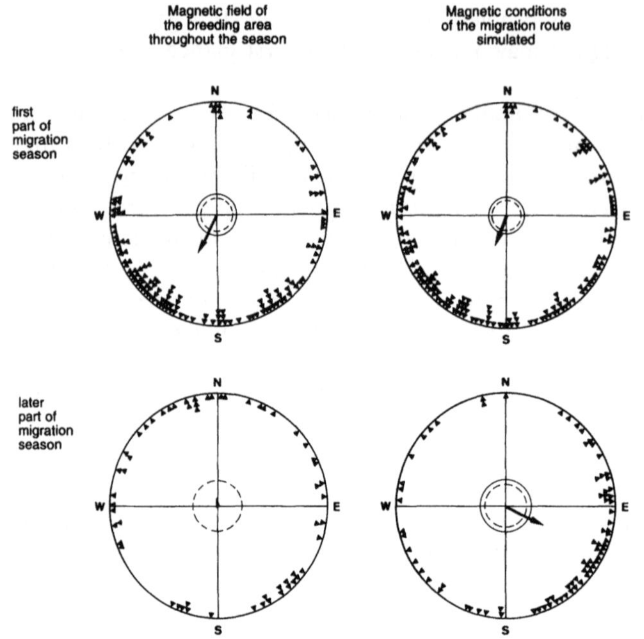

Fig. 6.18. Orientation of hand-raised Pied Flycatchers, *Ficedula hypoleuca*, during autumn migration. *Left diagrams* Birds tested in the local geomagnetic field of the breeding grounds (46 000 nT, 66 ° inclination) from late August to November; *right diagrams* birds tested in the local geomagnetic field of their breeding grounds from late August to 20 September, and in a field of 42 000 nT, 52 ° inclination from 21 September to 14 October (*upper right diagram*); in a field of 39 000 nT, 35 ° inclination from 15 October to 5 November, and in a field of 34 000 nT, 10 ° inclination after 6 November (*lower right diagram*). *Symbols* as in Fig. 5.4. (After Beck and Wiltschko 1988)

tests show that this change in direction is controlled by an endogenous time program (see Gwinner and Wiltschko 1978; Helbig et al. 1989; Munro et al. 1993). In nature, additional external factors may be involved to assure that the change occurs at the right location. Pied Flycatchers, *Ficedula hypoleuca* (Passeriformes: Muscicapidae), however, require specific magnetic conditions to alter their directional tendencies. Central European individuals first migrate on a southwesterly course to the Iberian Peninsula, then change to a south-southeasterly course to reach their wintering grounds south of the Sahara. Kept under the conditions of their breeding area, Pied Flycatchers first showed directional tendencies towards SW, which correspond to the first leg of their migration route. However, after mid-October, when their free-flying conspecifics leave the Mediterranean on a southeasterly course, these birds were no longer oriented. A second group of flycatchers was kept under magnetic conditions which roughly simulated those along the migration route. Total intensity and inclination were decreased in three steps from 46 000 nT, 66 ° inclination to 34 000 nT, 10 ° inclination. During mid-October, when these birds lived in a field of 39 000 nT, 35 ° inclination, as they would normally experience in northern Africa, they showed significant southeasterly tendencies (Fig. 6.18), which they maintained until the end of the migratory season (BECK and WILTSCHKO 1988). These findings show that a component of the magnetic field of northern Africa serves as a signal to activate the second part of the migratory program which guides Pied Flycatchers from Iberia towards their African wintering grounds.

This triggering mechanism works in connection with an endogenous time program. When the birds were transferred to the simulated magnetic conditions of northern Africa ahead of time at the beginning of migration in August, they were disoriented and continued to be disoriented during the entire season (BECK and WILTSCHKO 1988).

So far, this type of an external control of a change in migratory direction has been found in one species only. It is unknown how widespread it may be among migratory birds. Another specific magnetic condition, however, might be of general importance in eliciting a change in behavior. It concerns transequatorial migrants and has already been described in Section 4.1.3.2: at the magnetic equator, the *horizontal course of the field lines* serves as a trigger, causing the birds to cease flying 'equatorward' and begin flying 'poleward' (cf. Fig. 4.6). As the birds use an inclination compass, this change is crucial in order to maintain the same geographic heading beyond the the magnetic equator in the (magnetically) southern hemisphere (W. WILTSCHKO and WILTSCHKO 1992).

6.3.2 Control of Oceanic Migration in Young Sea Turtles

Recent data of LOHMANN and LOHMANN (1994b) appear to suggest a mechanism based on magnetic inclination controlling the migration of Loggerhead Turtles, *Caretta caretta* (Testudines: Cheloniidae). The authors tested newly emerged hatchlings from Florida beaches; young turtles from this population are assumed to spend the first years of their life in the North Atlantic gyre and the Sargasso Sea (CARR 1986). In control tests in the local geomagnetic field (inclination 57 °), the turtles headed in the usual eastward direction set by a dim light prior to testing (cf. Sect. 4.2.3.2). When tested at inclinations of 60 ° and 30 °, as they occur at the edges of the North Atlantic gyre, they preferred south-southwesterly and northeasterly headings, respectively. In fields with steeper or flatter inclinations, they were disoriented; the same was true, however, in fields of 45 ° inclination.

As the responses in fields of 60 ° and 30 ° inclination appear suitable to guide young turtles from the edges of the North Atlantic gyre back towards the center. LOHMANN and LOHMANN (1994b) discuss their data in view of the hypothesis that the inclination of the ambient magnetic field may elicit specific directional tendencies, even months before the young turtles could encounter the respective conditions in nature. Details are not yet known, however, and the disorientation at 45 ° inclination remains puzzling. Further analyses are in progress. In nature, additional factors might be involved in the mechanisms which keep marine turtles within a suitable habitat.

6.3.3 Immediate Control of the Migration Route

A theoretical model of how the route of migration could be controlled directly by the ambient magnetic field was forwarded by KIEPENHEUER (1984). The idea is based on the assumption that animals may possess a certain magnetic sensor which measures the apparent angle of dip. It is supposed to be fixed in such a way that it reads zero when the animal moves in its migratory direction. Any deviation would lead to a noticeable stimulus and thus evoke responses which bring the animal back on course. The model was devised for migrating birds and predicts that decreasing dip angles encountered on the way south would force the birds on routes which gradually shift from southwest (or southeast) towards south, being due south at the magnetic equator.

A main weakness of this model is the proposed intrinsic coupling of the magnetic sensor and the migratory course. Unlike a compass, the model allows only one specific direction and its reverse in a given field. This is in disagreement with experimental evidence regarding birds, where changing directions as a function of season have been observed under constant magnetic conditions (e.g. GWINNER and WILTSCHKO 1978; HELBIG et al. 1989; MUNRO et al. 1993). Also, in nature, the migration routes of many species do not follow the predictions of the model, e.g. changes in direction by many European migrants occur far north, those in the Australian honeyeater far south of the magnetic equator (cf. Fig. 5.9). The model, although an interesting concept, fails to describe the control of bird migration properly.

6.4 Theoretical Considerations

The examples given in this chapter concern different uses of magnetic information: magnetic factors as part of the navigational 'map' for determining position, direct guidance by the magnetic relief, and specific local magnetic values as triggers to initiate the next phase of the migratory program or elicit specific directional tendencies. However, they all somehow depend on the spatial distribution of magnetic parameters. So far, the respective findings have been discussed without considering in detail what magnetic information is available to the animals. The crucial question is whether or not the nature of the geomagnetic field allows the proposed kinds of orientation. Theoretical considerations soon lead to a number of unsolved problems, which involve the nature of the critical parameters, the required precision in detecting minute differences, and, as far as the 'map' and the relief hypothesis are concerned, also some aspects of the theoretical concepts.

6.4.1 What Factors Are Used?

Total intensity, horizontal and vertical intensity as well as inclination all have gradients running from the magnetic poles to the magnetic equator, i.e. these gradients show the same overall trend (cf. Figs. 1.3 and 1.4), even if regionally and locally there may be considerable differences between their directions. Indeed, the first 'map' concept proposed by VIGUIER (1882) was based on a combination of total intensity and inclination; YEAGLEY (1947), in contrast, suggested vertical intensity as one of the coordinates. Considering the general aspects of the various parameters, one might tend to favor total intensity, because it is sufficient to measure the variable itself. The use of inclination or the horizontal and vertical components seems to be more problematic, because these variables are defined with respect to a reference. In theory, such a reference is provided by gravity, but it is doubtful whether the accuracy in detecting the downward direction is sufficient to allow magnetic measurements of the required precision. The use of declination, also occasionally discussed, does not seem to be possible in view of the fact that this would require an independent means of detecting geographic North with highest precision.

Considering the potential role of magnetic information in the models described in this chapter, the nature of the crucial factor remains largely open. The relief hypothesis assumes the use of total intensity and spatial gradients of intensity. Other factors have seldom been considered, in part, because the necessary information on their distribution is more difficult to obtain. In regard to the navigational 'map' of pigeons, attempts to correlate the observed behavior with a variety of magnetic variables were only of limited success. Correlations with the K-values (cf. Sect. 6.1.1.2) were found, but this index is a summary of various kinds of changes involving various variables. LEDNOR (1982) was unable to find a relationship between the initial orientation around the loft at Ithaca, New York, and the distribution of any of the magnetic variables; GOULD (1985), on the other hand, suggested inclination as the most suitable cue in the Ithaca region. The multivariate analysis by DORNFELDT (1991) showed some relationship between the homeward component of the vanishing bearings and the absolute difference between the values of total intensity on the day of release and the preceding day, but this was also true for inclination. Attempts to correlate the initial orientation of pigeons from Frankfurt a.M., Germany, with magnetic variables yielded inconsistent results.

For young sea turtles, inclination was found to be the crucial factor eliciting the various directional tendencies, as all test fields had a similar intensity (LOHMANN and LOHMANN 1994b). In bird migrants, the same is true for the change in direction associated with crossing the magnetic equator, i.e. the horizontal field lines per se represent the crucial signal for the birds (W. WILTSCHKO and WILTSCHKO 1992). The nature of the factor acting as trigger for Pied Flycatchers in the Mediterranean, however, is still open, as total intensity as well as inclination were altered in the tests to simulate natural conditions. Here, it is also unclear whether the birds respond to absolute values or to differences from the values experienced in their breeding grounds.

6.4.2 Variability and the Detection of Minute Differences

An essential point to consider regarding the use of the spatial distribution of the geomagnetic field for orientation involves the necessity to detect differences between the local value at the present location and values perceived before, remembered or expected. In this respect, the trigger function in bird migration appears rather unproblematic, for the decrease in total intensity lies in the range of 10 to 15%, that of inclination around 40%. The differences in inclination to which young turtles were found to respond are likewise comparatively large. Use of magnetic 'map' factors and orientation along the magnetic relief, in contrast, requires the detection of minute differences, so that this aspect must be regarded as a crucial problem of these two orientation hypotheses.

6.4.2.1 Spatial Variability

It is generally assumed that pigeons do not make use of magnetic factors in the immediate vicinity of their loft, since the differences between local values and home values are still below threshold. The size of this area is unclear, however. An accuracy of about 5 to 10 km might be expected, as magnetic effects, i.e. differences in initial orientation between morning and noon (cf. Fig. 6.8), or permanent shift effects (cf. Fig. 6.9) were observed at distances of about 20 km from the home loft. Within this distance, the magnetic field shows only little change in intensity, inclination and derived parameters. The steepness of magnetic gradients varies greatly; pigeons would have to rely on the normal regional gradients, which are much smaller than those reported for anomalies (cf. Sect. 6.1.1). Gradients differ between regions. For example, in upstate New York around the Ithaca loft, the local intensity gradient is about 16 nT/km (see LEDNOR 1982), whereas it is only about 2.5 nT/km around the loft in Frankfurt a.M., Germany. Such tiny differences have to be detected against a background of 26 000 to 60 000 nT (cf. Fig. 1.4), depending on the local total intensity. Total intensity is used as an example here; similar considerations must be applied to other parameters.

This alone may not pose the main problem, since calculations have indicated that such differences might be detected by biological means (e.g. KIRSCHVINK and GOULD 1981; KIRSCHVINK and WALKER 1985; see Sects. 8.2.2 and 9.3). However, although total intensity shows a worldwide gradient, its local distribution may be rather irregular. Total intensity shows small-scale variations from the expected values practically everywhere; the true amount of these variations at ground level may be even greater than that reflected by commonly used aeromagnetic maps (cf. Fig. 2c of C. WALCOTT 1991). To make use of differences in intensity to determine position with respect to home and to derive the home course, the animals need detailed knowledge of the magnetic data of their home region. This is not necessarily an argument against the use of magnetic map information, because the 'map' is established by experience, and the magnetic topography of the home area should be familiar to the animals from previous excursions. Features of the magnetic relief might even be used actively as magnetic 'landmarks'. Unexpected irregularities met outside the familiar area would result in errors in position and thus in deviations from the true home course (cf. Fig. 6.1). Such deviations, called *release site bias* (KEETON 1973), are indeed observed when pigeons are released at weak anomalies. It is unclear whether some deviations found in seemingly magnetically undisturbed terrain might also be caused, at least in part, by small irregularities in the geomagnetic field.

For animals following features of the magnetic relief, an analogous problem lies in identifying local gradients. When magnetic minima are used as guidelines, the question is how far must an animal travel before it detects an increase or decrease in intensity? And how would the animal registering higher intensity distinguish between an increase associated with a magnetic 'ridge' and the general increase normally encountered when moving northward? These questions are still largely unanswered.

6.4.2.2 Temporal Variations

However, the most serious problem might arise from the fact that in large parts of the world, the spatial changes to be detected fall into the range of regular, daily variations (cf. Fig. 1.5). The fluctuations due to a moderate magnetic storm exceed them by far. This means that pigeons using magnetic map factors must be able to distinguish between changes caused by temporal variations and those caused by the spatial variation of the geomagnetic field, at the same time compensating for known local irregularities. This appears to be very difficult indeed. It is unclear whether the regular daily variations are taken into account; the findings described in Section 6.1.1.2 are inconsistent, even if the correlation of the vanishing direction with K_{12} seems to suggest that the amount of variation experienced in a critical period prior to release affects the behavior.

Animals using magnetic 'ridges' and 'valleys' as guidelines face a similar problem: they have to compensate for temporal variations in order to derive meaningful information from spatial variations. The observation that live strandings frequently occur after magnetic storms might be interpreted as indicating that strong, irregular fluctuations indeed interfere with orientation. The nature of the navigational errors is unclear, however. A simple explanation is provided by the assumption that atypical magnetic values that occur during a magnetic storm cause the whales to cross magnetic ridges unwittingly, thus leading them into valleys which later meet the coast.[8] The problem of how temporal and spatial variations might be distinguished is crucial with respect to the use of magnetic non-compass information; however, a solution is not in sight (cf. C. WALCOTT 1991).

---

[8] KLINOWSKA (1986, 1989) offered a different explanation: in analogy to traditional human seafaring, she assumes that temporal information is crucial for any kind of navigation, which is certainly not generally true for animals. She attributes the errors leading to stranding to a loss of time information (cf. Sect. 7.1.2.3) which, according to her concept, results in loss of position by dead reckoning. Such a strategy, however, seems alien to animals following magnetic 'valleys'; it is unclear why it should matter at what moment a specific point is passed.

6.4.3 Hypotheses with Open Questions

In contrast to the compass function of magnetic information, the use of spatial information from the geomagnetic field is not yet entirely clear. With the exception of the trigger function, which is fairly well understood, this applies to the data base as well as to the theoretical background. A number of findings, many of them correlations, suggest that magnetic information plays some role in homing of birds and possibly also of other vertebrates and in the migrations of cetaceans. This evidence cannot be dismissed, but it is not always consistent. The specific nature of the factors is also unknown, and important questions concerning the perception and the processing of the required information remain unanswered.

Even the theoretical background on the use of a 'map' and on magnetic 'valleys' as guidelines is not unproblematic. The traditional map hypothesis has a well-developed theoretical framework, which is consistent in itself (cf. WALLRAFF 1974; W. WILTSCHKO and WILTSCHKO 1982, 1987). Yet some findings on homing pigeons – magnetic effects manifesting themselves as deflections rather than directional tendencies – are not in agreement with predictions (cf. Sect. 6.1.1.2) and suggest that the role of magnetic parameters in homing may differ from the classical role of a map factor. With respect to the relief hypothesis, the situation is different insofar as the hypothesis itself is not well developed, and in its present form (KLINOWSKA 1987, 1988, 1989; KLIMLEY 1993) not without inconsistencies. The data base consists of correlative evidence only; critical experiments testing the hypothesis appear to be impossible owing to the sheer size of the animals in question and the distances involved. Thus, important questions remain unanswered, and it is unknown when we will finally understand how marine animals use magnetic information.

6.5 Summary

Several types of orientation responses, which are not based on the vector quality of the geomagnetic field, but instead use the spatial distribution of parameters like total intensity and inclination, have been described. The data supporting these effects consist mostly of correlative evidence, since effects of magnetic manipulation are mostly unpredictable.

The best-studied case involves the use of magnetic cues as components of the navigational 'map' of pigeons and other animals that home over considerable distances. The model assumes that the home course is determined by comparing values of navigational factors at the present location with the remembered home values. Findings supporting the hypothesis of magnetic 'map' factors are responses of displaced pigeons to natural magnetic phenomena, e.g. deviations from the home course and disorientation observed at magnetic anomalies, or differences in behavior associated with temporal variations of the magnetic field. The latter include correlations between the pigeons' mean headings and magnetic fluctuations in the 12-h period before release, differences between the mean headings observed in the morning and at noon, or between pigeons with different settings of their internal clock; magnets fixed on the pigeons' back suppressed these effects. Manipulations of the ambient magnetic field during release and homing flight usually resulted in a slight increase in scatter; treating pigeons with alternating fields *before* release produced a large increase in scatter up to disorientation or deviations from the mean of controls, depending on pulse shape and frequency. Some treatments with extremely strong fields or pulses resulted in deflections up to 30 ° and more; in other cases, there was no effect at all. One striking characteristic of these findings is the immense variability of effects on initial orientation observed on a spatial, temporal and individual level. Homing performance, on the other hand, is hardly affected. The evidence available so far does not form a consistent picture; some aspects appear to be conflicting and a number of problems remain unsolved.

The relief hypothesis assumes that extended magnetic 'ridges' and 'valleys' running roughly north-south in many parts of the ocean are used as guidelines by marine mammals during their migrations. The model is based on correlations between live strandings of whales and local magnetic topography in Britain and on the east coast of North America; a temporal relationship between British live stranding events and magnetic storms was also described. A similar strategy has been suggested for the orientation of hammerhead sharks near a seamount. The relief hypothesis also leaves important questions unanswered, in particular, how the animals determine the future course of magnetic 'ridges' and 'valleys'.

Specific magnetic conditons may serve as triggers, controlling the routes of extended migrations. For the Pied Flycatcher, a European bird migrant, magnetic fields as they are found in southern Iberia/northern Africa were necessary to initiate the second phase of migration and change the heading from SW to SE during autumn migration. At the magnetic equator, the horizontal course of the field lines causes transequatorial migrants to change their headings from 'equatorward' to 'poleward'. Inclinations as they are found at the northern and southern edge of the North Atlantic gyre are considered to elicit specific directional tendencies in Loggerhead Sea Turtles from the Caribbean population.

In regard to the magnetic 'map' and the use of the magnetic relief as guidelines, the nature of the crucial parameter is still open. Theoretical considerations seem to suggest total intensity, since it does not depend on a reference, in contrast to inclination. Both the map hypothesis and the relief hypothesis require the detection of minute differences in magnetic values. Their subsequent interpretation might also represent a problem, because the magnetic field is nowhere completely regular; such local anomalies would have to be considered. The most serious problem, however, arises from the fact that temporal fluctuations of the geomagnetic field may fall into the same range as the spatial variations. It is unclear how animals can distinguish between them and, at the same time, compensate for local irregularities.

Orientation by the spatial distribution of magnetic parameters is meanwhile supported by considerable evidence. Nevertheless, due to open questions concerning the detection and interpretation of minute differences, consistent interpretations of the various findings, and the theoretical background of the hypotheses, it remains a poorly understood phenomenon.

Chapter 7

Other Non-Compass Effects Associated with Magnetic Fields

Under the heading 'other non-compass effects', this chapter will cover a number of heterogeneous behavioral responses which all depend on the ambient magnetic field without being compass orientation or spatial orientation in connection with large-scale movements. They include responses to temporal variations of the earth's magnetic field which may be related to an 'orientation in time', effects of electromagnetic radiation emitted by technical facilities like radio transmitters and radar stations, and responses to a number of experimental manipulations. In most cases, the biological significance is far less obvious than that of the orientation responses described in the Chapters 4, 5 and 6. The phenomena in question do not represent orientation in the strict sense, but are indirectly related to orientation as they affect orientational responses and/or cause animals to modify their routes or move to other locations.

As in Chapter 6, most evidence is of indirect nature. Consequently, details of the responses and their precise relationship to specific magnetic parameters are not completely known. In most cases, the magnetic nature of the observed effect is documented, but interpretations are highly speculative. Therefore, we will concentrate on a few prominent cases, without going into too many details.

7.1 Effects Associated with Temporal Variations of the Geomagnetic Field

Daily variations of the geomagnetic field have been described in Section 1.1.3; at our temperate latitudes, they usually consist of some decrease in total intensity from sunrise to noon, followed by a corresponding increase from noon to sunset (cf. Fig. 1.5). In theory, these variations could provide information on the time of day, even though their regular course is frequently modified by irregular fluctuations and magnetic storms. Several of the reported effects are indeed related to periodic phenomena. The most prominent example, however, is the interaction with other orientation factors in the *waggle dance* of Honeybees, *Apis mellifera* (Hymenoptera), usually referred to as 'misdirection'.

7.1.1 'Misdirection' in the Waggle Dance of Honeybees

The waggle dance, its function, and misdirection have already been described in Section 3.3.2 (see von Frisch 1968; Lindauer 1975): successful foragers inform their nest mates about the direction to a rich food source. When the bees dance on the vertical comb, they transform the angle between food source and sun into an angle with respect to gravity. While indicating the direction, the bees make small 'errors', deviating from the true directions in an irregular fashion. The fact that these deviations, the '(residual) misdirection' *(Restmißweisung*, von Frisch 1948), disappeared when the ambient magnetic field was compensated to less than 5% of its intensity indicates the magnetic nature of this phenomenon (Lindauer and Martin 1968; see Fig. 7.2).

7.1.1.1 Curves of 'Misdirection' Recorded in the Geomagnetic Field

In the geomagnetic field (intensities range from ca. 42 000 to ca. 47 000 nT at the various test locations), the misdirection varies during the course of the day, leading to continuous curves with maximum deviations from the expected directions in the range of 10 ° to 15 ° (see Fig. 7.1). The shapes of the day curves of misdirection (cf. also Fig. 3.6) are highly variable, except for the *zero crossings* (i.e. the points where the misdirection becomes zero) regularly found when the dance direction runs parallel to the axis of the field lines or their projection (cf. Sect. 3.2.2). Curves recorded under the same conditions on different days may be similar, but they are not identical (Lindauer and Martin 1968). Curves of misdirection recorded simultaneously on the north and south side of the same comb on one occasion showed different shapes that seemed to be largely unrelated (Fig. 7 in Lindauer and Martin 1968,), on another occasion, the curves were identical (Fig. 3 in Martin and Lindauer 1977). The analyses of the misdirection phenomenon concentrated mainly on its relationship to the magnetic field; as a result, the relationship to other variables such as comb alignment, direction to the food, body axis of the dancing bee, etc. is not understood in detail. Also, it is unclear whether bees of different hives show identical curves of misdirection in identical situations or whether the behavior includes hive-specific components.

The distance to the food source was found to affect the size of the deviations: bees returning from distances of 3000 m or more show markedly smaller deviations than bees indicating food in the range of 400 m to 500 m. In particular, when bees from greater distances begin dancing, they first indicate the angle to the food source correctly, then gradually show increasing deviations and approach a normal level of misdirection in the course of a few minutes. Martin and Lindauer (1977) described this as an adaptation phenomenon. However, the bees usually experience similar magnetic conditions during their foraging flights and inside their hive – there seems to be no need to adapt to different magnetic conditions. The gradual increase in misdirection remains unexplained.

Being aware of the magnetic nature of the phenomenon, LINDAUER and MARTIN tried to correlate the day curves of misdirection with natural variations of the geomagnetic field. For combs oriented along a north-south axis, LINDAUER and MARTIN (1972) first suggested a relationship with the rate of change in total intensity

$$Mi = \log (\sin \alpha_s \cdot \Delta F),$$

where Mi = angle of misdirection, α_s = predicted dancing angle of the bee, and ΔF = variation of the magnetic intensity in units of time; however, the duration of the unit of time is not defined.

Later, after recording numerous additional curves, MARTIN and LINDAUER (1977), in an attempt to consider delayed responses to intensity changes (see Sect. 7.1.1.2), proposed the mathematical description

$$Mi = (eA_2 \pm v/2\,(eA_2 - eA_1)^2) \cdot \lambda, \text{ with } eA = \log(\Delta F + 1) \cdot \mu \cdot \sin \alpha_m\, d\alpha,$$

where v = Rayleigh constant, varying between 0.20 and 0.24 or 0.40 and 0.44, depending on temperature fluctuation at the time of the dances; ΔF = rate of change in field intensity, measured from the moment the curve passes zero, i.e. when the misdirection disappears; μ = magnetic permeability of a bee, roughly equal to 1; α_m = magnetic angle, which is 0 when the curve passes zero; λ = a scaling factor (not defined), and $d\alpha$ = angular steps (not defined) for calculating eA, the effective magnetic work (MARTIN and LINDAUER 1977). For transitions between a positive and a negative variation of F, another formula was suggested. Three slightly different versions of the formula have been published by LINDAUER (1976a,b, 1985).

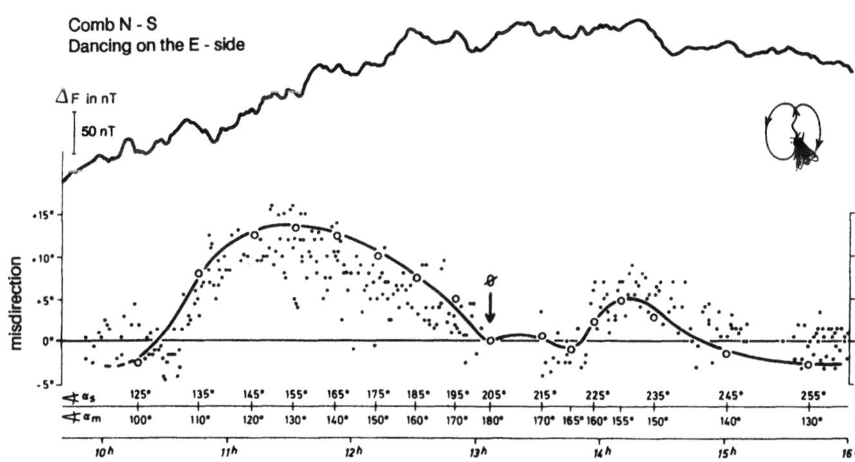

Fig. 7.1. Curve of 'misdirection' of the waggle dance of Honeybees in the course of a day, recorded on the east-facing side of a comb aligned N-S, together with simultaneously recorded variations of the ambient magnetic field (*upper solid line*). *Abscissae* α_s, expected angle of dancing (cf. Sect. 3.3.2.1) with respect to gravity; α_m, expected angle of dancing with respect to the magnetic vector; Ø marks the point where the dances are expected to run antiparallel to inclination; *below* time of day. *Ordinate* 'misdirection', i.e. the deviations of the dances from the expected values. The *open symbols* connected with the *lower solid line* indicate the curve calculated from magnetic variations using the formula published by MARTIN and LINDAUER (1977). For further explanations, see text. (After MARTIN and LINDAUER 1977)

Theoretical curves calculated according to these formulae appear to be in good agreement with the observed data on misdirection (Fig. 7.1). However, it must be pointed out that these curves were fit to the data after the event, choosing suitable values of *v* and other non-defined variables. GOULD (1980; TOWNE and GOULD 1985), criticizing this procedure, remarked that "*it would not seem unreasonable to disregard the formula altogether, were it not for its ability to track a variety of highly irregular misdirection curves so closely*" (TOWNE and GOULD 1985, p. 391). The merits of the formula seem to be indeed unclear, as a number of terms are assigned arbitrary values, which seems to allow a fitting of suitable curves to a wide variety of data, at the same time limiting the predictive value of the formula.

7.1.1.2 'Misdirection' in Magnetic Fields of Various Intensities

Observations of waggle dances under various conditions revealed a rather complex relationship between the size and the variation of misdirection and total intensity. Misdirection was always close to 0 ° in fields compensated to less than 5% of the normal intensity (Fig. 7.2). A small increase in intensity produced by an iron cage slightly increased the size of misdirection, altered the shape of the curve, and increased its scatter. Higher intensities up to 540 000 nT affected the maximal size of the deviations very little, but led to a marked increase in scatter (LINDAUER and MARTIN 1968; MARTIN and LINDAUER 1977). This proved to be a transient effect, however. Bees dancing for several days in such strong fields

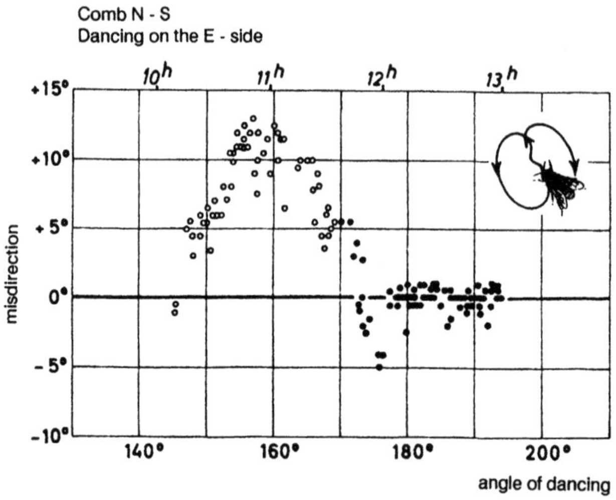

Fig. 7.2. Curve of 'misdirection' recorded from a bee dancing on an east-facing comb in the local geomagnetic field (*open symbols*) and, after 11:40, in a magnetic field compensated to less than 5% (*solid symbols*). *Abscissa* Expected angle of dancing (*below*) and time of day (*above*); *ordinate* deviation from the expected values. Note that 'misdirection' continues for about half an hour before the bees dance the angle more or less correctly as it is typical in compensated magnetic fields. (After LINDAUER and MARTIN 1968)

adapt in the sense that the misdirection decreases to a level below that observed in the geomagnetic field (KILBERT 1979).

Interestingly, responses to sudden changes in intensity occur with some delay. For example, when the field was compensated, the misdirection continued for 30 min to 1 h before the curve approached zero (Fig. 7.2)[1]. When the field returned to normal after increased intensities, delays of more than 1 h were observed. MARTIN and LINDAUER (1977) interpreted this phenomenon as 'adaptation', which they associated with hysteresis effects.

7.1.1.3 Problems and Open Questions

The 'misdirection' of the waggle dance is one of the most prominent, and, at the same time, one of the most enigmatic magnetic phenomena. It was documented with enormous efforts by von FRISCH, LINDAUER, MARTIN and their student KILBERT; MARTIN and LINDAUER (1977) claimed that 240 day curves were recorded between 1967 and 1976. Of these, 26 curves of bees dancing on a vertical comb in the geomagnetic field have been published; among them, 9 include recordings of the natural variations of total intensity and theoretical curves calculated from these recordings (LINDAUER 1976b; MARTIN and LINDAUER 1977). An independent replication confirming the findings on misdirection has not yet been presented[2].

In a theoretical discussion on the magnetic effects in bees, GOULD (1980) and TOWNE and GOULD (1985) critically reviewed the body of evidence on misdirection. They pointed out inconsistencies in some of the interpretations and a number of open questions. Although there is no doubt that the phenomenon itself depends on magnetic parameters, the nature of these factors and the specific relationship between them and misdirection are still far from being understood. In particular, the following aspects appear to require further explanations:

1. Misdirection is interpreted as a response to daily variation, i.e. to minimal changes in total intensity in the range of 50 to a few 100 nT. It seems that these small changes gain prominence when the static field is largely compensated. It is unclear why the misdirection is more pronounced in a natural geomagnetic field of about 45 000 nT, and what the delayed response to changes in intensity might mean.
2. It seems paradox that bees returning from greater distances start to indicate the angle to the food source correctly, but then gradually make mistakes of increasing size. If these errors are caused by conditions in the hive, it is unclear how these can be magnetic, since usually the magnetic field inside and outside the hive is similar.

---

[1] This may be the reason why VON FRISCH (1968) obtained negative results when he first tested bees in compensated magnetic fields, changing the intensity at frequent intervals.

[2] TOWNE and GOULD (1985) mentioned in parentheses that they once successfully repeated the compensation experiments, but no data were published.

3. It is unknown whether or not the bees picking up the information compensate for 'misdirections'. It has been argued that the errors must be automatically compensated, as dancing bees and the ones following them are subject to identical conditions, but this assumption has never been experimentally verified.
4. The crucial question concerning the biological significance of misdirection is still unanswered. In view of the function of the dances, namely, to inform nest mates about the direction to a food source, the occurrence of misdirection per se is unclear, as it seems to introduce an irregular error into an otherwise rather precise system. Whether or not misdirection is advantageous to the bees in a way not yet understood is still open. LINDAUER and MARTIN (1985) speculated that it might protect the dances against effects of other interfering factors such as light etc. (see below). But how one irregular 'error' might help to minimize others is difficult to understand.

Any attempt to describe the relationship between variations of the ambient magnetic field and the curve of misdirection must necessarily suffer from the limited understanding of the phenomenon in general. The relationship between the course of misdirection and the geometry of hive, food source etc. is not known in detail; also, the effect of distance has not been clearly determined. Other factors not yet considered may also be involved. In summary, with so many variables unclear, the relationship between magnetic field and misdirection is not understood well enough to allow a valid mathematical description. Moreover, it is still open whether misdirection is really based on active processing of magnetic information – it might still turn out to be a side effect of some other process which the bees cannot avoid.

7.1.1.4 Other Phenomena Associated with the Waggle Dance

The sound signals associated with the dances – sequences of vibrational pulses consisting of several single pulses (cf. Sect. 3.2.2) – were analyzed in detail by KILBERT (1979). They were found to be very closely correlated with the misdirection: the duration decreases as the misdirection increases (cf. Fig. 3.7). This effect is the same whether the deviations are to the right or to the left of the predicted dancing angle. Maximum duration of the sound signal is always observed when the misdirection is zero, i.e. at zero crossings and when dances are performed on a horizontal sheet of comb, but also, when bees returning from longer distances start dancing without misdirection. In fields of higher intensity, after an initial increase in scatter, the acoustic signals become rather constant with maximum duration, parallel to the observed decrease in misdirection (KILBERT 1979).

Various other magnetic effects associated with the waggle dance of Honeybees have been described. They include dances in fields changing between normal intensity and compensation on a short-term basis, in alternating magnetic fields (120 000 nT, 5 Hz) (LINDAUER and MARTIN 1968; MARTIN and LINDAUER 1977), and with artificial magnetic fluctuations (10 to 20 000 nT) periodically added (KORALL et al. 1988). Dances on inclined honeycombs and their relation to magnetic variation have been studied by MARTIN and LINDAUER (1977). LEUCHT and MARTIN (1990) observed dances on a horizontal comb oriented by the e-vector of polarized light from the natural sky. Normally, such dances are free of misdirection. However, when the geomagnetic field was compensated, misdirection up to 15 ° was observed; i.e., for reasons unknown, compensation affected horizontal dances in a way opposite to those on the vertical comb.

Findings on interactions of light stimuli with magnetic stimuli add yet another unexplained facet to the misdirection phenomenon. A light beam was found to affect the dances only during an interval lasting from approximately 30 min before to 2 h after a zero crossing. Oddly enough, this was true also for compensated magnetic fields where normal misdirection disappeared. This leads, among others, to the question how a zero crossing situation is detected when the magnetic field is no longer present. Ultraviolet light had a greater effect than blue or green light (LEUCHT 1984).

7.1.2 Interference with Biological Rhythms

Other effects of temporal variations of the magnetic field are related to periodic phenomena and the control of biological rhythms. F. BROWN (1965) was the first to point out a potential role of the geomagnetic field as *Zeitgeber*. Indications for a possible involvement of magnetic parameters in the control of rhythmicity emerge from successful attempts to interfere with diel rhythms through experimental manipulations of the magnetic field.

7.1.2.1 Breakdown of Rhythmicity Caused by Magnetic Manipulations

Earthworms. The intensity of the ambient magnetic field was found to interfere with the light-withdrawal reflex of earthworms, *Lumbricus terrestris* (Annelida, Clitellata, Lumbricida): when these worms withdraw from light, their reaction times vary during the course of the day, being 8.9 s at noon and 7.2 s at about 20:00 h. In a largely compensated field, the worms responded after 8.7 and 8.8 s, respectively (M.F. BENNETT and HUGUENIN 1969).

Honeybees. LINDAUER (1976a,b) observed that bees trained to visit a food source during a specific 2-h interval each day showed considerable scatter on magnetically disturbed days. Experiments with bees kept in a flight room under constant conditions without external *Zeitgebers* appeared to support a role of the normal daily variations of the geomagnetic field as *Zeitgeber*: bees could be trained to a specific feeding time, but their visits were scattered over the entire day when the magnetic field around the hive was increased to intensities between 145 000 and 500 000 nT (LINDAUER 1976b; MARTIN et al. 1983; KORALL 1987). After about 2 to 3 weeks, however, the bees seemed to adapt to the stronger fields and regained their sense of time (KORALL 1987). MARTIN et al. (1983) reviewed the findings that suggest a role of the geomagnetic field as *Zeitgeber* for bees.

GOULD (1980) obtained partly similar results, as only some of the bees appeared to be affected by the artificially increased magnetic field. Unfortunately, he just briefly summarized the outcome of these experiments, without giving data. A more extensive study by NEUMANN (1988), however, failed to produce any evidence for an influence of the magnetic field on the bees' sense of time. The bees scattered considerably, but this scatter was in the same range whether they were tested in the local geomagnetic field or in a stronger field.

7.1.2.2 Rhythmicity Based on Periodic Changes in Magnetic Intensity

Successful attempts to actively affect the periodicity of animals by experimentally introducing temporal changes in the ambient magnetic field have been reported for birds and mammals. A first study by BLISS and HEPPNER (1976) involved the circadian activity rhythm of House Sparrows, *Passer domesticus* (Passeriformes). The sparrows were kept in a LD 8:16 photoperiod; the controls were kept in the local geomagnetic field, whereas the experimental birds were subjected to a magnetic field that changed between 8 h of an intensity near zero and 16 h in the local geomagnetic field synchronous with the light cycle. When the birds were then transferred to constant darkness, maintaining their respective magnetic regime, the controls showed a prolongation of their cycle, while the activity period of the experimental birds did not shift significantly out of phase with the previous photoperiod.

F. BROWN and SCOW (1978) went one step further and tried to induce a cycle of different duration by periodic changes in total intensity. While subjecting Golden Hamsters, *Mesocricetus auratus* (Rodentia: Cricetidae), to a LD 12:12 photoperiod, they established a magnetic cycle with a 26 h period, consisting of 14 h of lower and 12 h of higher intensity. The respective differences varied from 800 to 26 000 nT in the various experiments. An analysis of the activity patterns of more than 800 'hamster days' revealed a clear 24-h cycle synchronous with the photoperiod; however, the magnetic cycles had modified the activity pattern, and a superimposed 26-h cycle in the range of about 14% of the mean activity became apparent.

*7.1.2.3 Magnetic Variations as **Zeitgeber**?*

The biological significance of a possible *Zeitgeber* role of magnetic parameters is evident, since it would help to synchronize biological rhythms with the natural day. This might be of particular value to animals spending considerable portions of their time in dark burrows, nesting places, etc., i.e. under conditions where they do not have ready access to a natural photoperiod. The idea was attractive and caused several authors to look for relationships between magnetic parameters and the behavior of animals. For example, STUTZ (1971) mentioned a possible correlation between magnetic horizontal intensity and the spontaneous activity of gerbils, *Meriones ungiculatus* (Rodentia: Gerbillidae). A correlation between magnetically disturbed days and the occurrence of cetacean live strandings (cf. Sect. 6.2.1.1) caused KLINOWSKA (1986, 1989) to speculate about a *Zeitgeber* function of daily variations in whales, but this relationship can also be explained in a different way (cf. Sect. 6.4.2.2).

Considerations regarding a *Zeitgeber* function of magnetic variations must take into account that the regular course of total intensity, inclination, etc. is often affected by various disturbances, which lead to atypical curves. Thus, they do not appear to be very reliable as *Zeitgeber*, which limits their usefulness considerably. Consequently, they should be expected to be of minor importance only. – Experimental evidence supporting a *Zeitgeber* role of magnetic variations is still insufficient. The reasons for the breakdown of the diel rhythm in fields of altered intensity are not entirely clear; in bees, such a breakdown seems to have been simulated by artifacts (NEUMANN 1988). The attempts to actively interfere with rhythms are more promising; in particular, the study by F. BROWN and SCOW (1978) seems to suggest an involvement of magnetic parameters in the control of the activity rhythm even under a periodic light regime. However, the animals were subjected to differences in total intensity that were in no way similar to those observed in nature, exceeding them many times. Also, the transitions between high field and low field conditions were abrupt instead of gradual. Hence, these experiments cannot be assumed to reflect a potential role of the natural changes realistically. In fact, the experiments by BLISS and HEPPNER (1976) clearly showed that the natural geomagnetic field, in contrast to the artificial intensity changes, was *not* able to keep the activity rhythm in phase.

Numerous studies involving animals of various systematic groups document that circadian rhythms become free-running in the absence of photoperiodic and temperature stimuli (e.g. ASCHOFF 1960). Since experimentalists usually did not interfere with the local geomagnetic field, these findings also argue against the idea that natural daily variations of the geomagnetic field provide an effective *Zeitgeber*. On the other hand, magnetic manipulations have been found to affect the production of melatonin, the pineal hormone that plays a crucial role in the control of daily rhythms in vertebrates (for summary, see REITER 1993b), but here, too, the fields involved differed markedly from the earth's magnetic field. On the whole, the available data seem to suggest that periodic changes in magnetic parameters involving non-physiologically large differences might act as *Zeitgeber*, whereas the slight variations of the natural geomagnetic field do not.

7.1.3 Effects of K-Variation and Magnetic Storms

The influence of temporal variations and irregular fluctuations of the magnetic field on orientation over long distances, e.g. homing in birds and migrations of whales, was already discussed in Chapter 6, since the respective effects are considered to be by-products of these animals' strategy of deriving navigational information from the spatial distribution of the geomagnetic field. Additionally, a number of other behavioral effects of irregular fluctuations of the magnetic field have been described.

Data on bird migration are inconsistent. F. MOORE (1977), observing night migrating birds in the light beam of a ceilometer, reported a positive correlation between the dispersion of flight directions and the size of magnetic fluctuations; scatter increased noticeably when K-values exceeded 4 (cf. Sect. 1.1.3). Previous radar studies (e.g. ABLE 1974; RICHARDSON 1974, 1976), however, did not indicate an influence of magnetic fluctuations on the flight directions of nocturnal migrants, and another ceilometer study (ABLE 1987) also failed to find such a relationship. SOUTHERN (1972b) observed that the young chicks of Ring-billed Gulls, *Larus delewarensis* (Charadriiformes), preferred a direction coinciding with their future migratory direction (cf. Sect. 4.1.3.1) when K-values were below 4; higher K-values associated with magnetic storms caused disorientation. Magnetic fluctuations have also been discussed as a cause for an increase in the scatter in the orientation of newts, *Eurycea lucifuga* (Urodela: Plethodontidae; PHILLIPS and ADLER 1978).

These effects are not easy to interpret, since the behaviors involved are usually considered to be compass responses, and at least the findings in birds (cf. Sect. 4.1) suggest that the compass is a rather robust mechanism which should not be affected by short-term fluctuations in the range of about 100 nT only. In adult homing pigeons, the same type of fluctuation did not cause disorientation, but a certain deflection of the preferred direction, or had no effect at all (cf. Sect. 6.1.1.2).

LEUCHT (1989) described an effect of magnetic fluctuations on the orientation of the body axis of larvae of the African toad *Xenopus laevis* (Anura: Pipidae). Tadpoles of this species usually maintain a head-down position at which their body axis forms an angle of 50 ° to 60 ° with respect to the horizontal, with fluctuations in the range of up to 4 °. On magnetically noisy days, these fluctuations were shown to be closely correlated with variations of the horizontal component of the geomagnetic field, whereas no such relationship was found on magnetically quiet days (LEUCHT 1989). Interestingly, the correlation was not with the ambient field, but with the field during the preceding 3 h. This finding has a parallel in responses of pigeons to K-variation (KEETON et al. 1974; KOWALSKI et al. 1988; cf. Sect. 6.1.1.2), although the duration of the crucial period in pigeons appeared to be 12 h. Attempts to correlate the long-term cycles in the spatial behavior of Goldfish, *Carassius auratus* (Cypriniformes: Cyprinidae) with magnetic data yielded inconclusive results (MATIS et al. 1974).

7.2 Effects of Transmitters, Radar Stations, etc. on Bird Orientation

Anthropogenic electromagnetic fields produced by various technical equipment add another type of parameter to the magnetic environment, namely, high frequency alternating fields in the kHz, MHz, and GHz range, produced by radio transmitters, radar stations and, more recently, by mobile telephones. These fields may have considerable strength in the immediate vicinity of the respective antennas.

Rumors that powerful stations may affect racing pigeons and migrating birds persist among pigeoneers and ornithologists, although tests whether birds could perceive such fields produced negative results (e.g. KRAMER 1951). Few studies have been performed that examine the possible effects of technical facilities on bird orientation; other animals have not been studied at all.

7.2.1 Migrating Birds

DROST (1949) and KNORR (1954) described effects of radar on bird migrants: when a flock of birds passed the beam of surveillance radar, the group broke up, but reassembled into a normal formation when the beam was no longer present. These observations were not confirmed by BUSNEL et al. (1956). EASTWOOD and RIDER (1964) performed an extended study on various types of radio frequencies ranging from 16 kHz to 10 GHz and did not observe unusual behavior; they concluded that any effect was of secondary importance, if it existed at all. As the power of the radiation in the study by EASTWOOD and RIDER (1964) exceeded that reported by KNORR (1954), the disturbing effects observed by DROST and KNORR remain unexplained.

In the following years, radar observations became a preferred method to record bird migration in Europe and North America. Tracking radars, which were later used to follow individual birds over considerable distances, are also not believed to affect the behavior of the target birds (e.g. ALERSTAM 1987).

R. LARKIN and SUTHERLAND (1977) tracked individual migrants flying over a large alternating-current antenna system which emitted a sinusoidal signal at 72 and 80 Hz. Birds were found to turn and change direction significantly more often when the antenna was operating than when it was inactive, suggesting that they may be able to sense the low-intensity electromagnetic fields. SOUTHERN (1975) examined the effect of the same antenna on the orientation of young chicks of Ring-billed Gulls, using the same design he had used to demonstrate an effect of K-variation. The birds were disoriented when the antenna was operating.

7.2.2 Homing Pigeons

Moreau and Pouyet (1968) looked for a possible effect of radio waves on pigeon homing by placing transportable army transmitters near the release site. Frequencies ranged from 1.5 to 2000 MHz, the power from 400 mW to 300 W. A general effect on behavior could not be observed; under overcast, however, the pigeons released while the transmitters were active tended to have slightly longer homing times. Occasionally, the vanishing intervals were also prolonged. The authors concluded that an effect of the transmitters could not be excluded.

A recent study examined the influence of a 7.5 to 21.8 MHz radio transmitter of 150 kW on initial orientation and homing of pigeons. Neither was found to be affected by the directional beam of the transmitter, independent of weather variables, training experience of the birds, etc. Thus, a general disruption of the orientational system could be largely excluded (Bruderer and Boldt 1994). However, vanishing intervals recorded in the various groups suggest that pigeons might be able to perceive the radiation of the transmitter. Boldt and Bruderer (1994) discussed the intriguing possibility that pigeons which had been repeatedly released when the transmitter was active may integrate radiation as a feature of the release site and use it actively as a 'landmark'.

Wagner (1972) studied possible effects of radar radiation with a frequency of 10 GHz on pigeon homing. A significant difference between experimental birds and controls did not emerge; however, the mean values of vector lengths and vanishing intervals of the experimental birds varied in such a way that a slight negative influence of radar waves could not be excluded.

In summary, while a clear and consistent effect could not be observed in any of these studies, the data do not altogether exclude some subliminal effect. Interestingly, there seem to be some parallels between the influence of electromagnetic waves and that of magnetic fluctuations as expressed by higher K-values. Both tend to increase the scatter in some of the behaviors, and both appear to have mostly negligible effects, except, maybe, under suboptimal weather conditions (cf. Moreau and Pouyet 1968; Schietecat 1988). Whether this is due to chance or whether this might possibly reflect a common origin of the observed effects, is unclear. It will remain unknown as long as it cannot be specified in what way natural magnetic fluctuations or anthropogenic electromagnetic fields affect the organisms.

7.3 Responses to Experimentally Induced Magnetic Changes

Altering the magnetic conditions of test animals, e.g. increasing intensity or introducing steep gradients, often caused various changes in behavior. We will briefly mention a few typical examples, without attempting to give a complete list.

7.3.1 Invertebrates

Many of the experiments in question date back to the 1960s and used strong magnets to alter the magnetic field locally. Animals were released into a test arena and subjected to various magnetic conditions. This led to deviations from a straight route. The specific effect depended on the direction of magnetic North. Such responses in connection with phototaxis were observed in the planarians *Dugesia dorotocephala* and *D. tigrina* (Turbellaria, Seriata: Planariidae; F. BROWN 1962, 1966; KARPENKO 1974), in the mud snail *Nassarius obsoletus* (Gastropoda, Prosobranchia, Neogastropoda; e.g. F. BROWN et al. 1959, 1960b, 1964) and in the fruit fly *Drosophila melanogaster* (Diptera; PICTON 1966). However, the change in path was only in the range of a few degrees when magnetic North was turned in the four cardinal compass directions; hence, the observed response was *not* a compass response, in contrast to the claims of some authors (e.g. F. BROWN et al. 1960a, 1964). F. BROWN (1962) also reported that *Paramecium* (Ciliata, Holotrichia) responded to changes in magnetic fields; PALINSCAR and DALE (1977) mentioned responses of the rotifer *Asplanchna brightwelli* (Rotatoria, Ploima).

The biological significance of such magnetically caused modifications of behaviors controlled by other factors, e.g. positive or negative phototaxis, is unclear. Why should the routes taken by various invertebrates when entering an arena be affected by the ambient magnetic field, in particular, when there is no constant relationship between the magnetic conditions and the observed differences from controls? An interesting aspect of these responses is that many of them varied during the course of the day and that they seem to be somehow related to the lunar cycle. In *Dugesia dorotocephala*, a reversal of the horizontal component of the magnetic field seemed to induce a phase shift of the lunar rhythm (F. BROWN and PARK 1965). Relationships between behaviors affected by magnetic conditions and lunar cycles have also been described in other contexts: the directions preferred by *Tritonia* (Gastropoda, Nudibranchia) and by the moth *Agrotis* (Lepidoptera: Noctuidae) with the help of a magnetic compass were found to vary in the course of a lunar month (LOHMANN and WILLOWS 1987; BAKER 1987b; cf. Sect. 4.3). The variations in the mean vanishing bearings of pigeons repeatedly released from the same site (KEETON et al. 1974; KOWALSKI et al. 1988), which were correlated with the K-values, also showed periodic changes in the course of the lunar month (T. LARKIN and KEETON 1978; KOWALSKI, unpubl.). SCHNEIDER (1963b) also discussed possible interrelations between magnetically affected behaviors and lunar and other cosmic cycles in the cockchafer *Melolontha melolontha* (Coleoptera: Scarabeidae).[3] However, knowledge on such relationships is far too limited to even speculate on what this might mean.

[3] MARTIN et al. (1983) described a similar interrelation between the behavior of bees and the constellations of sun and moon. However, NEUMANN (1988) pointed out weaknesses in their procedures; he attributed such relationships to artifacts.

The magnetotactic response observed in miracidia larvae of the eyefluke *Philophthalmus gralli* (Plathelminthes, Trematoda, Echinostomatida; cf. Sect. 3.1) became stronger when magnetic intensity was increased. The most pronounced preference for magnetic North was observed in fields between 300 000 and 7 500 000 nT; in stronger fields, the response decreased again to the level observed in the geomagnetic field (STABROWSKI and NOLLEN 1985). In chitons, *Chaetopleura apiculata* (Mollusca, Polyplacophora, Ischnochitonina), strong magnetic fields with intensities of 850 000 nT and 0.4 T were found to lower the general activity level significantly (RATNER and JENNINGS 1968).

SCHNEIDER (1957), who claimed a magnetic compass orientation in *Melolontha* (Coleoptera: Scarabaeidae; cf. Sect. 4.3.3.3), also subjected these beetles to various artificial magnetic fields. He described effects on orientation and activity which interacted with those of a simultaneously present electric field (SCHNEIDER 1960, 1963a,b). When exposed to a magnetic gradient, the marine isopod *Limnoria tripunctata* (Malacostraca, Isopoda: Sphaeromatidae) was attracted by higher intensities, but this attraction wore off after 5 to 6 days (G. BECKER 1980). In the strong (ca. 700 000 nT), inhomogeneous field of a magnet, bees spent significantly shorter intervals dancing than when they danced in the local geomagnetic field (TOMLINSON et al. 1980b). Exposure to a field of about 375 000 nT had little effect on the bees' activity; however, when conditions changed between a 375 000 nT field and the geomagnetic field every 10 or 15 min, the bees' activity oscillated in phase with magnetic field oscillations after a latency of ca. 40 to 60 min (HEPWORTH et al. 1980). Strong fields were also found to interfere with the nest-building activity of hornets, *Vespa orientatis* (Hymenoptera). In a magnetic gradient, the hornets built in the direction of decreasing intensity (KISLIUK and ISHAY 1977).

7.3.2 Vertebrates

Changes in the vertical component of the magnetic field, which caused considerable changes in the direction of the vector, but only comparatively small changes in intensity, did not affect turning behavior of elvers of the American Eel, *Anguilla rostrata* (Osteichthyes, Anguilliformes; MCCLEAVE and POWER 1978). Larger increases in intensity, however, were found to increase the activity of sticklebacks (Gasterosteiformes; KHOLODOV 1966), European Eels, *Anguilla anguilla* (Anguilliformes; BRANOVER et al. 1971; VASIL'YEV and GLEIZER 1973) and 3 of 12 Atlantic Salmon parr, *Salmo salar* (Salmoniformes; VARANELLI and MCCLEAVE 1974). When the magnetic field was locally distorted, animals tried to avoid these locations; such responses were observed in several members of the family Siluridae (Osteichthyes; LISSMANN and MACHIN 1963), European Eels (BRANOVER et al. 1971), the shark *Triakis semifasciata* (Elasmobranchii, Carcharhiniformes; KALMIJN 1978a), the newt *Notophthalmus viridescens* (Urodela: Salamandridae; PHILLIPS and ADLER 1978), and European Robins, *Erithacus rubecula* (Passeriformes: Turdidae; FIORE et al. 1984a,b). Both the increased activity and the tendency to move away from unusual magnetic conditions might reflect an attempt of the animals to avoid some unspecific interference with their general well-being.

The position of the body axis of tadpoles from African toads *Xenopus laevis*, normally forming an angle of 50°-60° with respect to the horizontal, changed when the inclination was altered from 65° to 90°, 35°, and 0°. The observed changes, however, ranged between +3° and - 5° only, i.e. they were much smaller than the changes in inclination, and they stayed mostly within the 9 ° range of positions recorded in the various control tests in the geomagnetic field (LEUCHT 1989, 1990).

7.4 Summary

A wide variety of effects caused by temporal variations of the geomagnetic field and experimental manipulations of the ambient magnetic field have been described. They do not represent orientation phenomena in a strict sense, but a number of them modify orientation in a way not yet completely understood, and others relate to 'orientation in time'.

One of the most prominent of these effects is the misdirection in the waggle dance of Honeybees, i.e. deviation in a range up to 15 ° from the expected dancing angle. When the field is compensated, the misdirection disappears after a transition period. In the natural geomagnetic field, the course of misdirection seems to be correlated with the small natural changes in total intensity. The same is true for associated phenomena like sound signals, etc.

Breakdowns of diel rhythms caused by magnetic manipulation and effects of periodic changes in field strength on activity rhythms were discussed, suggesting a role of daily variations as *Zeitgeber*. Experimental data, however, indicate that in most cases the natural variations of the magnetic field are not sufficient to synchronize the activity rhythm with the natural day; evidence is still insufficient to allow any definite conclusion.

A disorienting effect of magnetic storms is observed in young gulls and possibly also in newts; corresponding data on migrating birds are inconsistent. Larger magnetic fluctuations also affect the vertical orientation of the body axis in amphibian larvae. Radio transmitters and radar stations do not generally disrupt the orientation of migrating birds and homing pigeons; a negligible effect leading to a slight increase in scatter cannot be excluded, however.

Numerous responses to experimentally altered magnetic fields have been reported; they involve modifications of routes determined by other factors, preferred conditions of staying as well as increases or decreases in activity, interactions with lunar cycles, and a number of totally unexplained phenomena.

Chapter 8

Conditioning Experiments

The majority of conditioning experiments cannot be classified as orientation experiments, as they do not involve oriented behavior. Animals are usually trained to respond to a stimulus, either physiologically when anticipating a shock or a positive treatment, or by performing prescribed actions in order to obtain a reward or to avoid an adverse situation (Fig. 8.1). In the latter case, the stimulus is mostly used to indicate which of two alternatives is correct. Only very rarely is a specific orientation of the behavior to the direction of the stimulus required.

In magnetic orientation, however, conditioning experiments were often regarded as part of orientation tests. As man cannot perceive magnetic fields, at least not consciously (but cf. Sect. 4.2.4.3), demonstrating by conditioning that the animals in question could do so was considered an important prerequisite for any kind of magnetic orientation. Guided by the firm belief that "*we know of no case in which proven exteroreceptive sensitivity has remained impossible to demonstrate clearly with conditioning techniques*" (B. Moore et al. 1987, p. 116), authors have frequently argued against magnetic orientation or doubted respective findings of their colleagues on the ground that they themselves had not been able to condition animals to magnetic stimuli (e.g. Griffin 1952, 1982; Orgel and Smith 1954; Emlen 1970b; B. Moore et al. 1987). Conditioning experiments have also been performed by authors who basically accepted the

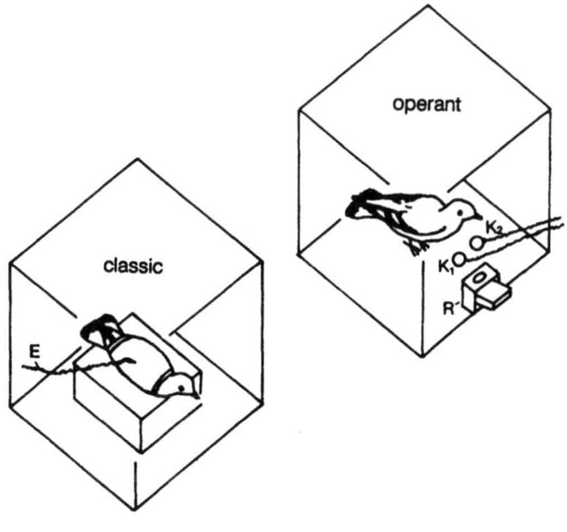

Fig. 8.1. Schematic views on apparatus for conditioning. *Left diagram* Classic conditioning: the animal is fixed in a harness; E = electrode for recording functions like heart rate, respiratory rate, etc., and for administering a slight shock as unconditioned stimulus. *Right diagram* Operant or instrumental conditioning: the animal is free to move. K_1, K_2 = alternative keys for pecking, R = apparatus distributing a reward if the right key is pecked. (After Delius and Emmerton 1978b)

premise that animals are able to perceive the magnetic field; in this case, the motivation for conditioning experiments was mostly the hope to find a clear, at all times available, response that could be used to address questions of magnetoperception.

The history of magnetic conditioning includes a considerable list of failures. The few successful cases, however, throw some light on the general conditions under which animals can be conditioned to magnetic stimuli.

8.1 Cardiac Conditioning

In cardiac conditioning, the animals are usually fixed in position (see Fig. 8.1, left). They learn to anticipate an adverse treatment following a warning stimulus; hence, the presentation of the warning stimulus is answered by a change in heart rate, which is recorded by electrodes. A response to light, sound, odor, and changes in electric fields could be elicited regularly when these stimuli signaled a forthcoming weak electric shock. Yet using an identical experimental design, numerous authors found it impossible to obtain responses to a variety of magnetic stimuli (see Table 8.1).

The only exception was Reille (1968) who tested pigeons, *Columba livia*; he reported positive responses which were only slightly smaller than the ones obtained with light stimuli. This was true when static or alternating fields with frequencies of 0.2 to 0.5 Hz and 300 to 500 Hz were presented (Fig. 8.2, above). Kreithen and Keeton (1974), in an attempt to repeat these experiments,

Table 8.1. Conditioning experiments: cardiac conditioning

Species	Type of stimulus	Result	Reference
Anguilla rostrata (Osteichtyes, Anguillif.)	dir+int, alt	neg.	McCleave et al. (1971)
Anguilla anguilla (Osteichtyes, Anguillif.)	dir+int, alt	neg.	Rommel and McCleave (1973)
Salmo salar (Osteichtyes, Salmoniformes)	dir, int, alt	neg.	» »
Columba livia (Aves, Columbiformes)	dir+int, alt	positiv	Reille (1968) [a]
» » » »	dir, alt	neg.	Kreithen and Keeton (1974) [a]
» » » »	dir+int	neg.	Beaugrand (1976)
Streptopelia turtur (Aves, Columbiformes)	dir	neg.	Beaugrand (1977)
Streptopelia decaocto (Aves, Columbiformes)	dir	neg.	» »
Anas penelope (Aves, Anseriformes)	dir	neg.	» »
Corvus monedula (Aves, Passeriformes)	dir	neg.	» »
Corvus corone (Aves, Passeriformes)	dir	neg.	» »
Turdus philomelos (Aves, Passeriformes)	dir	neg.	» »
Turdus viscivorus (Aves, Passeriformes)	dir	neg.	» »

Type of stimulus: dir = different direction or change in direction, int = different intensity or change in intensity; alt = alternating or pulsed magnetic fields.
[a] see Fig. 8.2.

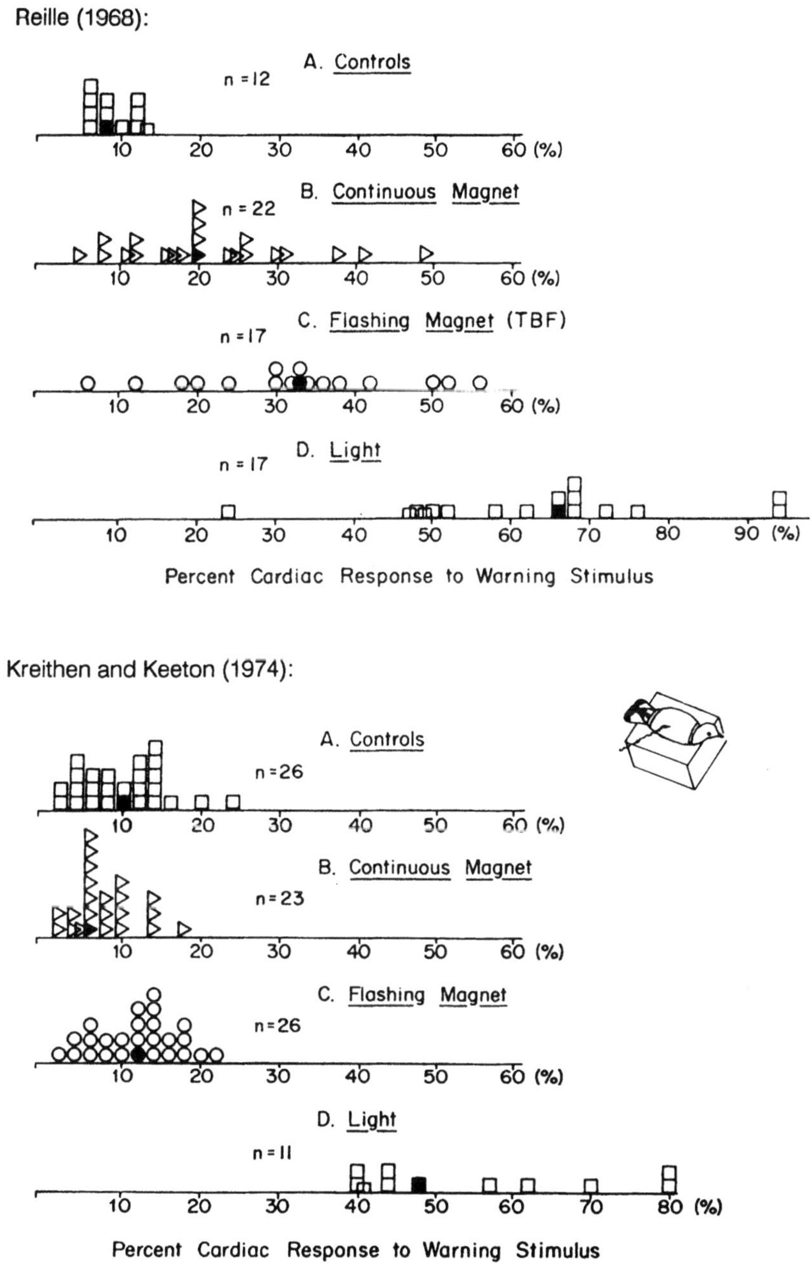

Fig. 8.2. Results of cardiac conditioning experiments with pigeons, *Columba livia*. REILLE (1968) found weak, but significant responses to a 10-s change in magnetic North by 93 ° ('continuous magnet') and to altering magnetic North between 0 ° and 93 ° over a period of 10 s ('flashing magnet'). KREITHEN und KEETON (1974) could not reproduce his findings. The response to a light stimulus is given for comparison. The median responses are indicated by a *solid symbol*. (KREITHEN and KEETON 1974)

duplicated his test arrangement, but in spite of considerable efforts, they were unable to replicate REILLE's positive findings (Fig. 8.2, below).

Positive results of heart rate studies with pigeons were reported by QUENTMEIER (1989) and HORNUNG (1993, 1994). Interestingly, these studies did not involve conditioning experiments, but *spontaneous* responses to magnetic stimuli similar to the ones used for conditioning. In QUENTMEIER's experiments, a 10-s inversion of the horizontal component was found to elicit a significant monophasic acceleration of heart rate within 4 s after stimulus onset; HORNUNG (1993), using the same stimulus, observed a response 6 to 8 s after stimulus onset. A sinusoidal change in the vertical component elicited a similar effect (HORNUNG 1994). With a 4% acceleration of heart rate compared to controls, the response to magnetic stimuli was smaller than the 6% acceleration observed when visual stimuli were presented. It might be important to note that these tests took place in a sound- and lightproof box in complete darkness. The birds were held in this almost stimulus-free situation for at least 30 min before testing began, and stimuli were presented only when the recordings showed that the heart rate was spontaneously decelerating.

8.2 Operant Conditioning

Attempts to demonstrate magnetic sensitivity by operant conditioning are more numerous (Table 8.2). Successful studies made use of two principles: they either used local distortions of the magnetic field to indicate which of two alternatives, presented simultaneously or successively, was correct, or they required the animals to look for a reward in a specific magnetic direction, thus making use of the compass quality of the magnetic field.

8.2.1 Directional Training

Successful conditioning of the latter type involved training spiny lobsters, *Palinurus argus* (LOHMANN 1985), and Round Stingrays, *Urolophus halleri* (KALMIJN 1978a,b), to look for food in a particular direction in their enclosure. In these cases, shifting the direction of magnetic North resulted in a corresponding shift of the preferred direction, indicating that the animals used magnetic information to locate the training direction. The stingrays, however, had to choose between two alternatives only so that it is unclear whether they responded to the change in magnetic direction or whether the inhomogeneity of the artificial field provided the crucial signal (cf. KIRSCHVINK 1989). Box turtles, *Terrapena carolina*, trained to look for land at the east or west end of a container, were seriously disturbed by magnets attached to their carapaces (MATHIS and MOORE 1988).

Table 8.2. Conditioning experiments: operant conditioning

Species	Type of stimulus	Result	Reference
Palinurus argus (Crust., Malac., Decapoda)	dir*	positive	LOHMANN (1985)
Apis mellifera (Insecta, Hymenoptera)	grad	positive	WALKER and BITTERMAN (1985, 1989a)
» » » »	grad, alt	neg.	WALKER et al. (1989)
» » » »	grad	positive	KISCHVINK and KOBAYASHI-KIRSCHVINK (1991)
» » » »	grad	neg.	» » » »
» » » »	dir*	positive?	» » » »
» » » »	alt	positive?	KIRSCHVINK et al. (1992)
Urolophus halleri (Chondricht., Myliobat.)	dir*?	positive	KALMIJN (1977)
Gymnarchus niloticus (Osteicht., Osteogloss.)	grad	positive	LISSMANN (1958)
Gymnotus carapo (Osteicht., Siluriformes)	grad	positive	» »
Apteronotus leptorhynchus (Osteicht., Silurif.)	dir	positive	SHERMAN (1992)
Carassius auratus (Osteichtyes, Cyprinif.)	int+grad	neg.	WALKER and BITTERMAN (1986)
Thunnus albacares (Osteicht., Perciformes)	int+grad	positive	WALKER (1984)
Chelonia mydas (Reptilia, Testudines)	int, dir	neg.	LEMKAU (1976)
» » » »	grad	neg.?	PERRY et al. (1985)
Caretta caretta (Reptilia, Testudines)	int, grad	neg.?	» » » »
Terrapene carolina (Reptilia, Testudines)	dir*	positive	MATHIS and MOORE (1988)
Larus ridibundus (Aves, Charadriiformes)	dir*	neg.	KATZ (1978)
Columba livia (Aves, Columbiformes)	alt	neg.	ORGEL and SMITH (1954)
» » » »	int	neg.	MEYER and LAMBE (1966)
» » » »	int+grad	positive	BOOKMAN (1977, 1978)
» » » »	dir+int	neg.	DELIUS and EMMERTON (1978a)
» » » »	int, dir	neg.	GRIFFIN (1982)
» » » »	grad, dir	neg.	ALSOP (1987)
» » » »	int	neg.	CARMAN et al. (1987)
» » » »	grad	neg.	MCISAAC and KREITHEN (1987)
» » » »	alt	neg.	B. MOORE et al. (1987)
» » » »	grad	neg.	COUVILLON et al. (1992)
Colibri serrirostris (Aves, Trochiliformes)	alt	neg.	IOALÈ and PAPI (1989)
Amazilia versicolor (Aves, Trochiliformes)	alt	neg.	» » » »
Passerina cyanea (Aves, Passeriformes)	dir	neg.	EMLEN (1970a)
Sturnus vulgaris (Aves, Passeriformes)	int, dir	neg	GRIFFIN (1982)
Monodelphis domesticus (Mamm., Marsup.)	dir*	neg.	MADDEN and PHILLIPS (1987)
Phodopus sungorus (Mammalia, Rodentia)	dir*	neg.	» » » »
Tursiops truncata (Mammalia, Cetacea)	grad	neg.	BAUER et al. (1985)

Type of stimulus: dir = different direction or change in direction; int = different intensity or change in intensity; grad = field with local gradiants in direction and intensity; alt = alternating or pulsed magnetic fields. * indicates directional training which made use of the compass quality of the magnetic field.

This experimental strategy, in contrast to most other conditioning experiments, tested magnetic sensitivity in a context where it can be expected to be used naturally, namely, in connection with directional tasks. This appears to be a biologically meaningful approach. There is a continuous transition from these tests to numerous other experiments demonstrating that animals use the magnetic field to remember and locate acquired directions (cf. Chap. 4). A similar test design, however, failed entirely for two species of mammals, the Bush Opossum, *Monodelphus domesticus*, and the Djugarian Hamster, *Phodopus sungorus*; the task was to look for food in the (magnetic) northern arm of a four-arm maze (MADDEN and PHILLIPS 1987). In these studies, the animals lived in a part of the test apparatus during the entire testing period and could move around freely. Thus, in contrast to most other training experiments, they experienced frequent changes in the magnetic field while staying in a familiar environment.

Attempts to condition Honeybees to leave a maze in a given magnetic direction yielded unclear results (KIRSCHVINK and KOBAYASHI-KIRSCHVINK 1991).

8.2.2 Discrimination of Two Alternatives

Most operant conditioning studies were of the more conventional type using magnetic stimuli to indicate a positive or negative target.

8.2.2.1 Arthropods

Successful conditioning experiments were performed with Honeybees, *Apis mellifera*, trained to use an artificial 'anomaly' as a guide to concentrated sucrose solutions (WALKER and BITTERMAN 1985): one of two feeders was marked by strong local magnetic distortion produced by coils placed directly beneath it. The bees experienced this anomaly upon approach, and they rapidly learned that they could suck sugar only in the distorted field and had to avoid the local field or vice versa (Fig. 8.3). When the initial intensity of 350 000 nT (added to the local intensity of 38 000 nT) was gradually reduced, bees were able to distinguish an additional field of about 260 nT from the earth's natural magnetic

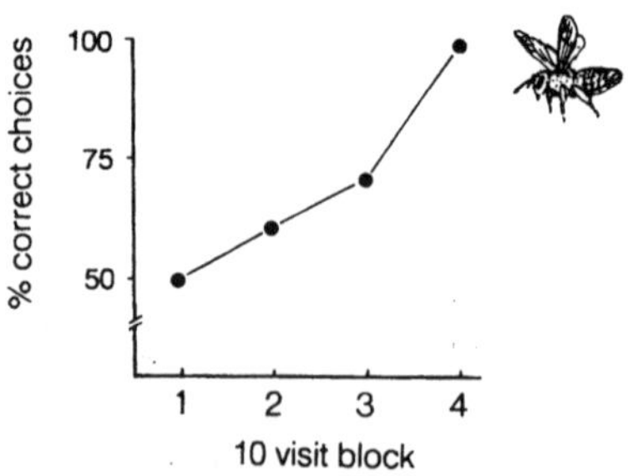

Fig. 8.3. Performance of an individual Honeybee, *Apis mellifera*, that learned to discriminate between two different magnetic situations in front of a pair of targets, given as portion of correct choices per ten visit block. A distorted field indicated a target providing a concentrated sugar solution; the ambient field indicated a target where the bee was shocked when it landed. (After WALKER and BITTERMAN 1985)

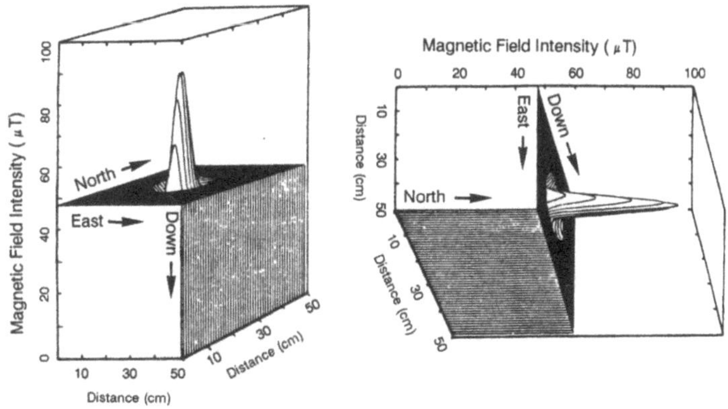

Fig. 8.4. Magnetic stimuli used in the conditioning experiments with bees by KIRSCHVINK and KOBAYASHI-KIRSCHVINK (1991). *Left* The upward directed anomaly, similar to the stimulus used earlier by WALKER and BITTERMAN (1985) produced positive results. *Right* When the same anomaly was presented horizontally, pointing northward, results were negative. (KIRSCHVINK and KOBAYASHI-KIRSCHVINK 1991)

field, with individual performance ranging from about 26 to 2 600 nT (WALKER and BITTERMAN 1989a, 1991; WALKER et al. 1990). It is unclear, however, whether the bees responded to the differences in intensity or to the steep local gradient.

These positive training results were successfully repeated by KIRSCHVINK and KOBAYASHI-KIRSCHVINK (1991), but only when the identical test arrangement was used. A seemingly small change in design, e.g. presenting the anomaly horizontally towards north instead of upward (Fig. 8.4) led to negative results. Also, time and circumstances of the stimilus presentation proved crucial: a stimulus to which bees readily responded when it was presented during the approaching flight was ineffective when presented to stationary bees sucking honey from a food dish (WALKER et al. 1989). Several other attempts to train bees to magnetic conditions or to changes in magnetic conditions also failed (WALKER et al. 1989; KIRSCHVINK and KOBAYASHI-KIRSCHVINK 1991).

8.2.2.2 Vertebrates

Operant conditioning experiments have been carried out in all major groups of vertebrates except amphibians.

Fish. Already in the 1950s, LISSMANN (1958) trained electric fishes, *Gymnarchus niloticus* and *Gymnotus carapo*, to take or not to take food, depending on the presence of a magnet behind the aquarium wall. This was the first successful conditioning experiment involving magnetic stimuli. WALKER (1984) trained Yellow-fin Tunas, *Thunnus albacares*, to discriminate between the earth's magnetic field and a non-uniform magnetic field of higher intensity. Yet the tunas could not distinguish two experimental fields which were produced by adding and subtracting the same ca. 10 000 nT field to the local geomagnetic field of 30 000 nT. This suggests that they responded to the inhomogeneity of the field rather than to its absolute intensity. Similar tests with Goldfish, *Carassius auratus*, yielded entirely negative results (WALKER and BITTERMAN 1986), although spontaneous reactions to magnetic fields have been reported for this species (cf. Sect. 3.4). SHERMAN (1992) trained the electric fish *Apteronotus leptorhynchus* to use a 90 ° deflection of magnetic North as a signal that food was available. The fish continued to respond when the deflection was reduced to 78 ° and 67 °, but when it was 45 ° or less, the number of correct responses decreased rapidly.

Thus conditioning studies with fish produced mainly positive results, the Goldfish being the only exception. It is striking, however, that the majority of the species studied were electric fish. Hence, it is not always clear whether the fishes really responded to changes in the magnetic field or to associated changes in the electric field. This question might not be significant, since it has been speculated that electric fish might normally use their electroreception system also to derive directional information from the geomagnetic field (KALMIJN 1978a; see Sect. 9.1). Yet it is unknown whether the fishes trained to respond to artificial magnetic stimuli are also able to use the natural geomagnetic field as a source of orientational information.

Other Vertebrates. In reptiles, mammals and birds, attempts to condition them to similar magnetic stimuli were largely unsuccessful. One Green Turtle, *Chelonia mydas*, seemed to learn the task, but after 5 days, the performance dropped to a random level (PERRY et al. 1985). In an apparatus similar to the one successfully used for Yellow-fin Tunas, BAUER et al. (1985) failed to obtain positive responses from Bottlenose Dolphins, *Tursiops truncata*. The longest list of unsuccessful conditioning experiments, however, comes from the studies with birds, although birds are known to orient with the help of information from the magnetic field. In particular with pigeons, authors tried several different designs, various apparatus, and various types of magnetic stimuli, but nine out of ten results were negative. Negative results were also obtained when the stimulus applied was the one that had been successfully used with bees (cf. Fig. 8.4, left; COUVILLON et al. 1992).

The notable exception were the experiments by BOOKMAN (1977) who succeeded in training pigeons to discriminate between a weak field of 2000 nT produced by the Mu-metal shielding of the room and a nonhomogeneous field of approximately earth's field strength produced by several coil systems. The birds (a pair was tested together, but only the first bird was counted) had to pass through a 3.5-m flight tunnel in which they experienced the test field which signaled which of two boxes at the end contained food. Interestingly, positive discrimination seemed to depend on the kind of movement: birds fluttering to the end of the tunnel scored much better than those that walked (Fig. 8.5). BOOKMANN emphasized the freedom of motion that his arrangement allowed the test birds. However, tests of the same design in a similar aviary failed when the geomagnetic field and a field in which the intensity had been reduced 50% served as stimuli, or when the birds were equipped with small magnets or brass bars (CARMAN et al. 1987; McISAAC and KREITHEN 1987).

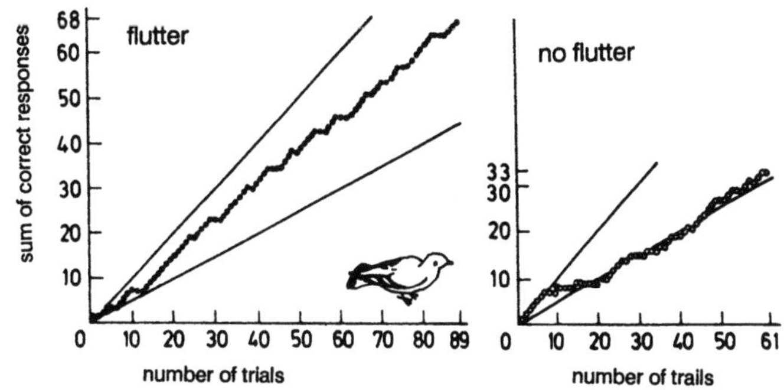

Fig. 8.5. Cumulative learning curve of pigeons trained to move along a flight tunnel of 3.5 m length; the magnetic conditions in that corridor signaled which of two boxes at the end contained a reward. The *graphs* show the number of correct responses as a function of the total number of trials, with the *two lines* indicating the 50% (chance) and the 100% (perfect) level of performance. The data are subdivided according to whether the birds fluttered (*left*) or walked (*right*). (Bookman 1978)

Altogether, the unsuccessful attempts outnumber the positive results. Being aware of the immense interest which magnetic sensitivity has met at times, and knowing that authors often hesitate to publish negative results, one can safely assume that there have been more unsuccessful attempts to condition animals to magnetic stimuli than have been documented in the literature.

8.3 Principal Differences Between Magnetic and Other Stimuli

A negative outcome of conditioning experiments per se does not allow one to draw a conclusion about the stimulus in question, even though some authors seem to be convinced that a failure to *respond* automatically means a failure to *detect* (e.g. B. Moore et al. 1987). Still, one cannot help wondering about the long list of negative results. This leads to the question why it is almost impossible to demonstrate the perception of the magnetic field, the use of which in orientation has been documented in numerous orientation experiments, by conventional conditioning techniques which have been highly successful with regard to almost all other sensory inputs.

There is no satisfactory answer as yet, but some aspects of the magnetic field may be considered with regard to this problem, namely,
1. its biological significance as an orientation cue;
2. the fact that, in nature, it never undergoes any rapid changes; and
3. the fact that it contains directional information.

8.3.1 Orientation Mechanisms in Training Experiments

One might expect that animals that have been shown to use a magnetic compass would readily respond to magnetic stimuli, at least when directional orientation is required. Compass mechanisms, however, are not always easily demonstrated by conditioning techniques. The sun compass provides a classic example: directional training was first believed to be a powerful method to study this mechanism (e.g. KRAMER and von SAINT PAUL 1950), but it proved rather difficult to train birds to select constant geographic directions with the help of the sun. They mostly chose constant angles to the sun, i.e. they used the sun, but not in connection with the time-compensating sun compass (e.g. RAWSON 1954; KRAMER 1957). The work of SCHMIDT-KOENIG (1958) is most significant: in spite of numerous training sessions at various times of the day, four out of six pigeons failed to learn the task of locating food by the sun compass. However, when these birds were released after their internal clock had been shifted, all of them showed the typical deviation, indicating the use of the sun compass (cf. Fig. 5.1). This indicates the crucial importance of the behavioral context in which the response is to be observed. Similar reasoning might be involved with regard to the negative results with Bush Opossums and Djugarian Hamsters by MADDEN and PHILLIPS (1987): the task was compass orientation, but the environment – a small enclosure of only 1 m diameter, very familiar to the animals – was one where reorientation was normally unnecessary, and compass orientation may not have been used.

The data from Honeybees also show that successful conditioning does not depend on the stimulus alone. Bees readily learned to recognize the experimental anomaly when they had to enter it before they approached the feeder. However, an anomaly with similar characteristics presented in a different direction so that the bees experienced it only just before reaching the feeder (KIRSCHVINK and KOBAYASHI-KIRSCHVINK 1991), or any attempts to condition bees at the feeder itself, yielded negative results. Timing and stimulus presentation proved to be crucial in these tests; WALKER and BITTERMAN (1985) were lucky that they had chosen an arrangement that worked. With respect to the negative results with stationary bees, WALKER et al. (1989) concluded that the bees might detect, but not process, magnetic stimuli in the training situation. In foraging Honeybees, various stimuli guiding them to a food source are integrated in a complex behavioral pattern, and the bees are alert for such stimuli only during specific phases of a foraging flight. This is true not only for magnetic stimuli: bees clearly responded to light when it was presented during the approaching flight, and seemed to ignore it when it was presented during food intake (WALKER et al. 1989).

8.3.2 Constraints of Learning

A second characteristic of the magnetic field that might contribute to the failure of conditioning experiments is that it always stays more or less constant. Rapid changes, which are typical for most other sensory inputs, and which the respective sensory organs are adapted to detect, do not occur in the geomagnetic field. In this sense, the stimuli and the situation in conditioning tasks are highly unnatural; thus, the negative results might reflect some biological constraint of learning (see e.g. SHETTLEWORTH 1972). Animals do not associate all stimuli, reinforcers and responses with equal ease. An impressive example is provided by the study of DELIUS and EMMERTON (1978b) who tried to condition pigeons to auditory and visual stimuli. They found that discrimination of pitch could be easily demonstrated in a classical heart-rate study with a shock as adverse stimulus, a method which failed to indicate discrimination of colors. Discrimination of colors, on the other hand, could be easily demonstrated using operant conditioning with a food reward, which, in turn, failed to indicate a discrimination of pitch. Discussing their results, the authors argued that colors might be easily associated with a food reward, because in nature, colors are important cues to identify food. Sound, in contrast, in its warning function, is associated with avoidance behavior, to which adverse heart-rate conditioning might be related. A later study showed indeed that the nature of unconditioned stimuli, and not cardiac versus operant conditioning, was crucial (KLINKENBERG et al. 1984).

In view of this, one would not necessarily expect that conditioning will make animals aware of a factor that normally never changes, and which, in nature, is significant to animals only in connection with orientation tasks. WALKER et al. (1986) also pointed out the importance of the behavior in connection to which a stimulus is tested. It is interesting that most of the positive conditioned responses in vertebrates come from fishes that use electric and magnetic fields to help them locate their prey and thus might be more alert for changes in their magnetic environment. Yet even such fishes may have problems with frequent changes in magnetic condition. KALMIJN (1978b) reported that frequent changes appeared to confuse the tested rays: in the training phase, the relationship between magnetic North and geographic North had been altered from day to day, whereas in the critical tests, magnetic North was reversed in a random order from trial to trial. The rays needed some time to adjust to the frequent changes in the magnetic direction and perform correctly again.

8.3.3 A Factor Containing Directional Information

Another possibility is suggested by an intrinsic property of the magnetic field which might prove crucial: the magnetic field, like the distribution of light and sound, contains directional information. Hence, any self-produced movement will cause changes in sensory input. This is obvious when compass information is recorded, but it might also apply when small changes in intensity, etc. are recorded, depending on the type of primary process involved (see Chap. 9).

Studies of the visual and auditory system showed that the problem associated with self-produced movements is solved by reafference processes: being aware of the effects of self-produced movements, the nervous system is able to distinguish self-induced changes in sensory input from changes caused by external events. Considerable parts of the brain are devoted to the task of keeping the impression of the environment constant while the animal moves, at the same time staying alert for events occurring in the outside world. This is vital, because changes in visual and auditory input might carry important information, and failure to respond in an appropriate way might be deadly.

With magnetic input, the situation is entirely different: the animals also face the problem that self-produced movements alter the sensory input, yet no changes caused by outside events have to be considered. Under these circumstances, the easiest way to cope with self-induced changes in magnetic input would simply be to shut out magnetic input and to call on it only when the situation demands it, i.e. when the animal has to orient in a strange environment. This leads to the conclusion that magnetic input might not be constantly processed.

This idea is highly speculative, and at the first glance, it may seem highly unlikely. Yet the nature of the magnetic field – providing directional information, never changing – does not rule out this possibility. Interestingly, some experiments with migratory birds seem to suggest that under certain conditions, the magnetic compass is only infrequently consulted. For example, European Robins, *Erithacus rubecula* (Passeriformes: Turdidae), responded to a shift in magnetic North only with some delay (cf. Fig. 5.6), and the same appears to be true for individual Bobolinks, *Dolichonyx oryzivorus* (Passeriformes: Icteridae). Perhaps this possibility ought to be included in further considerations.

8.4 Summary

Attempts to condition animals to magnetic stimuli were rarely successful; the failures outnumber the positive results by far. This is true for both cardiac and operant conditioning. With cardiac conditioning, attempts with fish and several bird species were negative. There is only one positive case: Reille (1968) reported that homing pigeons responded to changes in the direction of the magnetic field as well as to alternating magnetic fields. His findings could not be repeated, however.

Operant conditioning experiments used two types of test design: a few experiments were based on the compass quality of the magnetic field and required the animals to move in a certain magnetic direction. Lobsters, rays, and turtles learned the task, but studies with opossums and hamsters produced negative results. Most operant experiments, however, used magnetic stimuli to indicate which of two alternatives was correct. Positive results were reported from electric fish, tunas, and bees; for the bees, seemingly small details of the test situation proved to be crucial. Other fishes, marine turtles, several bird species and dolphins failed to learn the task. Especially in homing pigeons, one positive finding by BOOKMAN (1977) contrasts with nine negative attempts.

It is still largely unclear why magnetic stimuli mostly fail to elicit a response in conditioning tests. One of the reasons may lie in the normal use of magnetic information for orientation; other orientation mechanisms, like the sun compass, were also difficult to demonstrate by conditioning. Another reason may arise from the fact that the magnetic field never undergoes a rapid change in nature. Because of this, associating changing magnetic conditions with reward or punishment might face some constraints of learning that cannot be overcome. Moreover, the magnetic field contains directional orientation. The animals might evade the problem of self-induced movements thus changing the input by shutting out magnetic information most of the time, calling on it only when needed.

Chapter 9

Obtaining Magnetic Information

Traditionally, the question of how animals obtain the required information is an important issue in all considerations on magnetic orientation. Yet the search for receptors is handicapped in many ways. In contrast to the classical senses, it is not even clear in what part of the body magnetic information is perceived. The magnetic field, unlike light, sound and odors, penetrates living tissue with little modification. Hence, the sensory organs need not be on the surface of the body, but may lie within any structure inside. Sensory modalities like gravity and motion share this characteristic; the receptors for these stimuli are located in the head in the vicinity of the central nervous system. Thus one might spontaneously expect a similar arrangement for magnetic receptors, at least in vertebrates. This view is supported by the finding that coils around pigeons' heads could cause a reversal of orientation (C. WALCOTT and GREEN 1974; VISALBERGHI and ALLEVA 1979; cf. Fig. 4.7); the observation that magnets are most effective when placed at the temple or at the forehead just above the nose (BAKER 1989a,b) suggests that receptors are located in the frontal region of the head. The precise location is not yet known with certainty, however.

The search is further complicated by the fact that animals use various kinds of magnetic information, which might require receptors adapted to the different tasks. For example, animals like homing pigeons and bees use the magnetic field as a compass, and at the same time respond to minute intensity changes; hence, they must be expected to have two types of reception mechanisms (see KIRSCHVINK 1982; W. WILTSCHKO and WILTSCHKO 1988). Also, since there is more than one functional type of magnetic compass (cf. Sect. 4.4.1), there might be different receptor systems even for compass information. Thus, findings from one group of animals cannot necessarily be generalized.

The transducing mechanisms are likewise unclear. There have been many speculations on processes that might convey magnetic information to the organism. Almost all known physical interactions of magnetic fields were considered when magnetoreception was discussed. Summaries on how the magnetic field may be perceived, occasionally with a discussion of the pros and cons, have been published by themselves or are found in publications reporting or summarizing magnetic effects (e.g. BARNOTHY 1969; KEETON 1972; LEASK 1978; OSSENKOPP and BARBEITO 1978; GOULD 1984; LINDAUER and MARTIN 1985; LOHMANN and WILLOWS 1989; a.o.). Several specific transducer processes have been suggested, among them models based on biological superconductivity through JOSEPHSON junctions (COPE 1971, 1973), on semiconductor properties of neuroglia (RUSSO and CALDWELL 1971), and on streaming phenomena in electrolytes (BAMBERGER et al. 1978; GUNTER et al. 1978).

The current discussion focuses mainly on three principles, namely,

1. perception based on *induction*;
2. perception via *photopigments*; and
3. perception based on *magnetized particles*, in particular magnetite.

The respective concepts have been followed up by electrophysiological and behavioral studies.

Considerable research activities are currently engaged with these questions. Altogether, our knowledge is still rather limited, but recent findings have added to our understanding and modified some earlier ideas. In a field developing so rapidly, the present views will soon be outdated. Hence, this chapter does not intend to give an extended review of the present state of the art or details of the various hypotheses. Instead, we will just briefly mention the main concepts, summarize electrophysiological findings, and report results of orientation experiments which may throw light on the principles of magnetoreception.

9.1 Perception Based on Induction

The induction hypothesis assumes that magnetic information is perceived by electroreceptors. Not the magnetic field itself, but the corresponding electric field induced according to FARADAY's Law is recorded. The plausible primary process, together with the known existence of suitable sensory structures, made this model very attractive, although it is not applicable to all animals known to respond to magnetic stimuli.

9.1.1 Voltage Gradients Perceived by Electroreceptors

A number of fish have receptors specialized for the detection of electric fields. In their natural environment, these receptors help to locate prey, to avoid obstacles, etc. Electroreceptors are not restricted to 'active' electric fish producing their own electric signals, but are rather widespread among more primitive aquatic vertebrates like lampreys, sharks, rays, and members of other ancient groups. The receptors of these 'passive' electric fish are tonic, with a constant firing rate modulated by environmental electricity (for summary, see M.V. BENNETT 1971; BULLOCK and HEILIGENBERG 1986). The sensitivity is remarkable; a study by KALMIJN (1982) revealed thresholds of only 0.005 μV/cm in three species of elasmobranchs.

The possibility that these electroreceptors might additionally serve to obtain directional information by measuring the electric fields induced by the motion of the fish through the magnetic field was already considered by MURRAY (1962). When swimming forward, fishes cross the horizontal component of the geomagnetic field so that an electromotive force is induced, and a dorso-

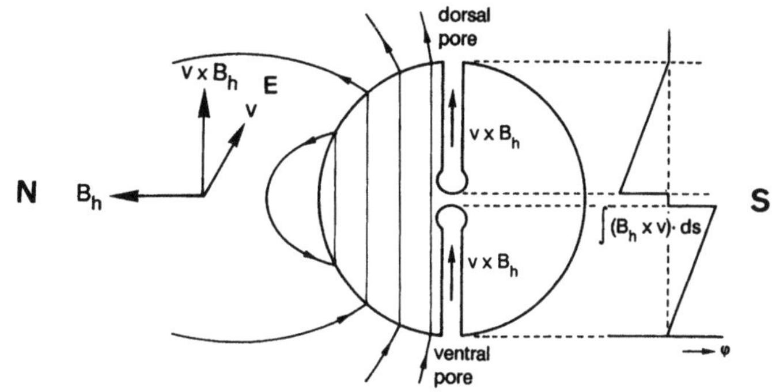

Fig. 9.1. Schematic cross section through a shark heading east, showing two *ampullae Lorenzini*, one with a dorsal and one with a ventral pore. While moving eastward (E) with the velocity v and crossing the horizontal component of the magnetic field B_h, a ventro-dorsal voltage gradient $v \times B_h$ is induced according to FARADAY's Law (see *left diagram*). Since the fish is virtually short-circuited by the highly conductive seawater, the induced voltage gradient is equalized. The same voltage gradient is induced in the *ampullae of LORENZINI*, but the *ampullae* act as high-ohmic voltmeters, preventing a flow of current. Hence the electromotive force induced along the lengths of the canals s can be measured. On the *right side of the diagram*, the *curve* indicates the potential distribution of ψ along the ampullary axes. (AFTER KALMIJN 1978a)

ventral voltage gradient will be generated (Fig. 9.1). This gradient is proportional to the swimming speed and varies according to the heading of the fish. Normally, it is counteracted by a current flowing through the fish; since the fish itself is virtually short-circuited by the seawater surrounding its body, the total potential difference is negligible. In electrosensitive organs like the *ampullae of LORENZINI* of elasmobranchs, the situation is different: the same voltage gradients are generated along the length of the ampullary canal, but the *ampullae* act as voltmeters with extremely high resistance at the canal walls. There is practically no compensating flow of current, and the induced voltage gradient can stimulate the sensory epithelia at the blind ampullary endings (Fig. 9.1). For a more detailed description of the physical principles, see KALMIJN (1974, 1978a,b).

At a constant swimming speed, the induced electric field is zero when fish head north or south, and it reaches maxima with different signs when the fish move east or west. Assuming a cruising speed of 1 to 2 m/s, KALMIJN (1978a) calculated voltage gradients up to 0.5 to 1.0 μV/cm, which are well above threshold. The response varies as a function of direction and thus could tell the fish the direction of movement[1]. The efficiency of this mechanism depends on the

[1] A similar mechanism would work when fish passively drift in water currents, since in this case they move with the water body through the geomagnetic field. The electromotive force induced by major ocean currents is considerably weaker than that induced by active motion, but might still be large enough to be detected by fish with specialized electroreceptors. Hence, an indirect magnetic effect providing information on the direction and strength of ocean currents and passive drift has been discussed (see KALMIJN 1974; SMITH 1985).

strength of horizontal intensity, which is highest near the magnetic equator; at lower and temperate latitudes, it could provide useful *compass* information.

The vertical component of the geomagnetic field would generate a similar, laterally oriented voltage gradient which could be perceived with horizontally oriented *ampullae*. In this case, the amount of the induced voltage gradient does not depend on the heading of the fish, but only on the vertical intensity. Since the vertical component of the geomagnetic field decreases towards the magnetic equator (cf. Fig. 1.3), such a mechanism might be used, at least theoretically, to determine magnetic latitude (KALMIJN 1978a).

9.1.2 Electrophysiological Recordings from the *Lateralis* System

In the 1970s, a group of Russian authors, studying functional characteristics of central neurons of the electroreceptor system, recorded electrophysiological responses from the *area octavo-lateralis* in the dorsolateral part of the *medulla*. The species studied were the Black Sea skates *Dasyatis pastinaca* and *Raja clavata* (Elasmobranchia: Rajiformes). In view of the suggested use of the *ampullae* for magnetoreception, they extended their study to magnetic stimuli. Almost all observed neurons responded to changing magnetic fields. The responses depended on the polarity of the field: some neurons were excited by one direction and inhibited by the other, while others responded in the opposite way (Fig. 9.2). Spike frequencies depended on the rate of change in the magnetic field; the smallest change causing a response was found to be around 200 000 nT/s. Responses to constant magnetic fields could not be recorded from stationary skates (ANDRIANOV et al. 1974; H. BROWN et al. 1974).

AKOEV et al. (1976) and H. BROWN and ILYINSKI (1978), recording from single fibers of the *nervus lateralis anterior* which connects the *ampullae of LORENZINI* with the central nervous system, confirmed these findings. The observed effects depended on the orientation of the canals: a stimulus that increased the spike frequency in canals pointing rostrally decreased the fre-

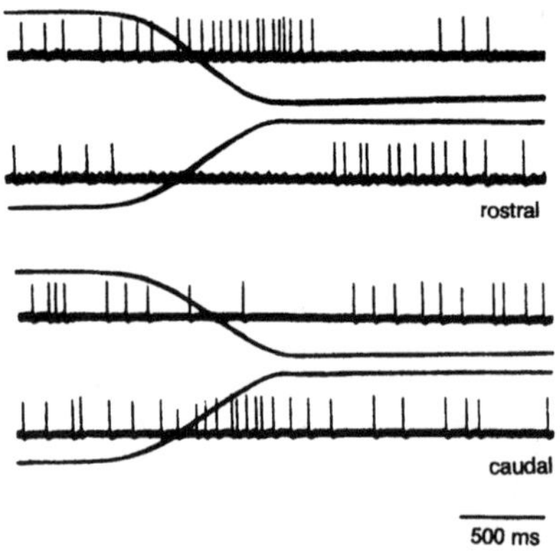

Fig. 9.2. Electrophysiological recordings from dorsal *ampullae of LORENZINI* of the left side of the skate *Dasyatis pastinaca*. *Above* Canal rostrally oriented; *below* canal caudally oriented. An *upward deflection* of the solid trace represents a northern, a *downward deflection* a southward direction of the magnetic field; the rate of change was approx. 93 000 nT/s. (After AKOEV et al. 1976)

quency in those pointing caudally (Fig. 9.2). Symmetrically located receptors on the right and on the left side of the skate's body gave reverse responses. *Ampullae* with differently oriented canals also differed in the intensity of responses to a standardized change in the magnetic field. The effect was inversely proportional to the length of the canals: the threshold of dorsal *ampullae* with a canal length of up to 12 cm was considerably lower than that of mandibular *ampullae* with their much shorter canals (AKOEV et al. 1976), as predicted by KALMIJN (1974). When the ampullary canal was surgically shortened, the spike rate decreased proportionally (H. BROWN and ILYINSKI 1978).

The recordings mentioned so far used changing magnetic fields as stimuli. Responses to a constant magnetic field could be recorded only when the fish were moved through the water; these responses were comparable to those obtained in changing fields (H. BROWN and ILYINSKY 1978). Although the authors used fields between 500 000 and 8 000 000 nT, which are considerably stronger than the geomagnetic field, both AKOEV et al. (1976) and H. BROWN and ILYINSKY (1978) pointed out that the observed sensitivity would be sufficient for the detection of the geomagnetic field.

Fluctuations of the magnetic field, as they occur during magnetic storms, are also reflected by corresponding electric fields. Responses to such fluctuations could be recorded in the *ampullae* of skates (H. BROWN et al. 1979), indicating that these variations, too, might be detected by electroreceptors.

9.1.3 A Special Mechanism of Some Marine Vertebrates

Taken together, the electrophysiological data are in general agreement with the suggested role of electroreceptors in magnetoperception. However, so far, it has only been shown that the *ampullae of LORENZINI* are capable of supplying the necessary information – a connection between specific sensory input and behavioral output has not yet been established. KALMIJN (1978a) claimed that he could train Round Stingrays, *Urolophus halleri*, to use the magnetic field for compass orientation (cf. Sects. 4.2.1.3 and 8.2.1), but it is unclear whether the required information indeed originates in the *ampullae*. Since the efficiency of magnetoperception via electroreceptors depends on the conductivity of the surrounding medium (KALMIJN 1978a, 1984), an easy means of verifying this possibility is provided e.g. by transferring individuals of suitable species, which were trained in seawater, to freshwater and test them there. Such critical tests have not been performed. Hence, it is still open whether sharks, rays, etc. indeed utilize information from their electroreceptors for magnetic orientation.

Magnetoperception via electroreceptors as described above is hardly applicable by animals in fresh water or on land (KALMIJN 1984). ROSENBLUM et al. (1985), considering signal-to-noise ratios and possible designs of sensory organs in terrestrial animals, pointed out that the required size of the organs per se would rule out induction-based magnetoreception for small species. They also concluded that the extreme sensitivity to intensity, e.g. required for the 'map' or suggested by the 'misdirection' of dancing bees, could not be achieved in this way. In summary, magnetoreception via induction appears to be restricted to primitive vertebrate groups living in saline water, and we must look for different mechanisms in other aquatic and terrestrial animals.

9.2 Perception Via Photopigments

First speculations regarding the involvement of photopigments in magnetoreception were based on estimates of diamagnetic anisotropy of rhodopsin (Hong 1977). Modern hypotheses involving photopigments propose a fundamentally different mechanism, namely, that macromolecules which have been elevated to an excited state function as transducers. Macromolecules in the ground state have a zero magnetic moment; certain excited states, however, may show various reactions which depend on the intensity and direction of the ambient magnetic field and could thus be used to obtain magnetic information. Since the molecules may reach excited states by photon absorption, the hypotheses were followed up by electrophysiological studies of the visual system and behavioral studies under various light conditions.

9.2.1 Excited Triplet State Macromolecules as Transducers

The possibility of magnetoreception via excited states of macromolecules was first suggested by Leask (1977, 1978). The model is based on optical pumping, i.e. the proposed process starts with a molecule being elevated by absorption of a photon to the singlet excited state. By excitation transfer it may then reach the lowest energy triplet state (Fig. 9.3). Molecules in this state have a magnetic moment and can interact with the magnetic field in various ways.

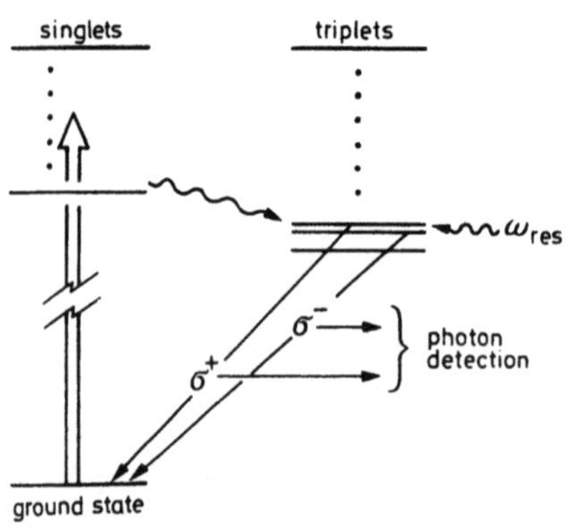

Fig. 9.3. Schematic diagram of the optical pumping process in a molecule having a singlet ground state. Photon absorption leads to the singlet excited states; excitation transfer may lead to triplet states. ω_{res}, resonance frequency; for further explanations, see text. (Leask 1978)

9.2.1.1 The Resonance Model

LEASK (1977) suggested a resonance process in the radio frequency range which would depend, among other things, on the alignment of the molecule relative to the ambient field. The organisms might obtain magnetic information by monitoring the decay rates from the triplet sublevels back to the ground state. The problem of magnetic field detection would then be a problem of photon detection (cf. Fig. 9.3). Magnetoreception according to this hypothesis would require an ordered array of molecules with specific properties. LEASK assumed that this condition was satisfied by the arrangement of photopigments in the retina. He suggested that magnetoreception might take place in the eye as a by-product of the normal visual process. For a detailed description of this hypothesis, see LEASK (1977, 1978).

The model was devised in view of the magnetic compass of birds and can indeed account for some of its unexpected functional characteristics (cf. Sect. 4.1.2): the axial sensitivity of the avian magnetic compass[2] could be attributed to the proposed resonance process being insensitive to polarity, the physiological window and its alterations could be explained by a change in the 'pattern of response', e.g. a shift of the resonance conditions to other regions of the retina (LEASK 1977). Yet one essential requirement of the hypothesis presents a crucial problem, namely, the existence of a *local, well-defined radio frequency* which will become resonant in the triplet states of the molecules. It is unclear how this could occur in the organism, and thus would seem to rule out resonance models of the type proposed by LEASK.

9.2.1.2 A 'Chemical Compass' Based on Biradicals

However, other interactions of macromolecules with magnetic moments appear possible. SCHULTEN (1982) and SCHULTEN and WINDEMUTH (1986) proposed a related model also based on excited triplet states of macromolecules. They suggested that a molecule in the excited singlet state may dissociate into a pair of radicals, which through hyperfine coupling may turn into a triplet pair. The latter process depends on the intensity as well as on the *direction* of the ambient magnetic field, with an axial directional dependence. The triplet pair can form a *triplet product* which is proposed to be chemically different from its precursor and might cause specific actions in membranes, etc.

SCHULTEN and WINDEMUTH (1986) also considered rhodopsin, iodopsin, or connected molecules as likely candidates. They pointed out that reactions as outlined above, involving known organic molecules, could take place in fields of earth-field strength with sufficient frequency to provide directional information and could be employed in a 'chemical compass'. For a more detailed description of the reaction steps and their dependence on the magnetic field, see BITTL and SCHULTEN (1986) and SCHULTEN and WINDEMUTH (1986).

[2] Magnetic compass responses based on the polarity of the magnetic field (cf. Table 4.8) had not yet been described when *LEASK* (1977) first published his model.

9.2.2 Electrophysiological Recordings from the Visual System

Since the eye had been suggested as a possible site of magnetoreception, it seemed promising to investigate the visual system for neuronal responses to changes in the magnetic field. Electrophysiological studies concentrated on animals known to use magnetic information; most were performed with birds, in particular pigeons, *Columba livia*.

Electrophysiological responses to changes in the magnetic field were recorded from the nucleus of the basal optic root (nBOR), a part of the accessary optic system receiving input from displaced ganglion cells distributed throughout the retina, and from the *stratum griseum et fibrosum superficiale* of the *tectum opticum* (Semm et al. 1984; Semm and Demaine 1986). The stimuli were gradual changes in direction at constant total intensity, either turning magnetic North to geographic South or changing inclination from 62 ° downward to 62 ° upward within 90 s.

On the basis of their visual responses, units of the nBOR may be classified into two major groups (Britto et al. 1981): movement-sensitive cells and direction-selective cells. The latter are excited by movements in a preferred direction and inhibited by motion in the opposite direction; only cells of this type responded to magnetic stimuli. Responses to magnetic stimuli could also be recorded from units of the *tectum opticum*, which showed directional selectivity to light stimuli. The portion of these cells found to respond to magnetic stimulation was about 70% in both the nBOR and the *tectum*. The majority showed increased spike frequencies (Fig. 9.4, left); about 15% of the cells were inhibited. In both cases, the responses depended on the presence of light and an intact retina and optic nerve. Units ceased to respond in total darkness, but resumed their responses when light was again presented. When the eyes were illuminated with light of various wavelengths, a peak of responsiveness to magnetic stimuli was indicated between 503 and 582 nm (Semm et al. 1984; Semm and Demaine 1986). These findings are in agreement with the excitation of molecules like rhodopsin by photon absorption, as suggested by Leask and Schulten.

Individual units showed distinct peaks of response at a particular phase of stimulation, i.e. to a particular alignment of the magnetic field, which varied between cells (Fig. 9.4, right). The rate of change in direction had no effect on the type of response. This is true for the nBOR as well as for the *tectum opticum*. In the *tectum*, the responses of individual units to the same stimulus differed when the pigeon's beak pointed in different geographic directions (Semm and Demaine 1986). These findings indicate that the information reaching the *tectum opticum* is direction-specific. The input of a number of units, processed collectively and integrated, would represent all directions in space and thus provide a suitable basis for compass orientation (Semm et al. 1984; Semm and Demaine 1986).

Similar responses to changes in the direction of the magnetic field were recorded from the *tectum opticum* of several passerines, such as Bobolinks,

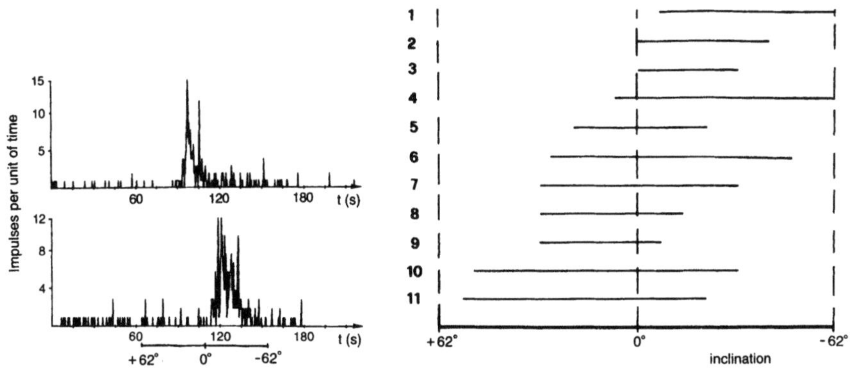

Fig. 9.4. Electrophysiological recordings from the visual system of pigeons, *Columba livia*. *Left side* Neuronal responses of direction-selective cells for the nucleus of the basic optic root (nBOR) to a gradual change of inclination from +62 ° downward (local value at the recording site) via a horizontal field (0 ° inclination) to - 62 ° upward (= vertical component inverted); total intensity and declination remained constant. *Abscissa* Time; *ordinate* spike rate, 1 unit of time corresponds to 300 ms. Note that the two cells respond at different phases of stimulation. *Right* When the stimulus was a gradual change of inclination from +62 ° downward to - 62 ° upward, different direction-selective cells in the nucleus of the basic optic root responded to different spatial directions of the magnetic vector. The *horizontal bars* indicate the range of augmentation of electrical activity of representative neurons. (Data from SEMM et al. 1984)

Dolichonyx oryzivorus (Icteridae), Starlings, *Sturnus vulgaris* (Sturnidae), and Zebra Finches, *Taeniopygia guttata* (Estrildidae). Here, too, responses could only be observed in the presence of light (BEASON and SEMM 1987; HOLTKAMP-RÖTZLER 1990; BEASON 1994). Processing of magnetic information in the *tectum opticum* appears to be widespread among birds.

9.2.3 Behavioral Tests for Light-Dependent Orientation Responses

In his papers, LEASK (1977, 1978) encouraged researchers to test his model by looking for effects of different illumination on magnetic orientation. Meanwhile, such experiments have been performed with a number of vertebrates and invertebrates. The results show a surprising variability and indicate considerable differences between the various animal groups.

9.2.3.1 Orientation in the Absence of Light

A basic question was whether or not magnetic orientation was possible in the absence of light. First data demonstrating magnetic compass orientation were obtained with night-migrating birds during migration. Unfortunately, this behavioral response could not be used to test the crucial role of light, because migratory birds cease activity in total darkness (GWINNER 1974); hence, direc-

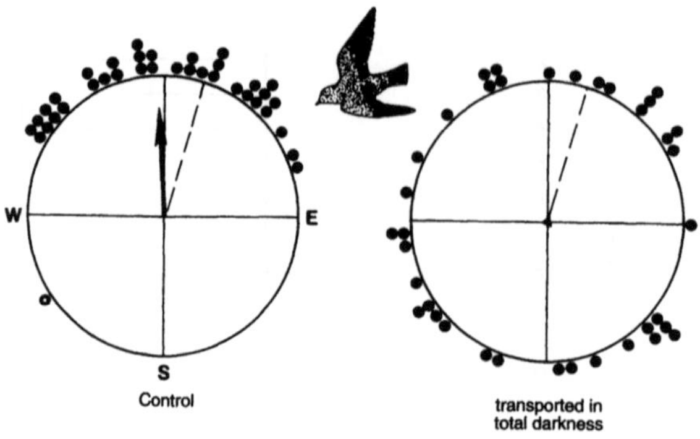

Fig. 9.5. Orientation of very young homing pigeons after displacement in total darkness; cf. Fig. 4.8. The home direction is indicated by a *dashed radius*; the *symbols at the periphery of the circle* indicate the vanishing bearings of individual birds, the *arrows* represent the mean vectors with respect to the radius of the circle. (W. WILTSCHKO and WILTSCHKO 1981)

tional tendencies cannot be recorded. However, the observation that young, inexperienced pigeons were disoriented after displacement in distorted magnetic fields (cf. Fig. 4.8) provided a possibility for such tests. Transportation in total darkness was indeed found to have a similar disorienting effect (Fig. 9.5; W. WILTSCHKO and WILTSCHKO 1981).

Meanwhile, more species have been studied in view of their capability to orient in the absence of light. The results listed in Table 9.1 indicate that disorientation in total darkness is the exception rather than the rule. Besides pigeons, only the newt *Notophthalmus* was found to be disoriented (PHILLIPS and BORLAND 1992a). The behavior studied was shoreward orientation, i.e. in newts, as in pigeons, an inclination compass was involved (cf. Sect. 4.2.2.1). The

Table 9.1. Magnetic compass orientation in the absence of light

Species	Possible?	Reference
Tritonia diomedea (Gastropoda, Nudibranchia)	[yes?]	LOHMANN and WILLOWS (1987)
Talitrus saltator (Crustacea, Amphipoda)	yes	ARENDSE (1978)
Tenebrio molitor (Insecta, Coleoptera)	yes	ARENDSE (1978)
Solenopsis invicta (Insecta, Hymenoptera)	[yes]	ANDERSON and VANDER MEER (1993)
Apis mellifera (Insecta, Hymenoptera)	yes?	SCHMITT and ESCH (1993)
Oncorhynchus nerka (Osteichtyes, Salmoniformes)	yes?	QUINN (1980); QUINN and BRANNON (1982)
Notophthalmus viridescens (Amphibia, Urodela)	no	PHILLIPS and BORLAND (1992a)
Caretta caretta (Reptilia, Testudines)	yes	LOHMANN (1991)
Dermochelys coriacea (Reptilia, Testudines)	yes	LOHMANN and LOHMANN (1993)
Columba livia (Aves, Columbiformes)	[no]	W. WILTSCHKO and WILTSCHKO (1981)
Apodemus sylvaticus (Mammalia, Rodentia)	[yes?]	MATHER and BAKER (1981)
Cryptomys sp. (Mammalia, Rodentia)	yes	MARHOLD (1995)
Homo sapiens (Mammalia, Primates)	yes?	BAKER (1989a)

With question mark: inferred from the description of the test arrangement, not explicitly stated by the author(s). [ ] refers to magnetic orientation indirectly demonstrated, cf. Chapter 4.

inclination compass of marine turtles (cf. Table 4.8), however, does not depend on the presence of visible light (LOHMANN 1991; LOHMANN and LOHMANN 1993), so that the findings do not form a pattern in this respect.

9.2.3.2 Orientation Under Light of Various Wavelengths

The optical pumping model implies that long wavelengths might not have sufficient energy to start the processes leading to magnetoreception. Hence, it became of interest in what way magnetic orientation might be affected by the various wavelengths of light.

Birds. Tests under monochromatic light of equal quantal flux at different wavelengths have been performed with passerine migrants. Silvereyes, *Zosterops lateralis*, migrating during twilight hours, were found to be disoriented under red light of 633 nm, while they were well oriented in their normal migratory direction under green light of 571 nm and under blue light of 443 nm (Fig. 9.6; W. WILTSCHKO et al. 1993b). Tests with night-migrating European Robins, *Erithacus rubecula* (Turdidae) under identical 'white', red and green lights produced similar results (W. WILTSCHKO and WILTSCHKO 1995). Further experiments indicated good orientation also under 443 nm blue light, whereas robins were disoriented under 588 nm yellow light (WILTSCHKO and WILTSCHKO, in prep.). Apparently, the transition between conditions that allow orientation and those that do not in the range between 571 nm green and 588 nm yellow light is very abrupt.

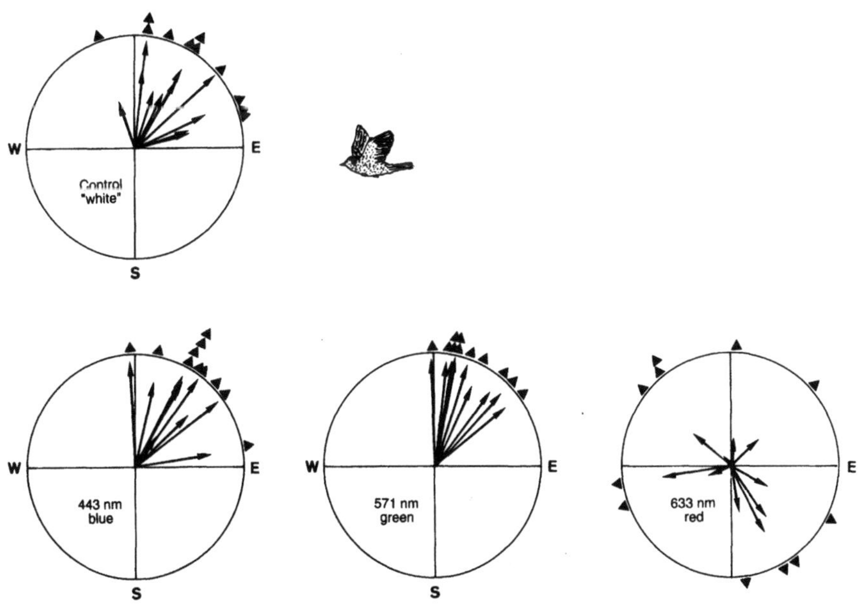

Fig. 9.6. Orientation of Australian Silvereyes, *Zosterops lateralis*, under light of various wavelengths. The *arrows* indicate the mean vectors of individual birds, based on 3 to 6 tests; the *symbols at the periphery of the circle* indicate the mean headings of these vectors. (Data from W. WILTSCHKO et al. 1993b)

The observed disorientation under long wavelengths raised the question of whether or not 588 and 633 nm light, clearly visible to humans, could be detected by the birds at all. Very little is known about the spectral sensitivity of passerines. The only species thoroughly studied so far is *Leiothrix lutea* (Timalidae); in the long wavelength range, a peak was found at 620 nm (MAIER 1992), which is beyond the 588 nm yellow light and close to the 633 nm red light under test conditions. Additionally, the observed normal activity levels argue against the assumption that the test birds did not detect the 588 and 633 nm light (W. WILTSCHKO et al. 1993b). In view of this, the data appear to suggest that red light allows normal activity, but somehow prevents the birds from obtaining orientational information, which would be in accordance with LEASK's predictions.

A recent study with Bobolinks suggests similar responses under long wavelength light, but possible differences under shorter wavelengths (BEASON 1994).

Amphibians. Wavelengths greater than 715 nm, which are beyond the visible range, caused disorientation in the newt *Notophthalmus viridescens* (Salamandridae; PHILLIPS and BORLAND 1992a). Tests under various other wavelengths indicated remarkable differences between the shoreward and homeward responses (Table 9.2).

Table 9.2. Orientation under visible light of various wavelengths in Eastern Red-Spotted Newts, *Notophthalmus viridescens*

	Shoreward orientation				Homeward orientation			
Wavelength	n	α	r	difference	n	α	r	difference
Full spectrum	85	5 °	0.55***		50	8 °	0.61***	
400 nm	9	359 °^{ax}	0.72**	−1 ° n.s.	13	335 °	0.48*	−7 ° n.s.
450 nm	18	7 °	0.68***	+15 ° n.s.	14	23 °	0.34 n.s.	(−14 °) n.s.
475 nm	17	97 °	0.07 n.s.	(+86 °) *				
500 nm	14	253 °	0.42^{+}	−95 ° *				
550 nm	12	268 °	0.49*	−106 ° *	14	166 °	0.22 n.s.	(+145 °) **
600 nm	7	241 °	0.82**	−109 ° *	7	131 °	0.17 n.s.	(+113 °) **

Both reference directions are set to 360 °. Control tests under full spectrum were run with each of the test series under different wavelengths; the pooled data are given in the first line. n, number of newts tested. α, r, direction and length of mean vector (^{ax} indicates axially bimodal behavior; the preferred end of the axis is given). Asterisks at r indicate significance by the Rayleigh Test. The difference to the respective control sample is given in parentheses when the experimental sample is not significantly oriented. Asterisks at the differences indicate significance by the Watson U-test. + = $p < 0.10$, * = $p < 0.05$, ** = $p < 0.01$, *** = $p < 0.001$; n.s. = not significant. (Data from Phillips and Borland 1992b,c, 1994)

When orienting shoreward, newts showed normal orientation at wavelengths of 450 nm and below, whereas their directional tendencies were deflected *counterclockwise* by approx. 90 ° at wavelengths of 500 nm and above. Under an intermediate wavelength of 475 nm, the newts were disoriented (PHILLIPS and BORLAND 1992b,c). In order to test whether this deflection was caused by properties of the spectral mechanisms involved in obtaining compass information, PHILLIPS and BORLAND (1992c) covered the outdoor housing tanks of a group of newts with a filter that blocked short wavelengths below

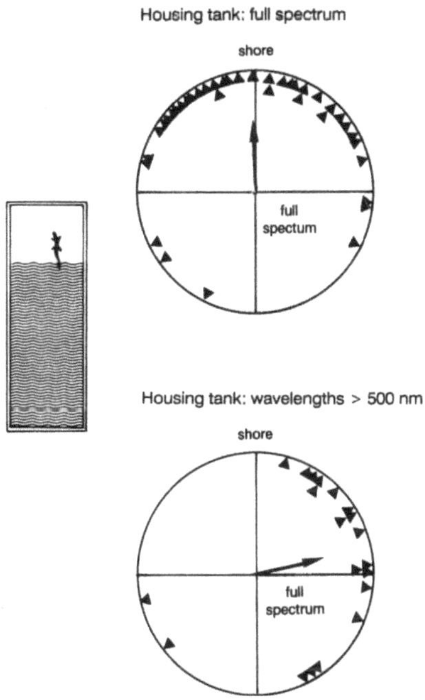

Fig. 9.7. Shoreward orientation of Red-spotted Newts, *Notophthalmus viridescens*, under full spectrum light and under light with wavelengths longer than 500 nm. *Upper diagrams* Newts kept under natural daylight; *lower diagrams* newts kept under a filter system that allowed only wavelengths longer than 500 nm to pass. (After PHILLIPS and BORLAND 1992c)

500 nm so that these newts experienced the directional relationship water-land solely under long wavelengths. When tested at the same long wavelengths, the newts preferred the true shoreward direction; under full spectrum light, however, they showed an approx. 90 ° *clockwise* deflection (Fig. 9.7). This suggests a receptor mechanism which provides similar information for 'north' under full spectrum light and 'west' under long-wavelength light.

Homeward orientation was also found to be affected by the various wavelengths of light, but in a different way (PHILLIPS and BORLAND 1994). Only at short wavelengths of 400 nm were the newts homeward oriented as under full spectrum light. At 450, 550 and 600 nm, in contrast, they showed a considerable increase in scatter, and significant tendencies could no longer be demonstrated (see Table 9.2). This disorientation at longer wavelengths suggests a response pattern similar to that observed in Silvereyes and European Robins, but with a different spectral range allowing orientation (see Fig. 9.8).

Insects. Exposing young fruitflies, *Drosophila melanogaster* (Insecta, Diptera), to light from a specific direction induces a preference for this direction, which is apparently memorized as a magnetic compass course (PHILLIPS and SAYEED 1993; cf. Sect. 4.3.3.4). This response was also found to be affected by the wavelength of light. When a preference was induced under ultraviolet light of

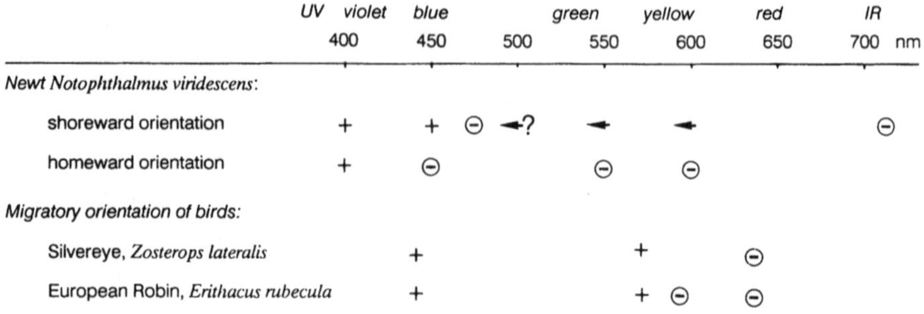

Fig. 9.8. Summary of vertebrate responses to light of various wavelengths. + indicates same response as under full spectrum light, ⊖ indicates disorientation; *arrows* indicate a shift in direction with respect to behavior under full spectrum light. (Data from PHILLIPS and BORLAND 1992c, 1994; W. WILTSCHKO et al. 1993b; W. WILTSCHKO and WILTSCHKO 1995 and unpubl.)

365 nm, the flies showed significant tendencies towards the induced direction when tested under the same wavelength; under 500 nm light, however, their tendencies were deflected *clockwise* approximately 90 °. This seems to suggest a mechanism similar to that involved in shoreward orientation of *Nothophthalmus*, but the deflection was to the other side.[3]

9.2.3.3 Two Patterns of Responses

Two types of patterns have been observed in the relationship of light-dependent orientation responses to light of increasing wavelengths: either disorientation at long wavelengths and oriented behavior at short wavelengths, or normal orientation at certain wavelengths and an approximately 90 ° deflection at others. Figure 9.8 summarizes the responses at the various wavelengths reported so far from newts and birds.

The migratory orientation of Silvereyes and European Robins (W. WILTSCHKO et al. 1993b; W. WILTSCHKO and WILTSCHKO 1995) follows the first pattern, with disorientation under 588 and 633 nm. These findings agree with LEASK's (1977) predictions that light above a certain wavelength might no longer reach the minimum threshold of energy required for magnetoreception. In the newt N*otophthalmus*, a similar response is observed in connection with homing. It has to be noted, however, that the spectral range of 'disorientation' seems to extend much further towards blue-green light. The newts were already disoriented at wavelengths of 550 nm, possibly even under 450 nm, i.e. in light where Silvereyes and European Robins showed excellent orientation (cf. Fig. 9.6) and where the neuronal responses suggested a peak of responsiveness in pigeons.

[3] Effects of various wavelengths of light on the 'misdirection' of the waggle dance in bees described by LEUCHT (1984) appear to be of a different nature, as they seem to be unconnected with processes of magnetoperception.

In connection with shoreward orientation, *Notophthalmus* showed the second type of pattern, with an approx. 90 ° counterclockwise deflection under long wavelength light. The reciprocal results obtained with newts trained under long wavelength light suggest a direct effect on the magnetoreception mechanism. PHILLIPS and BORLAND (1992c) attempted to explain these findings by a model which assumes two types of spectral mechanisms which have antagonistic effects.

Similar models might be evoked in order to explain the deflections observed in *Drosophila*. However, the findings on Silvereyes and European Robins can hardly be interpreted this way. The spectral curve of *Leiothrix lutea*, with peaks at 370, 460, 530 and 620 nm (MAIER 1992), suggests that the 571-nm green and the 443-nm blue light (cf. Fig. 9.6) stimulated different combinations of receptors, yet both produced the same orientation (W. WILTSCHKO et al. 1993b). This rather suggests an 'all-or-none' process, as predicted by LEASK. A plausible, common explanation for the various responses at different wavelengths is not yet available, and it is uncertain whether they represent effects of a homogeneous nature.

The same applies to the different spectral relationships found in shoreward orientation and in homing of *Notophthalmus*. Whether the two types of responses in the same species must be attributed to two underlying receptor mechanisms, is unclear. Initially, PHILLIPS (1986b), in view of the different responses to inversions of the vertical component of the magnetic field (cf. Fig. 4.11), suggested two separate magnetic pathways of compass information for the two behaviors. The findings summarized in Table 9.2 might be taken to support such an assumption. However, since homing additionally requires the determination of the home course as a first step (cf. Fig. 5.3), it also appears to be possible that certain wavelengths might interfere with a navigational mechanism based on magnetic information.

PHILLIPS and BORLAND (1994) speculate on such a model, suggesting that the compass mechanism described for shoreward orientation is used to align receptors necessary for the determination of the homeward course. The rotation under long wavelength light interferes with this alignment and thus prevents the newts from determining their home course. The advantage of this model is that the assumption of two separate compass systems in the same species is unnecessary, even though it does not account for the different responses to the inversion of the vertical component of the magnetic field. It predicts, however, that homeward orientation is possible as long as the light allows normal shoreward orientation. This is not necessarily supported by the data presented by PHILLIPS and BORLAND (1994; cf. Table 9.2): while the relationship between shoreward orientation and wavelengths is characterized by a change between two responses, separated by a phase of disorientation, the data on homing suggest a continuous increase in scatter with increasing wavelength. In particular, under blue light of 450 nm, *Notophthalmus* showed normal shoreward orientation, whereas homeward tendencies could no longer be demonstrated. Such an interpretation might be premature, however, as most samples have rather small sample sizes.

9.2.4 Compass Information from the Retina?

Experiments designed to test the hypotheses based on excited state macromolecules have so far focused on the aspect that magnetoreception might be associated with photoreception. This appears to be true only for a minority of animals. From these groups, however, an impressive body of evidence has been accumulated.

Although an association of magnetoreception and photoreception is suggested by the fact that light is necessary to obtain responses in birds and newts, a direct link between the visual process and the neuronal and behavioral responses has not yet been established. Recordings from avian photoreceptors face severe technical problems. Magnetic stimuli were found to affect the level of transmitters in the retina of some rodents. This effect, which depended on an intact retina and showed considerable inter- and intraspecific variability (for summary, see OLCESE 1990), has been discussed in connection with magnetoreception, although precise mechanisms are not known. The spectral dependence indicated by the data of SEMM and DEMAINE (1986) is consistent with an involvement of rhodopsin, which has peak absorption at 503 nm. Also, electrophysiological responses to magnetic stimulation from retinula cells of the blowfly *Calliphora* (Diptera; PHILLIPS 1987b) and from retinal ganglion cells and fibers of the optic tract in pigeons (SEMM and SCHNEIDER 1991) have been mentioned; this also suggests a link between photo- and magnetoreception.

The type of response observed in the avian visual system – neurons responding to changes in the *direction* of the magnetic field – indicates that this system provides *compass information*. An involvement in a magnetic 'map' seems to be unlikely, also, because the mechanisms proposed by LEASK and SCHULTEN are probably not powerful enough to detect small intensity differences in the range of 10 to 15 nT required for map use (KIRSCHVINK 1989). Both areas of the brain from which magnetic responses have been recorded are associated with processing visual directional information and general orientation in space. The nBOR projects to the vestibular system, which is intriguing in view of the possibility that magnetic information must be combined with gravity information in the birds' inclination compass. The *tectum opticum*, with its retinotopic organization, is an area where magnetic directional information might be integrated with visual information to form a homogeneous picture of the environment with respect to both types of sensory qualities. So far, the findings are consistent with the idea that a light-dependent mechanism in the retina provides directional information.

A number of other animals listed in Table 9.1 proved to be able to orient in the absence of visible light, which seems to rule out magnetoreception in association with visual processes. However, this does not necessarily exclude excited state macromolecules from being transducers. Optical pumping might be replaced by some kind of 'chemical pumping' to transfer molecules into excited states (LOHMANN 1993). A number of chemical reactions involving electron transfer, etc. might be considered in view of the required processes; a specific model has not been proposed, however. Presently, the discussion must remain open for all possibilities.

9.3 Perception Based on Ferromagnetic Particles

Magnetoreception based on ferromagnetic particles, which act as miniature magnets aligning themselves in the magnetic field, was frequently considered as an obvious possibility in the discussion on perception. Yet magnetic material of biogenic origin was unknown until LOWENSTAM (1962) described magnetite in the radula teeth of chitons. In the 1970s, technical progress made highly sensitive SQUID (superconducting quantum interference device) sensors widely available (see FULLER et al. 1985) so that minute quantities of magnetic material in tissues could be detected. Ferromagnetic particles were found in a wide variety of animals, often localized in areas where a connection with the central nervous system did not appear unlikely. This promoted hypotheses assuming a role of magnetite in magnetoreception and has led to intensive research on biogenic magnetic material. First results are impressively documented in the book *Magnetite Biomineralization and Magnetoreception in Organisms* edited by KIRSCHVINK et al. (1985a). Later publications added to these findings by reporting on electrophysiological and behavioral studies aimed at verifying an involvement of ferromagnetic material in orientation.

9.3.1 Magnetite Particles as Transducers

Magnetic material can be traced by its remanence which is detected with sensitive magnetometers. The magnetic moments observed were usually in the order of 10^{-10} to 10^{-8} A m^2/g (= 10^{-7} to 10^{-5} electromagnetic units (emu) per gram). MÖSSBAUER spectroscopy, comparisons of natural and induced remanence, and the relationship between magnetization and demagnetization curves at various temperatures provide information on the nature of the particles in question, their coercivity, and possible interactions (e.g. FRANKEL et al. 1985; FULLER et al. 1985; WALKER et al. 1985). Magnetic material found in animal tissue was mostly identified as the iron oxide *magnetite*, Fe_3O_4, whose general properties are described e.g. by BANERJEE and MOSKOWITZ (1985). Histological studies (e.g. B. WALCOTT 1985) served to detect the exact location and size of the particles, and to search for connections with the nervous system.

9.3.1.1 The Magnetic Properties of Magnetite Particles

The magnetic properties of ferromagnetic materials stem from the spin angular momentum of each atom. The magnetic behavior of particles depends strongly on their size, and to some extent on their shape, as shown in Fig. 9.9. Due to interactions, the spin axes of adjacent atoms show the same alignment, forming small regions, or *domains*, with all spins parallel. Large particles include multiple magnetic domains with spin axes oriented partially antiparallel (Fig. 9.9); hence, their magnetic moments tend to cancel one another, resulting in a low net remanence. If the particles are sufficiently small – for magnetite: approximately in the range between 0.04 and 0.12 μm – they consist of a single magnetic domain only; they have a stable magnetic moment that is constant in time. Very small particles also form single domains, but their magnetic moment fluctuates between several orientations as a result of thermal agitation. They can be easily aligned, however, by external fields, a property which makes them *superparamagnetic*. For more details, see e.g. BLAKEMORE and FRANKEL (1981) and KIRSCHVINK and GOULD (1981).

For magnetoreception, single domain particles with stable magnetization seemed to be the most appropriate candidates, as they can align themselves in the magnetic field. The particles found in magnetic bacteria are indeed single

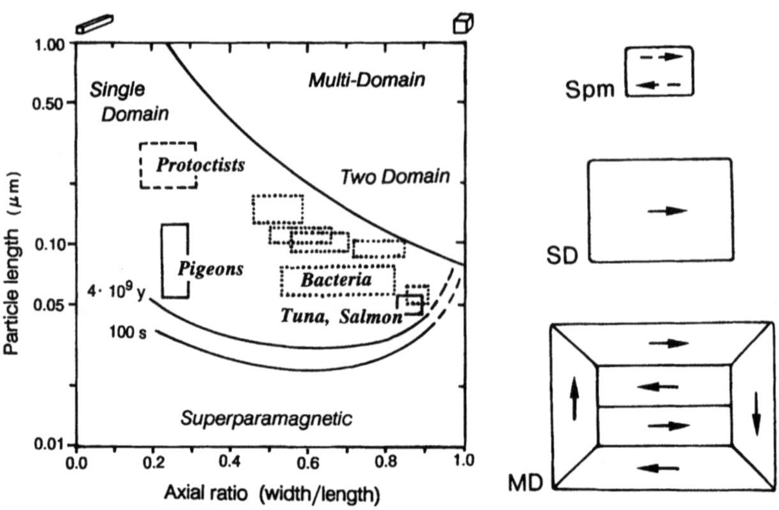

Fig. 9.9. Magnetic properties of magnetite particles of various sizes and shapes. *Left side* Domain stability field diagram; *right side* properties of particles of a given shape change with size. Small particles are superparamagnetic (*Spm*), having an instable magnetic moment. Particles above a critical size represent single domains (*SD*) with stable magnetic moment, while larger particles form more than one domain, becoming multidomains (*MD*) with magnetizations largely compensating each other so that the net magnetization is low. In the *left diagram*, the two lower lines for 100 s and 4×10^9 years of stable magnetization at room temperature indicate that the transition to stability is rather abrupt. Size and shape of magnetite particles found in various organisms are indicated. (After KIRSCHVINK 1989; BLAKEMORE and FRANKEL 1981)

domains (cf. Sect. 2.1.1). The particles identified in animals, however, are often in a range where the magnetic energy of orientation just barely exceeds the background thermal energy. This problem can be overcome by an orderly array of particles, e.g. they may be arranged in chains, clusters, or lattices where the single moments all point in the same direction, held in place by tissue structures. Adding their moments this way would increase their interactions with the geomagnetic field, leading to higher effectiveness. Superparamagnetic particles, on the other hand, might also be part of a receptor mechanism when arranged in a suitably ordered array or aligned by the local field of single domain particles.

Considerations by YORKE (1979, 1981, 1985) and KIRSCHVINK and GOULD (1981) indicated that ferromagnetic receptors could in principle account for the sensitivities indicated by the available behavioral evidence, provided a sufficient number of magnetic grains were available. While a few hundred single domain particles might be sufficient for obtaining information on the *direction* of the magnetic field, the detection of small *intensity changes*, to which homing pigeons and dancing bees were found to respond (cf. Sects. 6.1 and 7.1.1), would require the integration over a much larger number of particles. KIRSCHVINK and GOULD (1981), KIRSCHVINK and WALKER (1985), KIRSCHVINK et al. (1992), SEMM and BEASON (1990) and EDMONDS (1992, 1993) outlined a variety of ways in which magnetite particles of various size might form parts of receptors, interacting with membranes and modifying the activity of specialized cells. Interestingly, several of the possible interactions described are axial.

9.3.1.2 *Magnetic Material Is Widespread Among Animals*

The discovery of magnetotactic bacteria (BLAKEMORE 1975) was soon followed by reports on ferromagnetic material in bees (GOULD et al. 1978) and pigeons (C. WALCOTT et al. 1979), two species known to respond to magnetic stimuli. This started an almost systematic search for magnetite in other animals. Table 9.3 gives an overview; the species mentioned were selected mainly in view of documenting how widespread magnetic material is, with emphasis on species for which magnetic orientation was demonstrated or at least discussed. Further reports on magnetic material in various animals, including danaid butterflies, perciform fish, bats and cetaceans, are summarized by KIRSCHVINK et al. (1985a). Additional data on several species of birds and fish have been published by UEDA et al. (1982, 1986), EDWARDS et al. (1992) and HANSON and WESTERBERG (1987); no significant differences between migrating and non-migrating species were found. EDWARDS et al. (1992) described a correlation between the average intensities of remanence recorded in the heads and necks of birds and the birds' mean body mass.

Up to now, magnetite has been reported for a great number of animals, various parts of their bodies, and a wide variety of tissues. Magnetic material seems to be rather ubiquitous, at least in small quantities (KIRSCHVINK 1989). This raises the question about its biological significance. In chitons, magnetite

Table 9.3. Selected animal species with magnetic material found in their tissues

Species	Where?	Reference
Mollusca, Polyplacophora		
Cryptochiton stelleri (Acanthochitonida)	rad.	LOWENSTAM (1962); TOWE et al. (1963)
Lepidochitona hartwegii (Ischnochitonida)	rad.	NESSON and LOWENSTAM (1985)
Arthropoda, Crustacea		
Balanus eburneus (Cirripedia, Thoracica)	wh.b.	BUSKIRK and O'BRIEN (1985)
Penaeus aztecus (Malacostraca, Decapoda)	cth.	» » » »
*Palinurus argus** (Malacostraca, Decapoda)	cth.	LOHMANN (1984)
Arthropoda, Insecta		
*Apis mellifera** (Hymenoptera)	abd.	GOULD et al. (1978)
*Danaus plexippus**? (Lepidoptera)	hd.,th.	JONES and MACFADDEN (1982)
Vertebrata, Chondrichtyes		
Rhinobatos rhinobatos (Rajiformes)	hd.	O'LEARY et al. (1981)
Vertebrata, Osteichtyes		
Thunnus albacares (Perciformes)	hd.	WALKER et al. (1984)
*Anguilla anguilla** (Anguilliformes)	wh.b.	HANSON et al. (1984a,b)
Gnathonemus petersii (Osteoglossiformes)	wh.b.	HANSON and WESTERBERG (1987)
Clupea harengus (Clupeiformes)	»	» » » »
Cyprinus carpio (Cypriniformes)	»	» » » »
Zoarces viviparus (Gadiformes)	»	» » » »
*Oncorhynchus tschawytscha** (Salmoniformes)	hd.	KIRSCHVINK et al. (1985b)
*Oncorhynchus nerka** (Salmoniformes)	hd.	MANN et al. (1988); WALKER et al. (1988)
*Salmo salar**? (Salmoniformes)	l.l.	A. MOORE et al. (1990)
Vertebrata, Reptilia		
Chelonia mydas (Testudines)	hd?	PERRY et al. (1985)
Vertebrata, Aves		
*Columba livia** (Columbiformes)	hd.	C. WALCOTT et al. (1979)
Zonotrichia leucophrys (Passeriformes)	hd.,nk.	PRESTI and PETTIGREW (1980)
*Dolichonyx oryzivorus** (Passeriformes)	hd.	BEASON and NICHOLS (1984)
Colinus virginianus (Galliformes)	hd.	EDWARDS et al. (1992)
Chaetura pelagica (Apodiformes)	»	» » » »
*Sturnus vulgaris** (Passeriformes)	»	» » » »
Vertebrata, Mammalia		
Eptesicus fuscus (Chiroptera)	?	BUCHLER and WASILEWSKI (1985)
*Apodemus silvaticus** (Rodentia)	hd.	MATHER and BAKER (1981)
Macaca mulatta (Primates)	hd.	KIRSCHVINK (1980)
*Homo sapiens** (Primates)	ad.gl.	KIRSCHVINK (1981b)
» »	hd.	BAKER et al. (1983); KIRSCHVINK et al. (1992)
*Delphinus delphis**? (Cetacea, Odontoceti)	hd.	ZOEGER et al. (1981)
*Megaptera novaeangliae**? (Cet., Mystacoceti)	hd.	BAUER et al. (1985)

Species that were shown to use magnetic information for orientation are marked with *; *? indicates species for which magnetic orientation has been discussed. Under 'where?', it is indicated in what body region the particles were found, or, in case they occurred in various parts of the body, where the highest concentration was found: abd., abdomen; ad.gl., adrenal gland; cth., cephalothorax; hd., head; l.l., lateral line; nk., neck; rad., radula; th., thorax; wh.b., whole body.

covers the surface of radula teeth, obviously acting as a hardening agent to reduce wear. In various mud bacteria and algae, magnetite particles were found to be arranged in chains; their function is to align cells along the field lines for magnetotactic movements (cf. Sect. 2.1.1.). In most animals, however, their function is less obvious. A role in magnetoreception and orientation is usually suggested (e.g. GOULD et al. 1978; KIRSCHVINK 1983 and others), although such an assumption is not always supported by the quantity and/or the distribution of magnetic material (e.g. UEDA et al. 1982, 1986; HANSON et al. 1984a; HANSON and WESTERBERG 1987). Other functions, e.g. in the processes of ossification, or simply metabolic deposits, have also been considered, and they may indeed have been the precursers of a possible function in the perception of magnetic fields.

In view of magnetoreception, findings on the nature and the arrangement of magnetic material in species known or assumed to use magnetic information for orientation are of particular interest.

9.3.1.3 Search for Magnetite in Insects

In Honeybees, *Apis mellifera*, GOULD et al. (1978) located magnetic remanence in the front part of the abdomen. Eggs and larvae seemed to contain much less magnetic material than adult bees. A horizontal orientation of the magnetic moment perpendicular to the longitudinal axis was indicated. However, interindividual variability was very high, and there were individuals with no detectable natural remanence. The magnetization curves indicated superparamagnetic grains (GOULD et al. 1980). Histological, electron microscopy and analytical studies (KUTERBACH et al. 1982; KUTERBACH and WALCOTT 1986a) identified iron-rich granules in the trophocytes of the fat body in the abdomen near the segmental ganglions of post-eclosion adults. The size of the granules increased with increasing age at a rate which was directly related to the iron levels in the diet. Saturation values were normally reached about the time when the bees began showing foraging behavior, which was in agreement with a presumed role in magnetic orientation (KUTERBACH and WALCOTT 1986b). Iron was present as amorphous hydrous iron oxide, however. This form of iron is not magnetic and hence does not contribute to the remanence found by GOULD et al. (1978, 1980), but it is a frequent precursor of magnetite in biosynthesis. Only a small fraction (less than 1%) needs to be turned into magnetite to account for the reported magnetic moments (TOWNE and GOULD 1985). HSU and LI (1994) claimed that 10% of the granules contained small crystals of superparamagnetic magnetite. SCHIFF (1991), examining rigid hairs at the frontal part of the abdomen of bees, described small granules of electron dense material at their base, which he believes to be superparamagnetic crystals of magnetite. A few larger crystals were assumed to be single domains.

A connection to a sensory or nervous structure is unclear, however. The granules in the trophocytes were apparently randomly distributed within the cytoplasm, without being associated with any particular cellular organelle

(KUTERBACH and WALCOTT 1986a); however, HSU and LI (1993) described microtubule-like structures in close proximity to the granules attached to the membrane. Groups of granule-containing cells were electrically coupled, with a low coupling ratio; yet action potentials could not be excited. Moreover, although nearby nerves pass through the fat body, KUTERBACH and WALCOTT (1986a) failed to find the trophocytes innervated. HSU and LI (1994), on the other hand, mention a synaptic structure near one cell. The situation concerning the electron dense particles at the base of the hairs is even less clear, although SCHIFF (1991) speculates on possible receptor mechanisms.

In adult Monarch Butterflies, *Danaus plexippus*, magnetic material, probably magnetite, is concentrated in the front part of the body. As in bees, biosynthesis takes place during pupation and early adult life; eggs and caterpillars are largely free of iron-oxide particles (MACFADDEN and JONES 1985). This and the finding that much less magnetic material was found in four other, non-migrating danaid species were taken to suggest a function in connection with magnetic orientation during migration. Histological studies, on the other hand, did not reveal consistent concentrations of iron (B. WALCOTT and WALCOTT 1982).

9.3.1.4 Magnetite Found in the Head of Vertebrates

The mere size of most vertebrates makes it difficult to locate concentrations of ferromagnetic material in specific parts of the body. The search therefore centered on the head region and looked for orderly arrays as expected by magnetite hypotheses.

Fish. In cartilaginous fish, O'LEARY et al. (1981) found magnetite concentrations near the vestibular receptors of two species of guitarfish (gen. *Rhinobatos*). Magnetite particles formed a conspicuous curved band along the otoconia (calcite particles forming the statolithic mass of gravity receptors) in the ventral region of the *sacculus*, which in part coincided with a portion of the underlying hair cells of the *macula* (VILCHES-TROYA et al. 1984).

Several species of bony fish have been studied, including eels and salmons that orient with the help of the magnetic field. In eels, HANSON et al. (1984a) found magnetic material in practically all the bones examined; a concentration at a particular area in the head was not evident. Subsequent analysis revealed iron in the phases of magnetite, hematite and α-iron, with magnetite being a minority phase when present. WALKER et al. (1984, 1985) described a high content of magnetic material in the ethmoid region of the skull of several fish species, including tunas, marlins, and mackarels. X-ray diffraction allowed the identification of magnetite as the source of remanence in Yellow-fin Tunas, *Thunnus albacares* (Scombridae); size and shape of the crystals showed that they were single domains. In two species of Pacific salmon, magnetometric studies and electron microscopy indicated chains of single domain magnetite particles in the ethmoid tissue, held in place by organic structures (KIRSCHVINK et al. 1985b; MANN et al. 1988). Other tissues often also contained variable amounts of unidentified ferromagnetic material. The quantity of magnetite increased with increasing age; however, theoretical calculations revealed that the amount found in salmon fry may already be sufficient to mediate a compass response (WALKER et al. 1988). SAKAKI et al. (1990) described specific cells surrounded by chains of single domain magnetite, which they found throughout the forehead of Sockeye Salmons *Oncorhynchus nerka* (Salmonidae).

Birds. Initial reports on magnetic material in birds were somewhat confusing. C. WALCOTT et al. (1979) described ferromagnetic material, probably magnetite, in pigeons' heads between the *dura mater* and the skull or associated with the inner surface of the cranium. In this first study, magnetic measurements indicated permanent remanence in 40 % of the pigeons examined, while remanence could be induced in the other 60 %, suggesting superparamagnetic grains. A second study (B. WALCOTT and WALCOTT 1982) revealed no consistent permanent or induced remanence associated with the skull. Instead, iron-containing tissue was found in the harderian gland, at the base of the beak, and at a site just ventral to the olfactory bulbs. PRESTI and PETTIGREW (1980) found iron-rich permanently magnetic material throughout the head and neck region in a variety of birds, including pigeons and several migrants, but it appeared to be diffusely distributed. BEASON and NICHOLS (1984) described iron-rich particles in the sheaths around the olfactory nerve, in the nasal cavity as well as in a thin layer of tissue near the olfactory bulb in Bobolinks, *Dolichonyx oryzivorus* (Icteridae), and some other passerine species. This is in part in accordance with the findings of B. WALCOTT and WALCOTT (1982) in pigeons. The data suggested single domains, with the net natural remanence oriented horizontally from side to side (BEASON and BRENNAN 1986; BEASON 1989b). EDWARDS et al. (1992) found a similar orientation in additional bird species; however, the inter-individual variability was rather high.

Mammals. Magnetic iron-rich deposits, probably magnetite, are also indicated in the ethmoid region of the skull of wood mice and other rodents (MATHER and BAKER 1981; MATHER 1985). In various species of cetaceans, magnetic material was found in the *dura mater* (a tegument covering the brain; BAUER et al. 1985). Magnetite was also consistently found in the human *dura*, in the bones of the human ethmoid region, where ferric iron forms a thin continuous layer (BAKER et al. 1983), and in brain tissue of Rhesus monkeys, *Macaca mulatta* (Cercopithecidae), and humans, *Homo sapiens* (KIRSCHVINK 1981b; KIRSCHVINK et al. 1992).

9.3.1.5 Receptors

In view of the numerous histological studies locating iron-rich particles, often in orderly arrays, it is surprising that no sensory structure has yet been identified. If magnetite is indeed involved in magnetoreception, one would expect cell specializations, or at least some connection with the nervous system. The only one described so far is the localization of magnetite close to the hair cells of the saccular *macula* in guitarfish (VILCHES-TROYA et al. 1984). These findings have not been followed by similar reports from other species, so that they appear to represent an exceptional case among vertebrates. B. WALCOTT and WALCOTT (1982) and KUTERBACH and WALCOTT (1986a) explicitly state their inability to find a direct connection between the iron-containing cells and the nervous system in bees; HSU and LI (1994) mention a synaptic structure near one of the cells, without being able to describe a definite connection. In pigeons (HOLTKAMP-RÖTZLER and WILTSCHKO 1993) and Rainbow Trouts (WALKER et al. 1993), the search has been unsuccessful so far.

9.3.2 Electrophysiological Recordings

The failure to locate specific magnetoreceptors associated with magnetic material did not allow well-aimed electrophysiological studies. Yet the observation that accumulations of magnetite were repeatedly found in the same regions of the body seemed to suggest that recordings from the nerves innervating those areas might be worth investigating. In vertebrates, such recordings led indeed to the discovery of units responding to magnetic stimulation, while corresponding attempts in bees yielded ambiguous results.

9.3.2.1 Recordings from the Ophthalmic Nerve in Vertebrates

In fish as well as in birds and mammals, magnetite in orderly arrays was consistently found in the ethmoid region, which corresponds to the location of magnetic receptors suggested by the application of magnets in humans (BAKER 1989a). This part of the head is innervated by a branch of the *nervus trigeminus*, the ophthalmic nerve *or* the supraophthalmic nerve.

Birds. Recordings from the ophthalmic nerve of Bobolinks revealed that neurons respond with excitation, others with inhibition or with on-off responses, to shifting the horizontal or vertical component of the magnetic field in the course of 5 s (BEASON 1986, 1989b; BEASON and SEMM 1987). About 15 to 21% of spontaneously active units responded; formerly silent neurons were not found to respond to magnetic stimuli. The responses differed from those observed in the visual system insofar as they were not confined to a certain spatial direction; occasionally, their onset seemed to be delayed, and they often continued to be active after the end of stimulation. In a later study, SEMM and BEASON (1990) identified units which responded to intensity changes when the direction was held constant, which suggested that not changes in direction, but associated changes in intensity might have stimulated the ophthalmic system. A threshold was not determined exactly; some units, however, responded to changes as small as 200 nT. Extending the recordings to the trigeminal ganglion, the authors found that 9% of the spontaneously active units respond in a similar way. In some cells that were subjected to different stimuli, the number of spikes increased with increasing intensity changes (Fig. 9.10), suggesting a logarithmic receptor characteristic. SEMM and BEASON (1990) and BEASON and SEMM (1991) discussed their findings in view of small intensity changes as they might be involved in the navigational 'map'.

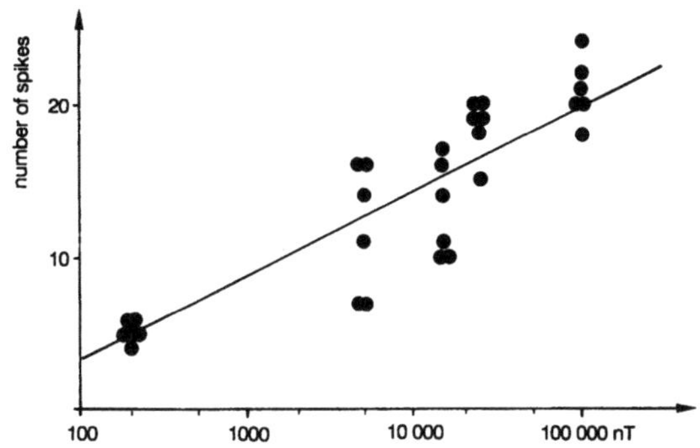

Fig. 9.10. Electrophysiological recordings from a trigeminal ganglion cell of the Bobolink, *Dolichonyx oryzivorus*, responding to different intensity changes of the magnetic field. (After BEASON and SEMM 1991)

Recordings from the ophthalmic nerve of Zebra Finches, *Taeniopygia guttata* (Estrilidae) and Starlings, *Sturnus vulgaris*, produced similar results (HOLTKAMP-RÖTZLER 1990; BEASON 1994). Recordings from pigeons, which are of particular interest owing to their responses to small variations of the magnetic field (cf. Sect. 6.1), have not been reported.

Fish. Recordings from the supraolphthalmic nerve of two salmonid species, Brown Trout, *Salmo trutta*, and Rainbow Trout, also produced responses to magnetic stimulation (WALKER et al. 1993). The portion of units involved was very low, however: only 5 of 179 units (about 3%) responded when a search stimulus, consisting of periodic changes in direction and intensity, was presented. The observed response was an increase in firing rate following the onset of the stimulus, with latency and duration in the range of 15 to 100 ms. When the components of the search stimulus were tested separately, the units seemed to respond to changes in intensity rather than to directional changes.

9.3.2.2 Recordings from Honeybees

In bees, magnetic measurements and histological studies (cf. Sect. 9.3.1.2) indicated accumulations of magnetic material in the front part of the abdomen; hence, recordings from that region seemed to be promising.

KORALL and MARTIN (1987) investigated the hair plate sensilla of the cervical and petiolar joints, which are gravity receptors involved in spatial orientation associated with comb building and dancing. Compensation of the geomagnetic field did not affect the firing rate; lasting intensity changes seemed to be followed by delayed responses; more frequent fluctuations did not cause consistent responses. Altogether, the results are rather unclear and do not support a role of the hair plate sensilla in magnetoreception. – SCHIFF (1991) made recordings from the second abdominal ganglion or the anterior or posterior commisures, using stimuli consisting of changes in intensity along the long body axis. The responses were just above the statistical noise and did not reflect the characteristics of the stimuli. It is unclear whether they were responses to changes in intensity or to associated changes in direction.

9.3.3 Treatments Designed to Affect Magnetite-Based Receptors

Attempts to verify a role of ferromagnetic material in magnetoreception were mostly aimed at interfering with the receptor, by altering the magnetization of the magnetite particles or by damaging its structure. The interpretation of any effect, however, is very difficult because of the general lack of knowledge on the structure and functional principles of the receptors in question. For example, a reversal of magnetization might grossly interfere with a receptor based on single domain particles fixed in specific positions, while it will have little effect on freely rotating particles or superparamagnetic grains. Also, non-specific effects of the rather rough treatments can never be excluded with certainty.

9.3.3.1 Honeybees

To substantiate an involvement of ferromagnetic material in magnetoreception, Gould et al. (1980) observed the dances of 'demagnetized' bees on a horizontal comb (cf. Sect. 3.3.2.2, Fig 3.8). Bees were treated with slowly decreasing alternating fields in order to randomize the magnetization of any magnetic particles; however, their dances in the geomagnetic field continued to show the usual preference for the cardinal compass directions. Bees that initially did not possess any detectable natural remanence were likewise well oriented. This seemed to exclude receptors based on single domain magnetite, at least as far as alignment responses are involved. The authors speculate on possible mechanisms based on superparamagnetic crystals. Kirschvink and Kobayashi-Kirschvink (1991) tried to train bees to leave a maze in a certain magnetic direction (cf. Sect. 8.2.1); they treated some of these bees with a short strong pulse of 1 ms duration and a peak intensity of 0.1 T. However, only three individuals were involved, and the performance of the bees was unclear.

An observation by Walker and Bitterman (1989b), on the other hand, indicated that receptors for magnetic information lie indeed in the front part of the abdomen where the iron-containing particles had been located. Trained bees (cf. Sect. 8.2.2.1) failed to discriminate between the positive and the negative target when they were equipped with small pieces of magnetic wire glued to the anterodorsal abdomen, whereas the same wires at the thorax or at the posterior part of the abdomen had no effect.

9.3.3.2 Homing Birds

The experiments designed to examine the role of magnetizable particles in magnetoreception of pigeons have already been described and discussed in connection with magnetic factors in the navigational 'map' (cf. Sect. 6.1.1.3). The results varied considerably, and their interpretation is not easy in view of the multifactorial nature of the 'map'.

C. WALCOTT et al. (1988) demagnetized pigeons by subjecting them to a 60-Hz oscillating field decreasing from 0.21 T. This treatment, as well as an attempt to align magnetic particles more accurately by 0.115 T and 0.145 T fields, had no effect on initial orientation. The findings seem to argue against receptors based on fixed ferromagnetic material for map information. However, it is unclear to what extent the particles in question might have rotated with the applied fields and thus kept their original magnetization. Superparamagnetic particles, on the other hand, would not be affected by the treatments. Yet exposing the pigeons' head to a strong gradient of 8 T/cm also failed to affect their orientation, although such a gradient would exert a translational force on magnetic grains and thus might have damaged the receptor.

W. WILTSCHKO and BEASON (1990) used a 0.5 T impulse of 4 to 5 ms duration – too short to overcome inertia, but strong enough to reverse the polarity of particles as they were found in birds. This treatment caused small deflections of variable size from the mean of controls. On the day of treatment, they were mostly in the range of 20 °, but occasionally reached 50 °. Their size appeared to be site-specific, and at a given release, their direction appeared to depend on the direction of the pulse (cf. Fig. 6.14). The effect tended to decrease in the course of the following days. Displaced Bank Swallows, *Riparia riparia* (Hirudinidae), on the other hand, were disoriented when treated with the same magnetic pulse (BEASON et al. 1994).

9.3.3.3 Migrating Birds

Passerine migrants were also treated with the short, strong magnetic pulse of 0.5 T used by W. WILTSCHKO and BEASON (1990). Immediately after treatment, Australian Silvereyes showed a ca. 90 ° clockwise deflection (Fig. 9.11). A similar tendency was also observed the following day. From the third day onward, with an intermediate period of disorientation, the birds slowly returned to their original headings (W. WILTSCHKO et al. 1994).

BEASON et al. (1995), using an identical pulse on Bobolinks, treated the birds in two different orientations, namely, north-anterior and south-anterior. The south-anterior treatment, which was identical to that used on pigeons and Silvereyes, caused a ca. 50 ° clockwise deflection, the north-anterior treatment led to a ca. 90 ° counterclockwise deflection. The shifted headings were maintained for at least 6 days. European Robins, treated with the same south-anterior pulse, showed a third response pattern: after treatment, their headings were scattered and did not show a common tendency. They were not completely random, however, as most of the birds continued to head in the same direction during the following tests; i.e. they showed significant directional preferences that differed from their behavior before treatment with the magnetic pulse, and did not agree between birds (W. WILTSCHKO and WILTSCHKO 1995). Thus, all three species tested during the migration phase responded to the magnetic pulse with a significant change in behavior, but their responses differed from

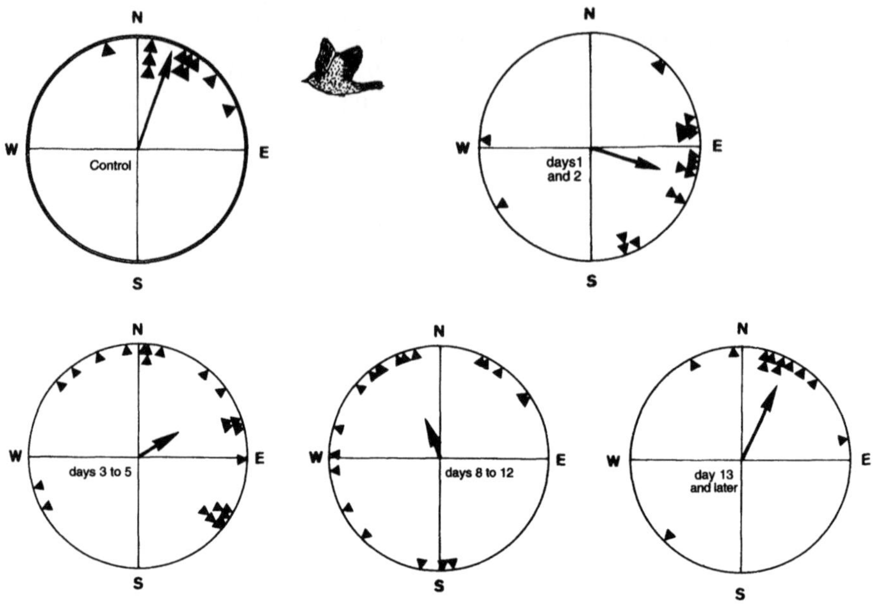

Fig. 9.11. Orientation behavior of Australian Silvereyes after treatment with a short, strong magnetic pulse of 0.5 T and 5 ms duration. The *symbols at the periphery of the circle* give the headings on the indicated days after treatment (days of treatment = day 1); the *arrows* represent the respective mean vectors proportional to the radius of the circle. (Data from W. WILTSCHKO et al. 1994)

one another. This is possibly due to different pre-experiences of the test birds (see W. WILTSCHKO and WILTSCHKO 1995 for discussion).

Recent findings seem to suggest that birds whose ophthalmic nerves had been locally anesthetized cease to be affected by the strong magnetic pulse (BEASON and SEMM, in prep.). This would mean that the information leading to the directional deflection indeed originates in the ophthalmic system.

9.3.3.4 The Interpretation of Behavioral Findings

The treatments, the effects of which have been described here, were specifically designed to affect receptors based on single domains in an orderly array. The negative findings with demagnetized bees (GOULD et al. 1980) and the results with pigeons reported by C. WALCOTT et al. (1988) do not support such a receptor type, but leave the possibility that magnetite in the form of superparamagnetic grains might be involved.

This contrasts to findings in which the treatment consisted of a short, strong magnetic pulse. The data from bees are not yet conclusive (KIRSCHVINK and KOBAYASHI-KIRSCHVINK 1991), but the responses of pigeons (W. WILTSCHKO and BEASON 1990) seem to suggest an involvement of single domains. The same is true for the responses of caged migrants. The decrease and final disappearance of the effect in the course of the following days observed in Silvereyes (cf. Fig. 9.11), however, is not easy to explain in view of single domain particles. Any new magnetization should be as stable as the original one; hence, a restoration of the original situation seems impossible. In pigeons that were allowed to fly freely and had the experience of homing after treatment, it is conceivable that the altered magnetic input is recalibrated and reintegrated with the other navigational factors. However, such an explanation is not plausible for the gradual return of the Silvereyes to their original headings, since the test birds had no access to other cues.

At the moment, some aspects of the treatments designed to affect magnetite remain unexplained. Therefore, a definite conclusion regarding the involvement of single domain or superparamagnetic magnetite is not yet possible on the basis of the available data.

9.3.4 Receptors Based on Magnetite Particles

Magnetite has been found in a wide variety of animals, but the question whether it is involved in magnetoreception, and if so, in what way, is still open. This applies to the nature of the information magnetite provides as well as to the question how the respective receptors in the animals might work.

9.3.4.1 Is the 'Map' Involved?

Although receptors based on magnetizable material could provide directional information as well as information on intensity, magnetite is usually discussed in view of the small intensity differences involved in the 'map' of birds and in the 'misdirections' of Honeybees. On the one hand, it seems to be the only mechanism known so far which, according to theoretical calculations (e.g. KIRSCHVINK and GOULD 1981; YORKE 1981, 1985; KIRSCHVINK 1989), is powerful enough to detect the small differences required. On the other hand, the few available electrophysiological recordings from neurons near deposits of magnetite in vertebrates suggest that these neurons respond to intensity changes rather than to directional changes.

The nature of the responses observed in pigeons after treatment with a short, strong pulse is in rough agreement with what one might expect if a receptor providing 'map' information is impaired: there is an effect, but this effect is small and variable, since only one factor in a multifactorial system is affected. The disorientation observed in displaced Bank Swallows might simply reflect a non-specific response of birds not so used to being handled as pigeons are. On the other hand, Bank Swallows are migrants that return to their breeding colony every year over distances of thousands of kilometers; hence, it seems possible that their response to the treatment indicates an increased importance of magnetic map factors in birds that regularly cover great distances.

The effects on caged migrants are more difficult to explain. Since migratory orientation is usually considered as compass orientation with an innate set direction (cf. Sect. 5.2.2), a magnetite-based compass seems to be suggested. However, the fact that the magnetic compass of birds is an inclination compass makes this interpretation unlikely. Normally, particles can be magnetized only along their long axis; hence, any change in magnetization must be expected to be a reversal. This should not affect a mechanism that does not use polarity. Also, it is unclear how a treatment affecting compass mechanisms could lead to such a variety of responses. On the other hand, a map-based effect cannot be excluded, because in all three cases, the birds were experienced migrants and were not tested during their first migration; i.e. they were not necessarily flying innate courses, but were heading for a region where they had been before. This may involve navigation (cf. Sect. 5.2.2) and thus includes the possibility that the magnetic pulse has simulated a 'displacement' on the 'map'. In this case, the differences in response might be related to differences in experience: the Silvereyes had been captured at their wintering ground, the Bobolinks on their breeding grounds, and the European Robins were young birds caught as transmigrants during autumn migration, but tested the following spring.

9.3.4.2 Single Domains or Superparamagnetic Particles?

Remanence measurements, spectroscopy, and histological data indicate single domains as well as superparamagnetic particles in animal tissues, but even in the same species, data obtained with the various methods (and by the various authors) do not always agree. In several cases, the nature of the particles and/or their arrangement are poorly defined, and original reports had to be modified by later studies. On the whole, these findings were difficult to interpret.

The history of magnetite research in bees may illustrate the confusing state of matters: the initial reports in 1978 indicated considerable natural remanence in some individual bees, but practically none in others (GOULD et al. 1978, 1980). Inspired by the arrangement of magnetite in magnetic bacteria and their 'magnetotactic' behavior (cf. Sect. 2.1.1), this gave rise to speculations about single domains. Subsequent calculations seemed to confirm that magnetically controlled behavior, such as dancing on the horizontal comb, is compatible with receptors based on such particles (e.g. KIRSCHVINK 1981a; YORKE 1985; but see Sect. 3.3.2.2). When behavioral studies with demagnetized bees failed to support an involvement of single domains (GOULD et al. 1980), attention turned to superparamagnetic particles. Histological studies, on the other hand, identified granules containing amorphous hydrous iron oxide, not magnetite, in trophocytes in the abdomen of bees. A connection to the nervous system could not be established, and KUTERBACH and WALCOTT (1986a) concluded that a function as part of a magnetoreceptor, although not totally impossible, appeared to be rather unlikely. Another histological work (SCHIFF 1991) described electron dense material at the base of abdominal hairs, which was discussed to be superparamagnetic *and* single domain particles of magnetite, without verifying whether it was magnetite at all. A later report suggested that small quantities of magnetite form superparamagnetic grains in the trophocytes, and mentioned elements of the cellular skeleton and a synaptic structure in the vicinity (Hsu and Li 1994), yet without definite information on receptor structures. – With birds, the situation is similar, as the reports on the location and the nature of magnetite did not always agree, and behavioral evidence so far does not allow a definite conclusion whether single domains or superparamagnetic material are involved.

9.3.4.3 Open Questions

So far, only few findings support the concept of receptors based on magnetite; especially behavioral data are still rare. In none of these cases has a link between magnetic particles and magnetoreception been established. The initial euphoria following the discovery of magnetite has long vanished in view of the tedious search for actual evidence of its involvement in magnetoreception. Direct proof is difficult to obtain. Failure to find natural or induced remanence, on the other hand, does not per se exclude an involvement of magnetite in magnetoreception, as the minimum quantities of single domains required for compass orientation – a few hundred or thousand – are below the threshold or just at the threshold of detection of the usual SQUID sensors. Likewise, superparamagnetic particles are too small to be identified under normal microscopes, and electron microscopy search requires much time and effort.

A crucial problem of any magnetite hypothesis remains the lack of knowledge on receptive structures. In theory, a wide variety seems to be possible (see e.g. Kirschvink and Gould 1981; Yorke 1985); specific concentrations of magnetite have repeatedly led authors to develop ideas on corresponding sensors. For example, Presti and Pettigrew (1980), after finding magnetite in the neck muscles, suggested magnetoreception by modified stretch receptors in the muscle spindles; Sakaki et al. (1990), describing specific cells surrounded by chains of magnetite, outlined a corresponding receptor. Semm and Beason (1990), who obtained responses in the avian *trigeminus* system which is associated with pressure receptors, suggested magnetite embedded in modified pressure receptors such as Herbst or Grandry corpuscles, and Edmonds (1992, 1993), inspired by the presence of magnetite in the vestibular system (O'Leary et al. 1981), proposed a magnetoreceptor based on magnetically loaded hair cells associated with the gravity receptors. Schiff (1991), after describing magnetite in the bases of hairs of bees, also speculated about its possible function in a receptor mechanism. However, an actual receptor mechanism has not been identified as yet, and with the exception of the *maculae* in guitarfish, no direct connection between magnetite and a sensory organ or the central nervous system has been described, despite extensive search. Considering that first reports on ferromagnetic material in animals responding to magnetic fields date back to the late 1970s, one begins to wonder.

9.4 Magnetoreception in the Light of the Available Evidence

Although behavioral evidence indicates widespread use of magnetic information for orientation, our knowledge on how animals obtain this information remains insufficient. The evidence on magnetoreception available so far results in a somewhat confusing picture. The number of data sets is still rather limited, and a wide variety of animals is involved. There is as yet no single case in which the path of magnetic information from sensory input to behavioral output can be continuously followed up. In this situation, speculations replace missing facts, thus advancing hypotheses on how magnetoreception might work. The probability of any of these hypotheses is difficult to assess. However, most of the findings are fairly new and many have not yet been replicated. So additional studies might clarify the situation by resolving apparent discrepancies and adding to our still limited knowledge.

9.4.1 Open Questions About Transducers

Perception of the magnetic field via electroreceptors is restricted to certain groups of marine vertebrates because of functional limitations (e.g. JUNGERMAN and ROSENBLUM 1980; ROSENBLUM et al. 1985); the other two mechanisms – perception via excited state macromolecules and via magnetite – are free from such restrictions. Moreover, while all three processes could mediate compass information, theoretical considerations so far indicate that only receptors based on magnetite would be sufficiently powerful to mediate information on minute intensity differences indicated to affect behavior, i.e., 10 to 15 nT for the 'map' of pigeons (cf. Sect. 6.4.2) and 1 to 10 nT or rates of change of 1 to 3 nT/min for dancing bees (TOWNE and GOULD 1985; KIRSCHVINK 1989). Each of the three hypotheses in discussion is accompanied by some electrophysiological and/or behavioral data which are in agreement with the suggested transducing process, even though a direct link with the transducer has not been established as yet. So far, the proposed mechanisms are more or less plausible possibilities only.

The crucial point is the necessity to demonstrate specific transducer mechanisms. Previous attempts are ambiguous in the sense that they are in agreement with proposed mechanisms, without being able to exclude other interpretations. The best-documented case is represented by light-dependent receptors associated with the visual process in birds: they are indicated, on the one hand, by electrophysiological responses to changes in the direction of the ambient magnetic field in the visual system, in particular in the *tectum opticum* and, on the other hand, by behavioral effects of various light conditions on compass orientation. The crucial questions regarding the specific transducing processes and receptive structures are still unanswered, however.

With respect to magnetite-based receptors, the situation is more difficult, as non-invasive experimental techniques which selectively affect this type of receptor, e.g. the lights of different wavelengths used to test for a role of photopigments, are not possible. It is practically impossible to interfere with magnetite without affecting other types of receptors. Possible side effects of the treatment can hardly be assessed. The short, strong magnetic pulse might alter the magnetization of single domains; yet it is not completely certain that subsequent effects are actually based on magnetite. The mere fact that there were aftereffects of the treatment argues for a magnetite-based receptor, since it would seem that the other transducing mechanisms suggested so far are affected only when the external field is present. However, transient aftereffects of treatments with alternating fields, too weak to affect the magnetization of magnetite, have been described in pigeons (e.g. PAPI et al. 1983; IOALE and TEYSSEDRE 1989; cf. Sect. 6.1.1.3), so that it might be premature to take the existence of aftereffects per se as proof for an involvement of magnetite. The gradual return to the original orientation observed in Silvereyes is also hard to reconcile with the single domain hypothesis, since a newly acquired magnetization would be as stable as the original one. Demonstrating an involvement of superparamagnetic magnetite grains seems to be impossible by similar means, at least as long as the site and structure of the sensor remain unknown.

Birds are the only group that has been tested in view of both light-dependent processes and magnetite. Treatments designed to interfere with both types of mechanisms were found to affect orientation in different ways. In pigeons, the disorientation of young birds after displacement in total darkness suggested a light-dependent compass, while the deviations of vanishing bearings observed under sunny conditions after treatment with a short magnetic pulse seemed to indicate a magnetite-based 'map'. This is in accordance with available electrophysiological data (for summary, see BEASON and SEMM 1994), and was considered to be a first step in narrowing down the two receptor types expected on the basis of different uses of magnetic information. Both treatments also affected the migratory orientation: long wavelength light caused disorientation, which is in agreement with a light-dependent compass mechanism (W. WILTSCHKO et al. 1993b), while the pulse caused changes in the preferred direction (W. WILTSCHKO et al. 1994). This need not necessarily disagree with the pigeon data: since the test birds were experienced migrants, the effect might also be attributed to a magnetite-based 'map', even if the true role of navigational processes during migration is not entirely clear (see HELBIG 1992). However, other interpretations cannot be excluded as long as magnetoreception in general is so little understood. At first glance, the light-dependent effect and the pulse effect on migratory orientation suggest two mechanisms of different nature; yet it is unclear whether the two different treatments affected the behavior at different levels, namely, one affecting the compass and the other the 'map', whether they affect different components of the same complex mechanism, or the same mechanism in different ways.

9.4.2 Evolutionary Considerations

Magnetic orientation is widespread among animals; this and the discovery of magnetite in organisms ranging from bacteria and algae to arthropods and vertebrates seemed to suggest a common origin of magnetoreception. Magnetoreceptors developed at an early stage and passed on through evolution would seem to be an attractive idea. However, bacteria, arthopods, and vertebrates belong to very different lineages which diverged at a point far ahead of the reconstruction of paleontological history. A comparison of other receptor systems among organisms reveals that usually ecological requirements rather than phylogenetic relations determine the development of sensory structures which utilize the most suitable transduction processes. For example, von Salvini-Plawen and Mayr (1977) concluded that photoreceptors originated independently in at least 40 different phylogenetic lines. This is remarkable in view of the general advantage offered by visual information. On the other hand, maintaining receptors for a sensory function which does not pay off by increasing the chances of survival and reproduction means an unnecessary expenditure of energy and hence will not be favored by selection. Thus, magnetoreceptors would be reduced when the information they provide ceases to be of importance. Therefore, one can hardly expect a uniform type of magnetoreceptor within the animal kingdom.

Among vertebrates, the situation is different insofar as they represent a monophyletic group; their sensory organs are basically homologous. However, a closer look reveals amazing adaptations to specific needs, e.g. in the mechanisms for focusing the eyes, for frequency analysis in the ears, etc. When magnetic material was repeatedly found in or around the ethmoid sinus in different groups of vertebrates, Kirschvink (1983), Walker et al. (1984), and Baker (1989b) speculated about a possible ancestral magnetoreceptor based on deposits of magnetite in that region of the head. However, a variety of mechanisms is indicated, as some groups require light for compass orientation, while others do not (cf. Table 9.1). A similar diversity is implied by the findings on inclination compass and polarity compass (cf. Table 4.8). There is not even a consistent relationship between the type of compass and the ability to orient in the absence of light. Thus the findings do not support a common origin of magnetoreception in vertebrates.

Assuming that the primitive groups with specialized electroreceptors indeed use these receptors to obtain magnetic compass information, as Kalmijn (1978b, 1984) proposed, means that the ancestors of modern vertebrates gave up one type of receptor and developed another which works on a different principle. Electroreception seems to be one of the ancestrial senses of ancient vertebrates; Agnatha, Chondrichthyes, and groups close to the roots of tetrapods, like crossopterygians and lungfishes, as well as some amphibians, possess ampullary organs today. Most bony fishes, on the other hand, including marine fishes living in the same environment as Chondrichthyes, lack specialized electroreceptors; some groups living in muddy freshwater have developed a new type of electroreceptor with slightly different functional properties (Bullock et al. 1983). In regard to magnetic orientation, this would mean that both the ancestors of bony fishes and the common ancestors of reptiles, birds, and mammals lost one way of perceiving magnetic fields together with electroreception. It is hard to believe that animals would give up information so useful for orientation. However, the fact that magnetoreception based on induction requires motion may have proved disadvantageous, and for freshwater fishes and terrestrial animals, this type of magnetoreception might have proved to be insufficient so that selection favored alternative mechanisms.

The new receptors differ between groups. The light-dependent mechanism indicated in the avian magnetic compass is in accordance with the fact that birds are a primarily diurnal group, mostly living in an open environment and having a highly developed visual system. Likewise, it is not surprising that the magnetic compass of a predominantly nocturnal group like mammals does not require light, especially when fossorial forms are considered. A light-dependent mechanism in amphibians seems to be odd, however, as salamanders are mostly active during twilight or at night and generally prefer shady habitats. The light-dependency of the magnetic compass suggests parallels between birds and amphibians, on the one hand, and between reptiles and mammals, on the other, while the inclination compass suggests parallels between birds and reptiles, with fishes and mammals having a different type of mechanism and amphibians apparently having both types. Thus, even if the elasmobranchs had additional magnetoreceptors besides the *ampullae of Lorenzini*, magnetoreceptors can hardly be considered homologous among vertebrates.

9.5 Other Electrophysiological Responses to Magnetic Stimuli

This section briefly summarizes miscellaneous electrophysiological recordings which produced responses to magnetic stimuli in the range of the geomagnetic field. They are not so closely associated with specific hypotheses as the ones mentioned so far, but document a widespread receptivity for magnetic cues.

9.5.1 Recordings from Vertebrates

The first electrophysiological responses to magnetic stimulation were reported from the pineal gland of the Guinea Pig, *Cavia aperea* (Mammalia, Rodentia): about 20% of the pineal cells changed their spontaneous activity when the intensity and/or the direction of the magnetic field were altered (Semm et al. 1980). Changes in spontaneous activity could also be observed in the *cerebellum* and in the habenular nuclei. When the studies were extended to pigeons, about 30% of the pineal cells were found to respond to a change in the magnetic field with either excitation or inhibition, while a second change within the next

30 min had little effect (Semm 1983). However, the responses did not appear to reflect any of the obvious characteristics of the stimulus like the amount or rate of change. Pineal cells continued to respond after the optic nerves and the pineal stalk had been cut, indicating that these cells do not receive magnetic information from other cells, but are sensitive to changes in the magnetic field (Demaine and Semm 1985).

The pineal gland developed from the 'third eye' of ancient vertebrates and is phylogenetically part of the visual system. In birds, cells of the pineal gland maintained their sensitivity to light (Semm and Demaine 1983). Hence, the documented sensitivity of pineal cells to magnetic changes is another indication for a connection between magnetoreception and the visual process. At the same time, this sensitivity is intriguing in view of the role of the pineal hormone *melatonin* in maintaining the diel rhythm and the suggested effects of magnetic changes on this rhythmicity (cf. Sect. 7.1.2). For a summary of magnetic effects on pineal activity, see Olcese (1990).

Responses to magnetic stimulation could also be recorded from cells of the *vestibulo-cerebellum* and the vestibular nuclei of pigeons. The neurons responded only when the animal was tilted out of the horizontal plane so that the vestibular system was activated, suggesting some interaction between magnetic and gravitational information (Semm et al. 1984). At first glance, this seems to be in agreement with receptors in the vestibular systems, as proposed by Edmonds (1993). However, the responses depended on light and an intact contralateral eye, indicating that the information originated in the visual system. The responses in the *vestibulo-cerebellum* and the vestibular nuclei did not reflect the direction of the magnetic field in the way the majority of responses recorded from units in the nBOR or the *tectum opticum* had done (cf. Fig. 9.4).

9.5.2 Recordings from Invertebrates

Electrophysiological responses to directional changes in the magnetic field were recorded from the neuron PE5 of the pedal ganglia of the marine slug *Tritonia diomedea* (Gastropoda, Opisthobranchia, Nudibranchia), a species known to use the magnetic field for orientation. The cell PE5 can be readily identified by its position and size; its function is unknown (Lohmann et al. 1991). The applied stimuli were clockwise rotations of the ambient magnetic field by 20 °, 60 °, and 90 °, either presented for a series of 1-min intervals, each followed by a 1-min interval in the geomagnetic field, or continuously for 30 min. The cells responded with a marked increase in spike frequency to both types of stimuli about 6 to 16 min (!) after stimulus onset, which mostly, but not always, persisted for several minutes after the end of stimulation (Lohmann and Willows 1989; Lohmann et al. 1991). The biological significance of these findings is unclear; even considering that *Tritonia* is a rather slow moving animal, the observed latencies are too long to provide meaningful orientational information.

Neuronal responses to magnetic stimulation were also reported from terrestrial snails, *Helix pomatias* (Gastropoda, Pulmonata, Stylommatophora), but they involved high frequency alternating fields about three times that of the earth's field intensity (KULLNICK 1991). Recordings from various neurons of arthropods also produced responses to alternating fields which were mostly of unphysiological strength (see e.g. SADAUSKAS and SHURANOVA 1984; UENO et al. 1986).

9.6 Summary

The search for magnetoreceptors is handicapped by the lack of knowledge regarding the location of sensory organs and the functional principles. There are presently three main hypotheses concerning transducing mechanisms: (1) induction, (2) excited state macromolecules, in particular photopigments, and (3) ferromagnetic material such as magnetite.

The induction hypothesis assumes that marine vertebrates obtain magnetic compass information by recording the electric fields produced by their own motion in various directions. The sensitivity of the *ampullae of LORENZINI* is sufficient to allow detection of directions; electrophysiological recordings from the *lateralis* system are in agreement with the hypothesis. Because a highly conductive external medium is required, magnetoperception via electroreceptors would be restricted to marine animals with highly specialized electroreceptors.

The second hypothesis assumes photopigments as transducers. Through photon absorption, these macromolecules may reach excited states; excited triplet states have magnetic moments and could thus interact with the ambient magnetic field. Magnetoreception was suggested to take place in the eye as a by-product of the visual process. Electrophysiological recordings revealed neuronal responses to magnetic stimuli in the accessory optic system and in the *tectum opticum* of birds, both of which depended on the presence of light. The responses were direction-specific; they varied between neurons in a way suggesting that the responses of numerous cells processed collectively and integrated might provide compass information. Behavioral tests revealed light-dependent compass responses in birds, salamanders and flies. Their behavior under light of various wavelengths indicated different response patterns, namely, disorientation under long wavelengths and normal orientation under short wavelengths, as found in birds and salamanders orienting homeward, or a deflection of orientation at specific wavelengths, as found in salamanders orienting shoreward, and in flies. Beetles, fishes, reptiles and mammals, on the other hand, appear to be capable of magnetic orientation in the absence of light.

The third hypothesis assumes magnetoreception based on ferromagnetic material, in particular magnetite forming single domains or superparamagnetic particles. Various types of receptors have been suggested. Ferromagnetic material has been identified in a wide variety of animals; in vertebrates, it was consistently found in the ethmoid region of the head. Honeybees have iron-rich particles in the front part of their abdomen. Neurons of the ophthalmic nerve, a branch of the *nervus trigeminus*, of birds and fishes responded to changes in magnetic intensity. Behavioral studies with bees failed to support a role of single domain magnetite in magnetoreception, which left the possibility of a receptor based on superparamagnetic grains. In birds, the results of treatments designed to alter the magnetization of magnetite ranged from no effect at all to 90 ° deviations from the mean of untreated controls. The observed effects are mostly compatible with receptors based on fixed arrangements of single domain magnetite particles, although effects on other types of receptors cannot be excluded at the present stage. The findings are usually discussed in regard to the navigational 'map'.

The available evidence on magnetoreception does not yet provide a complete picture. In particular, a direct link between the proposed transducing mechanisms and the electrophysiological and behavioral findings could not be established. The best-documented case so far is that of light-dependent magnetoreception in birds. A role of magnetite is difficult to demonstrate, as there is no treatment which will selectively affect magnetite. Data on pigeons and migrants are in agreement with the assumption of a light-dependent compass and a magnetite-based 'map' in birds; however, other interpretations cannot be excluded yet.

The variety of compass mechanisms observed among vertebrates – inclination compass and polarity compass, light-dependent and light-independent – speaks against a common principle of magnetoreception, and suggests that different types of magnetoreceptors might have developed independently among the various groups.

Neurons responding to magnetic stimuli were also found in the pineal gland of rodents, the *cerebellum*, the habenular nuclei, the pineal and the vestibular nuclei of pigeons, and the neuron PE5 of the pedal ganglion of a marine slug.

Part 5: Outlook

Chapter 10

Research on Magnetic Phenomena in Biology

Today, the scientific public is aware of the magnetic field as a factor that must be taken into account in animal orientation. This view was accepted only very recently, however, which is surprising considering the usefulness of magnetic information in spatial tasks and man's own use of the technical magnetic compass. General evolutionary considerations also should have encouraged research on magnetic orientation: the geomagnetic field has always been an integral part of the earth's geophysics, which means that the evolution of sensory systems took place under conditions of a magnetic field providing directional and spatial information. Animals are very versatile and have developed complex receptive organs and mental processing units to make use of an astounding variety of environmental cues in order to promote their fitness. In view of this, it would be really surprising if animals had not taken advantage of a useful, reliable cue like the geomagnetic field.

10.1 Slow Acceptance of Magnetic Orientation

The reasons why the role of the magnetic orientation cues was not recognized before are rather complex. Spatial orientation, like animal behavior in general, does not belong to the classic fields of biological sciences, e.g. systematics, anatomy and biogeography. Hence, it does not share their long tradition. Regular studies of behavior began only in the first half of this century. The spatial behavior of most animals was very poorly known; aside from the periodic occurrence of some commercially exploited fish, homing of carrier pigeons and bird migration were the only exceptions. Thus, it is not surprising that two early orientation hypotheses attempted to explain the orientation of birds.

Interestingly, both hypotheses considered the magnetic field as orienting factor: von Middendorff (1859) suggested that migrating birds orient along magnetic meridians, which means, in modern terms, that they use a magnetic compass, and Viguier (1882) proposed a navigational 'map' for homing birds, based on total intensity and inclination (cf. Sect. 6.1.1). However, first attempts to verify an involvement of the magnetic field in avian orientation by equipping displaced birds with magnets failed to yield positive results (cf. Table 6.2). As a consequence, the magnetic field was no longer regarded as a relevant orientation factor. Later hypotheses proposing the use of magnetic cues like Yeagley's (1947) met with heavy criticism.

The discovery of the sun compass by KRAMER (1950) in birds and VON FRISCH (1950) in bees initiated a new age in orientation research. The control of spatial orientation received considerably more attention than ever before, but most interest focused on the newly described mechanism and other celestial cues. Observations that animals could orient also in the absence of visual cues were repeatedly questioned, because researchers hesitated to accept the existence of a further orientation factor. The general attitude in the 1950s and early 1960s is expressed in a typical way by the quotation given below; the respective letter was written in December 1961 by a well-known orientation researcher to a colleague who had demonstrated to him that migratory birds were able to find their migratory direction in closed rooms, and had mentioned the magnetic field as possible cue:

"I still find it very difficult to believe that there is really an entirely new sensory channel between these birds and their environment, though I do not see any obvious defects in the experiment."

Two reasons played a most important role in this reluctance: firstly, a third compass mechanism besides the sun compass and stellar compass seemed superfluous. It seemed to violate *Occam's razor*, a traditional principle of science, namely, to prefer the most parsimonious explanation. Scientists were not yet aware of the complex, partly redundant nature of orientational systems. Only in the 1970s and 1980s was it realized that animals very often have more than one option. Secondly, psychological barriers hindered the acceptance of an ability so alien to humans in animals. In their attempt to understand the remarkable orientation feats, authors often clung to explanations based on the more 'conventional' senses, yet mostly without worrying about details, e.g. the nature of the required information or the precise mechanisms. Altogether, this was a rather anthropocentric attitude, but it is perhaps understandable in the light of the still limited knowledge of the animals' sensory world at a time when many of the specific sensory qualities of animals that exceed our human senses had not yet been discovered.

A certain reluctance to accept a 'new' sensory quality was not the only reason why magnetic orientation was not easily accepted, however. The animals' ability to perceive magnetic stimuli was generally doubted, because crucial questions concerning magnetoreception remained unanswered. While anatomy and physiology of the sensory organs for the more conventional orientation stimuli were familiar, the primary processes of magnetoreception, the locations of the receptors, etc. were completely unknown. Moreover, there was a long list of unsuccessful attempts to condition animals to magnetic stimuli. This was in sharp contrast to conditioning experiments with visual, acoustic, chemical, etc. stimuli and continued to throw doubts on the animals' ability to perceive the magnetic field, in spite of an increasing body of behavioral evidence indicating a role of magnetic cues in orientation. Possible reasons for the negative conditioning results have been discussed in Section 8.3: experimentalists largely ignored the specific functions of magnetic information as orientation cue and disregarded the natural behavioral context in their test designs.

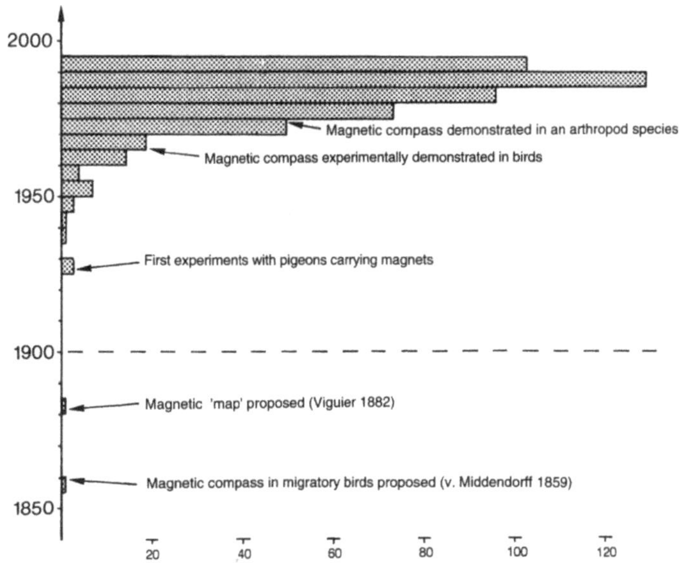

Fig. 10.1. Number of articles on magnetic orientation and magnetoreception published within 5-year periods

From 1960 onward, the number of papers on behavioral effects of the magnetic field increased steadily (see Fig. 10.1), even though at that time, the respective findings still met with considerable scepticism. During a symposium in 1971, magnetic orientation was among the topics of a panel discussion on 'Unconventional theories of orientation' (F. BROWN 1971). The use of magnetic information for direction finding was first demonstrated in 1966 in the European Robin, a passerine bird; magnetic compass orientation in non-avian vertebrates and arthropods was not described until the late 1970s. From then onward, magnetic orientation gradually won acceptance. More and more experimentalists examined whether their experimental animals oriented by magnetic cues. About three-fourths of the papers listed in Tables 4.2, 4.4 and 4.6 as documenting magnetic compass orientation were published in the 1980s and 1990s. During the last decade, non-compass uses of magnetic information have also been considered more often for various animals in connection with various behaviors.

10.2 Present and Future Research

The research on magnetic phenomena is no longer a homogeneous field of biology, but develops along various lines. Presently, most research efforts concern the effects of electromagnetic radiation and fields of anthropogenic origin, which became part of basic medical research. Other, more zoologically oriented research focuses on the geomagnetic field and its natural uses, studying behavioral questions and the physiology of magnetoreception.

10.2.1 Bioelectromagnetic Research

Research on the biological effects of electromagnetic radiation developed into a rapidly growing field of its own. The increased awareness of human impact on the environment and potential hazards to human health led to great public concern about possible adverse effects of electromagnetic fields produced by various kinds of technical equipment. These include low frequency alternating fields from common household installations, radiation with frequencies used for broadcasting, television and mobile telephones, up to high frequencies originating from microwaves, radar stations, etc. The term '*electromagnetic smog*' reflects the omnipresence of such radiation as well as the growing uneasiness about it. At the same time, modern medical treatments involve fields exceeding the strength of the geomagnetic field more than 100 times, so that safeguards against possible side effects become a necessity.

Although disturbing effects of electromagnetic radiation on bird orientation were already described in the middle of this century (e.g. DROST 1949; KNORR 1954), such effects did not receive very much attention at first. During the last decades, however, increasing efforts were devoted to questions concerning the biological effects of electromagnetic radiation, which became a most active field of research. This is reflected by a rapidly increasing number of publications, specialist journals devoted to these topics, etc. Research followed up these effects on all levels, from the biochemical and cellular level to questions concerning responses of the whole organism, e.g. general effects on growth and development, neuroendocrine control, behavior, etc. A number of recent reviews report progress in this field. For example, BERNHARD (1992) summarizes research on electromagnetic smog, MISAKIAN et al. (1993) review the effects of extremely low frequency magnetic fields and a series of papers edited by AMEMIYA (1994) review the effects of electromagnetic fields ranging from extremely low to ultra-high frequencies. Reviews on a wide variety of effects of electric and magnetic fields have been published, for example, by WADAS (1991) and CARPENTER and AYRAPETYAN (1994). Owing to the continuous development of new technical devices producing electromagnetic radiation and new and different applications of strong magnetic fields, together with growing public concern about their possible harmfulness, this field of research will continue to undergo very rapid development. One can expect to obtain deeper insights into the interactions of electromagnetic radiation with biological processes, which hopefully will help to avert potential dangers to our health and well-being.

However, the nature of the man-made fields studied in bioelectromagnetic research is fundamentally different from the geomagnetic fields animals use as a source of orientational information, just as the effects on cellular biochemistry, tissue interactions, etc. are very different from the normal processes which make magnetic information available to the animals. Thus these studies, although they might give hints to possible mechanisms of receptive processes, do not contribute directly to the questions of magnetic orientation.

10.2.2 Magnetoreception and Magnetic Orientation

Parallel to bioelectromagnetic research, studies concerned with the biological effects of the geomagnetic field, in particular with the animals' use of magnetic information, continue to produce intriguing new results. Numerous efforts are currently devoted to magnetoreception, a pressing open question that must be answered to fully understand magnetic orientation phenomena. Here, recent progress in the understanding of the theoretical background may prove helpful. While scientists previously tended to search for one mechanism which can detect the direction of the magnetic field *and*, at the same time, can record minute changes in intensity, it is now largely realized that the various types of magnetic information might require different types of mechanisms: for example, hearing and feeling, both being mechanical senses, have very different receptors, each adapted to the specific nature of their stimuli. This understanding allows more specific considerations on how receptors might work.

Several hypotheses on magnetoperception have been proposed, some of which are meanwhile supported by first experimental evidence (cf. Chap. 9). Even though the present data base is still extremely limited and leaves essential questions unanswered, it means an entry and thus helps to direct research efforts. In the search for receptors, we are no longer completely groping in the dark; we have at least tentative ideas as to what we might be looking for and where it might be worth looking. For example, electrophysiological responses to changes in the magnetic field recorded in specific neurons may help to locate the receptor sites. Histological and neurochemical studies may follow up this track in order to find out about sensory structures, receptor mechanisms, etc. This might involve a lot of tedious work, however. The present state of the search for magnetite-based receptors (cf. Sect. 9.3.4.2) warns us not to expect fast answers. Nevertheless, one may be optimistic and hope that within the next few years we will learn to understand the principles of magnetoreception.

Behavioral studies continue to elucidate the important role of magnetic information in the control of spatial behavior. Orientation phenomena like bird migration or the zonal orientation of littoral arthropods have been thoroughly analyzed, and the role of magnetic information in these multifactorial systems is fairly well understood. The orientation mechanims of eels and salmons on their migrations have been outlined; use of magnetic information is indicated, but many details have to be explained by future studies. The orientation of marine turtles and the role of magnetic information are currently under investigation.

Many other groups, however, have hardly been studied at all. Animals known to perform oriented movements would be most promising subjects for the examination of magnetic orientation. Yet the spatial behavior of most species is still poorly known, so that a first step would have to be the search for an adequate behavior as a basis for orientation tests. This is especially true for the majority of invertebrate species. The apparently widespread use of magnetic information suggested by the taxonomic diversity of species that have been shown to use a magnetic compass indicates that such studies may be rewarding.

Altogether, by collecting the existing findings on magnetic orientation, we hope to encourage other researchers to engage themselves in this field, and we look forward to new, fascinating results in the years to come.

References

Able KP (1974) Environmental influences on the orientation of free-flying nocturnal bird migrants. Anim Behav 22:224–238

Able KP (1982) Skylight polarization patterns at dusk influence migratory orientation in birds. Nature 299:550–551

Able KP (1987) Geomagnetic disturbances and migratory bird orientation: is there an effect? Anim Behav 35:599–601

Able KP (1989) Skylight polarization patterns and the orientation of migratory birds. J Exp Biol 141:241–256

Able KP (1991a) Common themes and variations in animal orientation systems. Am Zool 31:157–167

Able KP (1991b) The development of migratory orientation mechanisms. In: Berthold P (ed) Orientation in birds. Birkhäuser, Basel, pp 166–179

Able KP (1993) Orientation cues used by migratory birds: a review of cue-conflict experiments. Trends Ecol Evol 8:367–371

Able KP (1994) Magnetic orientation and magnetoreception in birds. Prog Neurobiol 42:449–473

Able KP, Able MA (1990a) Ontogeny of migratory orientation in the savannah sparrow, *Passerculus sandwichensis:* calibration of the magnetic compass. Anim Behav 39:903–913

Able KP, Able MA (1990b) Calibration of the magnetic compass of a migratory bird by celestial rotation. Nature 347:378–389

Able KP, Able MA (1990c) Ontogeny of migratory orientation in the savannah sparrow, *Passerculus sandwichensis*: mechanisms at sunset. Anim Behav 39:1189–1198

Able KP, Able MA (1993a) Magnetic orientation in the savannah sparrow. Ethology 93:337–343

Able KP, Able MA (1993b) Daytime calibration of magnetic orientation in a migratory bird requires a view of skylight polarization. Nature 364:523–525

Able KP, Able MA (1994) The development of migratory mechanisms. J Ornithol 135:372

Able KP, Able MA (1995) Interactions in the flexible orientation system of a migratory bird. Nature (in press)

Able KP, Cherry JD (1986) Mechanisms of dusk orientation in white-throated sparrows (*Zonotrichia albicollis*): clock-shift experiments. J Comp Physiol A 159:107–113

Adey WR (1981) Tissue interactions with non-ionizing electromagnetic fields. Physiol Rev 61:435–514

Akoev GN, Ilyinsky OB, Zadan PM (1976) Responses of electroreceptors (ampullae of Lorenzini) of skates to electric and magnetic fields. J Comp Physiol 106:127–136

Alerstam T (1987) Bird migration across a magnetic anomaly. J Exp Biol 130:63–86

Alerstam T, Högstedt G (1983) The role of the geomagnetic field in the development of bird's compass sense. Nature 306:463–465

Alsop B (1987) A failure to obtain magnetic discrimination in the pigeon. Anim Learn Behav 15:110–114

Altmann G (1981) Untersuchung zur Magnetotaxis der Honigbiene, *Apis mellifica* L. Anz Schädlingskd Pflanzenschutz Umweltschutz 54:177–179

Amemiya Y (ed) (1994) Special issue on biological effects of electromagnetic fields. IEICE Trans Commun E77-B:683–769

Anderson JB, Vander Meer RK (1993) Magnetic orientation in the fire ant, *Solenopsis invicta.* Naturwissenschaften 80:568–570

Andrianov GN, Brown HR, Ilyinsky OB (1974) Response of central neurons to electrical and magnetic stimuli of the Ampullae of Lorenzini in the black sea skate. J Comp Physiol 93:287–299

Arendse MC (1978) Magnetic field detection is distinct from light detection in the invertebrates *Tenebrio* and *Talitrus.* Nature 274:358–362

Arendse MC (1980) Non-visual orientation in the sandhopper *Talitrus saltator* (Mont). Neth J Zool 30:535–554

Arendse MC, Barendregt A (1981) Magnetic orientation in the semi-terrestrial amphipod *Orchestia cavimana*, and its interrelationship with photo-orientation and water loss. Physiol Entomol 6:333–342
Arendse MC, Kruyswijk CJ (1981) Orientation of *Talitrus saltator* to magnetic fields. Neth J Sea Res 15:23–32
Arendse MC, Vrins JCM (1975) Magnetic orientation and its relation to photic orientation in *Tenebrio molitor* L (Coleoptera, Tenebrionidae). Neth J Zool 25:407–437
Aschoff J (1960) Exogenous and endogenous components in circadian rhythms. Cold Spring Harbor Symp Quant Biol 25:11–28
August PV, Ayvazian SG, Anderson JTG (1989) Magnetic orientation in a small mammal, *Peromyscus leucopus*. J Mammal 70:1–9
Baker RR (1978) The evolutionary ecology of animal migration. Holmes and Meier, New York
Baker RR (1980) Goal orientation by blindfolded humans after long-distance displacement; possible involvement of a magnetic sense. Science 210:555–557
Baker RR (1987a) Human navigation and magnetoreception: the Manchester experiments do replicate. Anim Behav 35:691–704
Baker RR (1987b) Integrated use of moon and magnetic compass by the heart-and-dart moth, *Agrotis exclamationis*. Anim Behav 35:94–101
Baker RR (1989a) Navigation and magnetoreception by horses and other non-human land mammals. In: Orientation and navigation – birds, humans and other animals. Proc Int Conf Royal Inst Navig, Cardiff, Paper 12
Baker RR (1989b) Human navigation and magnetoreception. Manchester University Press, Manchester
Baker RR, Mather JG (1982) Magnetic compass sense in the large yellow underwing moth, *Noctua pronuba* L. Anim Behav 30:543–548
Baker RR, Mather JG, Kennaugh JH (1983) Magnetic bones in human sinuses. Nature 301:78–80
Balda RP, Wiltschko W (1991) Caching and recovery in scrub jays: transfer of sun-compass direction from shaded to sunny areas. Condor 93:1020–1023
Baldaccini NE, Benvenuti S, Fiaschi V, Ioalè P, Papi F (1979) Initial orientation of homing pigeons: effects of transport in an artificially altered magnetic field. Monit Zool Ital (NS) 13:196
Bamberger S, Valet G, Storch F, Ruhenstroth-Bauer G (1978) Electromagnetically induced fluid streaming as a possible mechanism of the biomagnetic orientation of organisms. Z Naturforsch 33c:159–160
Bams RA (1976) Survival and propensity for homing as affected by presence or absence of locally adapted paternal genes in two transplanted populations of pink salmon (*Oncorhynchus gorbuscha*). J Fish Res Board Can 33:2716–2725
Banerjee SK, Moskowitz BM (1985) Ferrimagnetic properties of magnetite. In: Kirschvink JL, Jones DS, MacFadden BJ (eds) Magnetite biomineralization and magnetoreception in organisms. Plenum Press, New York, pp 17–41
Barnothy MF (1969) Biological effects of magnetic fields, vol II. Plenum Press, New York
Batschelet E (1981) Circular statistics in biology. Academic Press, London
Bauer GB, Fuller M, Perry A, Dunn JR, Zoeger J (1985) Magnetoreception and biomineralization of magnetite in cetaceans. In: Kirschvink JL, Jones DS, MacFadden BJ (eds) Magnetite biomineralization and magnetoreception in organisms. Plenum Press, New York, pp 489–507
Beason RC (1986) Magnetic orientation and magnetically sensitive material in migratory birds. In: Maret G, Boccara N, Kiepenheuer J (eds) Biophysical effects of steady magnetic fields. Springer, Berlin Heidelberg New York, pp 167–172
Beason RC (1987) Interaction of visual and non-visual cues during migratory orientation by the bobolink (*Dolichonyx oryzivorus*). J Ornithol 128:317–324
Beason RC (1989a) Use of an inclination compass during migratory orientation by the bobolink (*Dolichonyx oryzivorus*). Ethology 81:291–299
Beason RC (1989b) Magnetic sensitvity and orientation in the bobolink. In: Orientation and navigation – birds, humans and other animals. Proc Int Conf Royal Inst Navig, Cardiff, Paper 7

Beason RC (1992) You can get there from here: responses to simulated magnetic equator crossing by the bobolink (*Dolichonyx oryzivorus*). Ethology 91:75–80
Beason RC (1994) Potential neural mechanisms of avian magnetic perception. Abstr XXI Int Congr Ornithol, J Ornithol 135:412
Beason RC, Brennan WJ (1986) Natural and induced magnetization in the bobolink, *Dolichonyx oryzivorus* (Aves: Icteridae). J Exp Biol 125:49–56
Beason RC, Nichols JE (1984) Magnetic orientation and magnetically sensitive material in a transequatorial migratory bird. Nature 309:151–153
Beason RC, Semm P (1987) Magnetic responses of the trigeminal nerve system of the bobolink (*Dolichonyx oryzivorus*). Neurosci Lett 80:229–234
Beason RC, Semm P (1991) Two different magnetic systems in avian orientation. In: Bell BD, Cossee RO, Flux JEC, Heather BD, Hitchmough RA, Robertson CJR, Williams MJ (eds) Acta XX Congr Int Ornithol. New Zealand Ornithol Congr Trust Board, Wellington, pp 1813–1819
Beason RC, Semm P (1994) Detection of and receptors for magnetic fields in birds. In: Carpenter DC, Ayrapetyan S (eds) Biological effects of electric and magnetic fields, vol I: Sources and mechanisms. Academic Press, New York, pp 241–260
Beason RC, Semm P (in prep.) Does the avian ophthalmic nerve carry magnetic navigational information?
Beason RC, Harper J, McNulty S, Dussourd N, Freas J (1994) Magnetic effects on homing bank swallows (*Riparia riparia*). J Ornithol 135 (Special issue):88
Beason RC, Dussourd N, Deutschlander M (1995) Behavioral evidence for the use of magnetic material in magnetoreception by a migratory bird. J Exp Biol 198:141–146
Beaugrand JP (1976) An attempt to confirm magnetic sensitivity in the pigeon, *Columba livia*. J Comp Physiol 110:343–355
Beaugrand JP (1977) Test of magnetic sensitivity in seven species of European birds using a cardiac nociceptive conditioning procedure. Behav Proc 2:113–127
Beck W, Wiltschko W (1981) Trauerschnäpper (*Ficedula hypoleuca* Pallas) orientieren sich nicht-visuell mit Hilfe des Magnetfelds. Vogelwarte 31:168–174
Beck W, Wiltschko W (1982) The magnetic field as reference system for the genetically encoded migratory direction in pied flycatchers, *Ficedula hypoleuca*. Z Tierpsychol 60:41–46
Beck W, Wiltschko W (1983) Orientation behaviour recorded in registration cages: a comparison of funnel cages and radial perch cages. Behaviour 87:145–156
Beck W, Wiltschko W (1988) Magnetic factors control the migatory direction of pied flycatchers, *Ficedula hypoleuca*. In: Ouellet H (ed) Acta XIX Congr Int Ornithol. University of Ottawa Press, Ottawa, pp 1955–1962
Becker G (1963) Magnetfeld-Orientierung von Dipteren. Naturwissenschaften 50:664
Becker G (1964) Reaktion von Insekten auf Magnetfelder, elektrische Felder und atmospherics. Z Angew Entomol 54:75–88
Becker G (1965) Zur Magnetfeld-Orientierung von Dipteren. Z Vergl Physiol 51:135–150
Becker G (1974) Einfluß des Magnetfelds auf das Richtungsverhalten von Goldfischen. Naturwissenschaften 61:220–221
Becker G (1975) Einfluß von magnetischen, elektrischen und Schwerefeldern auf den Galeriebau von Termiten. Umschau 75:183–185
Becker G (1976) Chemische und physikalische Einflüsse auf Aktivität und Richtungsverhalten beim Galeriebau von Termiten. Mitt Dtsch Entomol Ges 35:99–104
Becker G (1980) Reaktion der Holzbohrassel *Limnoria tripunctata* auf Magnetfeld-Intensitätsgradienten. Naturwissenschaften 67:42–43
Becker G, Speck U (1964) Untersuchungen über die Magnetfeldorientierung von Dipteren. Z Vergl Physiol 49:301–340
Becker M, Füller E, Wiltschko R (1991) Pigeon orientation: daily variation between morning and noon occur in some years, but not in others. Naturwissenschaften 78:426–428
Bell GB, Marino AA (1989) Exposure systems for production of uniform magnetic fields. J Bioelectricity 8:147–158
Bellrose F (1967) Radar in orientation research. In: Snow DW (ed) Proc XIV Int Ornithol Congr. Blackwell, Oxford, pp 281–309

Bellrose F, Graber RR (1963) A radar study of the flight directions of nocturnal migrants. Proc XIII Int Ornithol Congr, Ithaca, pp 362–389
Bennett MF, Huguenin J (1969) Geomagnetic effects on a circadian difference in reaction times in earthworms. Z Vergl Physiol 63:440–445
Bennett MVL (1971) Electroreception. In: Hoar WS, Randall DS (eds) Fish Physiology 5. Academic Press, New York, pp 493–574
Benvenuti S (1986) Pigeon homing: the effect of an altered magnetic field prior to release. Am Racing Pigeon News, Oct 1986:17–19
Benvenuti S, Ioalè P (1988) Initial orientation of homing pigeons: different sensitivity to altered magnetic fields in birds of different countries. Experientia 44:358–359
Benvenuti S, Baldaccini NE, Ioalè P (1982) Pigeon homing: effect of altered magnetic field during displacement on initial orientation. In: Papi F, Wallraff HG (eds) Avian navigation. Springer, Berlin Heidelberg New York, pp 140–148
Bernhard JH (1992) Non-ionizing radiation safety: radiofrequency radiation, electric and magnetic fields. Phys Med Biol 37: 807–844
Berthold P (1988) The control of migration in European warblers. In: Ouellet H (ed) Acta XIX Congr Int Ornithol. University of Ottawa Press, Ottawa, pp 215–294
Berthold P, Helbig AJ, Mohr G, Querner U (1992) Rapid microevolution of migratory behaviour in a wild bird species. Nature 360:668–669
Bingman VP (1981) Savannah sparrows have a magnetic compass. Anim Behav 29:962–963
Bingman VP (1983a) Magnetic field orientation of migratory savannah sparrows with different first summer experience. Behaviour 87:43–53
Bingman VP (1983b) Importance of earth's magnetism for the sunset orientation of migratory naive savannah Sparrows. Monit Zool Ital (NS) 17:395–400
Bingman VP (1987) Earth's magnetism and the nocturnal orientation of migratory European robins. Auk 104:523–525
Bingman VP, Wiltschko W (1988) Orientation of dunnocks (*Prunella modularis*) at sunset. Ethology 77:1–9
Bingman VP, Beck W, Wiltschko W (1985) Ontogeny of migratory orientation: a look at the pied flycatcher, *Ficedula hypoleuca*. In: Rankin MA (ed) Migration: mechanisms and adaptive significance. Contrib Mar Sci (Port Aransas) 27:544–552
Bittl R, Schulten K (1986) Study of polymer dynamics by magnetic field-dependent biradical reactions. In: Maret G, Kiepenheuer J, Boccara N (eds) Biophysical effects of steady magnetic fields. Springer, Berlin Heidelberg New York, pp 90–98
Blackbourn DJ (1987) Sea surface temperature and pre-season prediction of return timing in Fraser River sockeye salmon (*Oncorhynchus nerka*). Can Spec Publ Fish Aquat Sci 96:296–306
Blakemore RP (1975) Magnetotactic bacteria. Science 19:377–379
Blakemore RP, Frankel RB (1981) Magnetic navigation in bacteria. Sci Am 245:58–65
Bliss VL, Heppner FH (1976) Circadian activity rhythm influenced by near zero magnetic field. Nature 261:411–412
Bochenski Z, Dylewska M, Gieszczykiewicz J (1960) Homing experiments on birds. Part XI: Experiments with swallows *Hirundo rustica* L. concerning the influence of earth magnetism and partial eclipse of the sun on their orientation. Zesz Nauk Uniw Jagiellonk 33:125–130 (in Polish)
Boldt A, Bruderer B (1994) Anfangsorientierung von Brieftauben im Einflußbereich eines Kurzwellensenders. Ornithol Beob 91:111–123
Bookman MA (1977) Sensitivity of the homing pigeon to an earth-strength magnetic field. Nature 267:340–342
Bookman MA (1978) Sensitivity of the homing pigeon to an earth-strength magnetic field. In: Schmidt-Koenig K, Keeton WT (eds) Animal migration, navigation, and homing. Springer, Berlin Heidelberg New York, pp 127–134
Bovet J (1992) Combining V-test probabilities in orientation studies: a word of caution. Anim Behav 44:777–779
Bovet J, Dolivo M, George C, Gogniat A (1988) Homing behavior of wood mice (*Apodemus*) in a geomagnetic anomaly. Z Säugetierkd 53:333–340

Brabyn MW, McLean IG (1992) Oceanography and coastal topography of herd-stranding sites for whales in New Zealand. J Mammal 73:469–476
Brannon EL (1967) Genetic control of migrating behavior of newtly emerged sockeye salmon fry. Int Pacific Salmon Fish Comm Prog Rep 16
Brannon EL (1972) Mechanisms controlling migration of sockeye salmon fry. Int Pacific Salmon Fish Comm Bull 21, New Westminster, Canada
Brannon EL (1982) Orientation mechanisms of homing salmonids. In: Brannon EL, Salo EO (eds) Proc Salmon and Trout Migratory Behavior Symp, University of Washington, Seattle, pp 219–227
Brannon EL (1984) Influence of stock origin on salmon migratory behavior. In: McCleave JD, Arnold GP, Dodson JJ, Neill WH (eds) Mechanisms of migration in fishes. Plenum Press, New York, pp 103–111
Brannon EL, Quinn TP, Succetti GL, Ross BD (1981) Compass orientation of sockeye salmon fry from a complex river system. Can J Zool 59:1548–1553
Branover GG, Vasil'yev AS, Gleizer SI, Tsinober AB (1971) A study of the behavior of the eel in natural and artificial magnetic fields and an analysis of its reception mechanism. J Ichthyol 11:608–614
Breiner S (1973) Application manual for portable magnetometers. Geometric, Sunnyvale, CA
Britto LRG, Natal CL, Marcondes AM (1981) The accessory optic system in pigeons: receptive field properties of identified neurons. Brain Res 206:149–154
Brown AC, McLachlan A (1990) Ecology of sandy shores. Elsevier, Amsterdam
Brown AI, Lednor T, Bernstein N (1986) The „K" factor: a story that continues. Am Racing Pigeon News 102:26–31
Brown FA (1962) Responses of the planarian, *Dugesia*, and the protozoan, *Paramecium*, to very weak horizontal magnetic fields. Biol Bull 123:264–281
Brown FA (1965) A unified theory for biological rhythms. In: Aschoff J (ed) Circadian clocks. North-Holland, Amsterdam, pp 231–261
Brown FA (1966) Effects and after-effects on planarians of reversals of the horizontal magnetic vector. Nature 209:533–535
Brown FA (1971) Unconventional theories of orientation: panel discussion. Ann NY Acad Sci 188:331–358
Brown FA, Park YH (1965) Phase-shifting a lunar rhythm in planarians by altering the horizontal magnetic vector. Biol Bull 129:79–86
Brown FA, Scow KM (1978) Magnetic induction of a circadian cycle in hamsters. J Interdiscip Cycle Res 9:137–145
Brown FA, Bennett MF, Brett WJ (1959) Effects of imposed magnetic fields in modifying snail orientation. Biol Bull 117:406
Brown FA, Bennett MF, Webb HM (1960a) A magnetic compass response of an organism. Biol Bull 119:65–74
Brown FA, Webb HM, Webb WJ (1960b) Magnetic response of an organism and its lunar relationship. Biol Bull 118:382–392
Brown FA, Webb HM, Barnwell FH (1964) A compass directional phenomen in mud-snails and its relation to magnetism. Biol Bull 127:206–220
Brown HR, Ilyinsky OB (1978) The ampullae of Lorenzini in the magnetic field. J Comp Physiol 126:333–341
Brown HR, Andrianov GN, Ilyinsky OB (1974) Magnetic field perception by electroreceptors in black sea skates. Nature 249:178–179
Brown HR, Ilyinsky OB, Muravejko VM, Corshkov ES, Fonarev GA (1979) Evidence that geomagnetic variations can be detected by Lorenzinian ampullae. Nature 277:648–649
Bruderer B, Boldt A (1994) Homing pigeons under radio influence. Naturwissenschaften 81:316–317
Bruderer B, Jenni L (1988) Strategies of bird migration in the area of the Alps. In: Ouellet H (ed) Acta XIX Congr Int Ornithol. University of Ottawa Press, Ottawa, pp 2150–2161
Buchler ER, Wasilewski PJ (1985) Magnetic remanence in bats. In: Kirschvink JL, Jones DS, MacFadden BJ (eds) Magnetite biomineralization and magnetoreception in organisms. Plenum Press, New York, pp 483–487

Buck T (1988) Untersuchungen zur Biologie der Erdkröte *Bufo bufo* L. PhD Thesis, Universität Hamburg, Hamburg
Bullock TH, Heiligenberg W (eds) (1986) Electroreception. John Wiley, New York
Bullock TH, Bodznick DA, Northcutt RG (1983) The phylogenetic distribution of electroreception: Evidence for convergent evolution of a primitive vertebrate sense modality. Brain Res Rev 6:25–46
Burda H, Marhold S, Westenberger T, Wiltschko R, Wiltschko W (1990) Evidence for magnetic compass orientation in the subterranean rodent *Cryptomys hottentotus* (Bathyergidae). Experientia 46:528–530
Burda H, Marhold S, Wiltschko W, Beiles A, Nevo E, Simon S (1991) Spontane magnetische Richtungspräferenz bei Blindmäusen, *Spalax ehrenbergi* superspecies (Rodentia, Mammalia). Verh Dtsch Zool Ges 84:339
Buskirk RE, O'Brien WP Jr (1985) Magnetic remanence and response to magnetic fields in crustacea. In: Kirschvink JL, Jones DS, MacFadden BJ (eds) Magnetite biomineralization and magnetoreception in organisms. Plenum Press, New York, pp 365–383
Busnel RG, Giban J, Gramet P, Pasquinelly F (1956) Absence d'action des ondes du radar sur la direction de vol de certains oiseaux. C R Soc Biol 150:18–20
Carey FG, Scharold JV (1990) Movements of blue sharks (*Prionace glauca*) in depth and course. Mar Biol 106:329–342
Carman GJ, Walker MM, Lee AK (1987) Attempts to demonstrate magnetic discrimination by homing pigeons in flight. Anim Learn Behav 15:124–129
Carpenter DC, Ayrapetyan S (eds) (1994) Biological effects of electric and magnetic fields. Vol 1: Sources and mechanisms. Vol 2: Beneficial and harmful effects. Academic Press, New York
Carr A (1986) Rips, FADs and little loggerheads. BioScience 36:92–100
Carr A, Switzer WP, Hollander WF (1982) Evidence for interference with navigation of homing pigeons by a magnetic storm. Iowa State J Res 56:327–340
Casamajor J (1926) Le mystérieux „sens de l'espace", chez le pigeon voyageur. La Nature (Paris) 54:366–367
Casamajor J (1927) Le mystérieux „sens de l'espace". Rev Sci 65:554–565
Chapman S, Bartels J (1940) Geomagnetism. Clarendon, London
Chernosky EJ, Fougere PF, Hutchison RO (1966) The geomagnetic field. In: Valley SL (ed) Handbook of geophysics and space environments. McGraw-Hill, New York, pp 11–61
Chew GL, Brown GE (1989) Orientation of rainbow trout (*Salmo gairdneri*) in normal and null magnetic fields. Can J Zool 67:641–643
Collett TS, Baron J (1994) Biological compasses and the coordinate frame of landmark memories in honeybees. Nature 368:137–140
Cope FW (1971) Evidence from activation energies for superconductive tunneling in biological systems at physiological temperatures. Physiol Chem Phys 3:403–410
Cope FW (1973) Biological sensitivity to weak magnetic fields due to biological superconductive Josephson Junctions? Physiol Chem Phys 5:173–176
Cosens D, Toussant N (1985) An experimental study of the foraging strategy of the wood ant *Formica aquilonia*. Anim Behav 33:541–552
Couvillon PA, Asam AM, Bitterman ME (1992) Further efforts at training pigeons to discriminate changes in the geomagnetic field. J Exp Biol 173:295–299
Deelder CL, Tesch FW (1970) Heimfindevermögen von Aalen (*Anguilla anguilla*), die über große Entfernungen verpflanzt worden waren. Mar Biol 6:81–92
DeJong D (1982) Orientation of comb building by honeybees. J Comp Physiol 147:495–501
Delius JD, Emmerton J (1978a) Sensory mechanisms related to homing in pigeons. In: Schmidt-Koenig K, Keeton WT (eds) Animal migration, navigation, and homing. Springer, Berlin Heidelberg New York, pp 35–41
Delius JD, Emmerton J (1978b) Stimulus-dependent asymmetry in classical and instrumental discrimination learning by pigeons. Psychol Rec 28:425–434
Demaine C, Semm P (1985) The avian pineal gland as an independent magnetic sensor. Neurosci Lett 62:119–122
Deoras PJ (1962) Some observations on the termites of Bombay. In: Termites in the humid tropics. Proc New Delhi Symp 1960, UNESCO, Paris, pp 101–103

Dodson JJ (1988) The nature and role of learning in the orientation and migratory behavior of fishes. Environ Biol Fishes 23:161–182
Dornfeldt K (1977) Statistische Auswertungen der Wettfluggeschwindigkeiten von Preistauben. Brieftaube 1977:904–906
Dornfeldt K (1991) Pigeon homing in relation to geomagnetic, gravitational, topographical, and meteorological conditions. Behav Ecol Sociobiol 28:107–123
Drost R (1949) Zugvögel perzipieren Ultrakurzwellen. Vogelwarte 15:57–59
Drury WH Jr, Nisbet JCT (1964) Radar studies of orientation of songbird migrants in southeastern New England. Bird-Banding 35:69–119
Duelli P, Duelli-Klein R (1978) Die magnetische Nestausrichtung der australischen Kompaßtermite *Amitermes meridionalis*. Mitt Schweiz Entomol Ges 51:337–342
Dyer FC, Gould JL (1981) Honeybee orientation: a back-up system for cloudy days. Science 214:1041–1042
Eastwood E, Rider GC (1964) The influence of radio waves upon birds. Br Birds 57:445–458
Edmonds DT (1992) A magnetite null detector as the migrating bird's compass. Proc Royal Soc Lond B 249:27–31
Edmonds DT (1993) A model avian compass based upon a magnetite null detector. In: Orientation and navigation - birds, humans and other animals. Proc Int Conf Royal Inst Navig, Oxford, Paper 11
Edwards HH, Schnell GD, DuBois RL, Hutchison VH (1992) Natural and induced remanent magnetism in birds. Auk 109:43–56
Eloff G (1951) Orientation in the mole-rat *Cryptomys*. Br J Psychol 428:134–145
Emlen ST (1967a) Migratory orientation in the indigo bunting, *Passerina cyanea*. Part I: Evidence for use of celestial cues. Auk 84:309–342
Emlen ST (1967b) Migratory orientation in the indigo bunting, *Passerina cyanea*. Part II: Mechanisms of celestial orientation. Auk 84:463–489
Emlen ST (1970a) Celestial rotation: its importance in the development of migratory orientation. Science 170:1198–1201
Emlen ST (1970b) The influence of magnetic information on the orientation of the indigo bunting, *Passerina cyanea*. Anim Behav 18:215–224
Emlen ST (1975) Migration: orientation and navigation. In: Farner DS, King JR (eds) Avian biology 5. Academic Press, New York, pp 129–219
Emlen ST, Emlen JT (1966) A technique for recording migratory orientation of captive birds. Auk 84:361–367
Emlen ST, Wiltschko W, Demong NJ, Wiltschko R, Bergman S (1976) Magnetic direction finding: evidence for its use in migratory indigo buntings. Science 193:505–508
Enright JT (1972) When the beachhopper looks at the moon: the moon compass hypothesis. In: Galler SR, Schmidt-Koenig K, Jacobs GJ, Belleville RE (eds) Animal orientation and navigation. NASA SP-262, US Government Printing Office, Washington, DC, pp 523–555
Ercolini A, Scapini F (1972) On the non-visual orientation of littoral amphipods. Monit Zool Ital (NS) 6:75–84
Esch H (1961) Über die Schallerzeugung beim Werbetanz der Honigbiene. Z Vergl Physiol 45:1–11
Etienne AS, Maurer R, Portenier V, Saucy F, Teroni E (1986) Short-distance homing of the golden hamster under conditions of darkness and light. In: Beugnon G (ed) Orientation in space. Privat, Toulouse, pp 33–44
Evans WE (1974) Radiotelemetric studies of two species of small odontocete cetaceans. In: Schevill E (ed) The whale problem: a status report. Harvard University Press, Cambridge, MA, pp 385–394
Ferguson DE (1967) Sun-compass orientation in anurans. In: Storm RM (ed) Animal orientation and navigation. Oregon State University Press, Corvallis, pp 21–32
Fiore L, Greppetti L, Mela P (1984a) Short term effects of horizontal geomagnetic field deflections on the oriented behavior of caged robins, *Erithacus rubecula* L. Monit Zool Ital (NS) 18:123–131
Fiore L, Greppetti L, Mela P (1984b) Effects of intensity changes in the horizontal geomagnetic field component on the oriented behaviour of caged robins *Erithacus rubecula* L. Monit Zool Ital (NS) 18:337–345

Frankel RB (1986) Magnetite and magnetotaxis in bacteria and algae. In: Maret G, Kiepenheuer J, Boccara N (eds) Biophysical effects of steady magnetic fields. Springer, Berlin Heidelberg New York, pp 173–179

Frankel RB, Blakemore RP, Wolfe RS (1979) Magnetite in freshwater magnetotactic bacteria. Science 203:1355–1356

Frankel RB, Blakemore RP, Torres de Araujo FF, Esquivel DMS, Danon J (1981) Magnetotactic bacteria at the geomagnetic equator. Science 212:1269–1270

Frankel RB, Papaefthymiou GC, Blakemore RP (1985) Mössbauer spectroscopy of iron biomineralization products in magnetotactic bacteria. In: Kirschvink JL, Jones DS, MacFadden BJ (eds) Magnetite biomineralization and magnetoreception in organisms. Plenum Press, New York, pp 269–287

Frei U (1982) Homing pigeons' behaviour in the irregular magnetic field of Western Switzerland. In: Papi F, Wallraff HG (eds) Avian navigation. Springer, Berlin Heidelberg New York, pp 129–139

Frei U, Wagner G (1976) Die Anfangsorientierung von Brieftauben im erdmagnetisch gestörten Gebiet des Mont Jorat. Rev Suisse Zool 83:891–897

Fromme HG (1961) Untersuchungen über das Orientierungsvermögen nächtlich ziehender Kleinvögel (*Erithacus rubecula, Sylvia communis*). Z Tierpsychol 18:205–220

Füller E (1986) Die individuelle und tageszeitliche Variation im Orientierungsverhalten von Brieftauben (*Columba livia*). PhD Thesis, Universität Frankfurt, Frankfurt a.M.

Fuller M, Goree WS, Goodman WL (1985) An introduction to the use of SQUID magnetometers in biomagnetism. In: Kirschvink JL, Jones DS, MacFadden BJ (eds) Magnetite biomineralization and magnetoreception in organisms. Plenum Press, New York, pp 103–154

Gaibar-Puertas C (1953) Variacion secular del campo geomagnetico. Observ del Ebro 11, Tortosa, Spain

Gleizer SI, Khodorkovsky VA (1971) Experimental determination of geomagnetic reception in the European eel. Dokl Akad Nauk SSSR Ser Biol 201:964–967 (in Russian)

Gordon DA (1948) Sensitivity of homing pigeons to the magnetic field of the earth. Science 108:710–711

Gould JL (1980) The case for magnetic sensitivity in birds and bees (such as it is). Am Sci 68:256–267

Gould, JL (1982) The map sense of pigeons. Nature 296:205–211

Gould JL (1984) Magnetic field sensitivity in animals. Annu Rev Physiol 46:585–598

Gould JL (1985) Are animal maps magnetic? In: Kirschvink JL, Jones DS, MacFadden BJ (eds) Magnetite biomineralization and magnetoreception in organisms. Plenum Press, New York, pp 257–268

Gould JL, Kirschvink JL, Deffeyes KS (1978) Bees have magnetic remanence. Science 201:1026–1028

Gould JL, Kirschvink JL, Deffeyes KS, Brines ML (1980) Orientation of demagnetized bees. J Exp Biol 86:1–8

Graue LC (1965) Initial orientation in pigeon homing related to magnetic contours. Am Zool 5:704

Griffin DR (1940) Homing experiments with Leach's petrels. Auk 57:61–74

Griffin DR (1952) Bird navigation. Biol Rev Camb Philos Soc 27:359–400

Griffin DR (1955) Bird navigation. In: Wolfson A (ed) Recent studies in avian biology. University of Illinois Press, Urbana, pp 154–197

Griffin DR (1973) Oriented bird migration in or between opaque cloud layers. Proc Am Philos Soc 117:117–141

Griffin DR (1982) Commentary: ecology of migration: is magnetic orientation a reality? Q Rev Biol 57:293–295

Grigg GC (1973) Some consequences of the shape and orientation of magnetic termite mounds. Austr J Zool 21:231–237

Grigg GC, Underwood AJ (1977) An analysis of the orientation of magnetic termite mounds. Austr J Zool 25:87–94

Grigg GC, Jacklyn P, Taplin L (1988) The effect of buried magnets on colonies of *Amitermes* spp. building magnetic mounds in Northern Australia. Physiol Entomol 13:285–289

Groot C (1965) On the orientation of young sockeye salmon (*Oncorhynchus nerka*) during their seaward migration out of lakes. Behaviour Suppl 14
Groot C (1982) Modification on a theme – a perspective on migratory behavior of Pacific salmon. In: Brannon EL, Salo EO (eds) Proc Salmon Trout Migratory Behav Symp. University of Washington, Seattle, pp 1–21
Groot C, Quinn TP, Hara TJ (1986) Responses of migrating adult sockeye salmon (*Oncorhynchus nerka*) to population-specific odours. Can J Zool 64:926–932
Gunter RC Jr, Bamberger S, Valet G, Crossin M, Ruhenstroth-Bauer G (1978) The trajectories of particles suspended in electrolytes and the influence of crossed electric and magnetic fields. Biophys Struct Mech 4:87–95
Gwinner E (1974) Endogenous temporal control of migratory restlessness in warblers. Naturwissenschaften 61:405
Gwinner E, Wiltschko W (1978) Endogenously controlled changes in the migratory direction of the garden warbler, *Sylvia borin*. J Comp Physiol 125:267–273
Gwinner E, Wiltschko W (1980) Circannual changes in the migratory orientation of the garden warbler, *Sylvia borin*. Behav Ecol Sociobiol 7:73–78
Hansen LP, Doving KB, Jonsson S (1987) Migration of farmed adult Atlantic salmon with and without olfactory sense, released on the Norwegian coast. J Fish Biol 30:713–721
Hansen LP, Jonsson N, Jonsson B (1993) Oceanic migration in homing Atlantic salmon. Anim Behav 45:927–941
Hanson M, Westerberg H (1987) Occurrence of magnetic material in teleosts. Comp Biochem Physiol 86A:169–172
Hanson M, Karlsson L, Westerberg H (1984a) Magnetic material in European eel (*Anguilla anguilla* L.). Comp Biochem Physiol 77A:221–224
Hanson M, Wirmark G, Öblad M, Strid L (1984b) Iron-rich particles in European eel (*Anguilla anguilla* L.). Comp Biochem Physiol 79A:311–316
Harden-Jones FR (1968) Fish migration. Arnold, London
Hartwick RF (1976) Beach orientation in talitrid amphipods: capacities and strategies. Behav Ecol Sociobiol 1:447–458
Hasler AD, Scholz AT (1983) Olfactory imprinting and homing in salmon. Springer, Berlin Heidelberg New York
Hasler AD, Wisby WJ (1951) Discrimination of stream odors by fishes and its relation to parent stream behavior. Am Nat 85:223–238
Helbig AJ (1990) Are orientation mechanisms among migratory birds species-specific? Trends Ecol Evol 5:365–367
Helbig AJ (1991a) Inheritance of migratory direction in a bird species: a cross breeding experiment with SE- and SW-migrating blackcaps (*Sylvia atricapilla*). Behav Ecol Sociobiol 28:9–12
Helbig AJ (1991b) Dusk orientation of migratory European robins, *Erithacus rubecula*: the role of sun-related directional information. Anim Behav 41:313–322
Helbig AJ (1992) Ontogenetic stability of inherited migratory directions in a nocturnal bird migrant: comparison between the first and second year of life. Ethol Ecol Evol 4:375–388
Helbig AJ, Wiltschko W (1989) The skylight polarization patterns at dusk affect the orientation behavior of blackcaps, *Sylvia atricapilla*. Naturwissenschaften 76:227–229
Helbig AJ, Orth G, Laske V, Wiltschko W (1987) Migratory orientation and activity of the meadow pipit (*Anthus pratensis*): a comparative observational and experimental field study. Behaviour 103:276–293
Helbig AJ, Berthold P, Wiltschko W (1989) Migratory orientation of blackcaps (*Sylvia atricapilla*): population-specific shifts of direction during autumn. Ethology 82:307–315
Hennig R (1931) Die Frühkenntnis der magnetischen Nordweisung. Beitr Gesch Technik Ind 21:25–42
Hepworth D, Pickard RS, Overshott KJ (1980) Effects of the periodically intermittent application of a constant magnetic field on the mobility in darkness of worker honeybees. J Apic Res 19:179–186
Herrnkind WF (1983) Movement patterns and orientation. Biol Crustacea 7:41–105

Herrnkind WF, Van Der Walker J, Barr L (1975) Population dynamics, ecology and behavior of the spiny lobster, *Panulirus argus*, of St. John, US Virgin Island: habitation and pattern of movements. Bull Nat Hist Mus LA County 20:31–45

Heusser H (1960) Über die Beziehung der Erdkröte (*Bufo bufo* L) zu ihrem Laichplatz. Behaviour 16:93–109

Hiramatsu K, Ishida Y (1989) Random movements and orientation in pink salmon (*Oncorhynchus gorbuscha*) migration. Can J Fish Aquat Sci 46:1062–1066

Holtkamp-Rötzler E (1990) Neurobiologische Untersuchungen am Nervus ophthalmicus des Zebrafinken (*Taeniopygia guttata*) unter besonderer Berücksichtigung seiner Magnetfeldempfindlichkeit. Diplomarbeit, Universität Frankfurt, Frankfurt a.M.

Holtkamp-Rötzler E, Wiltschko W (1993) Orientation of impulse-magnetized homing pigeons (*Columba livia*). Animal and Cell Abstracts A 1.2, Soc Exp Biol, Canterbury meeting, p 1

Hong FT (1977) Photoelectric and magneto-orientation effects in pigmented biological membranes. J Colloid Interface Sci 58:471–497

Hornung U (1993) Perception of earth strength magnetic fields in pigeons (*Columba livia*). In: Orientation and navigation – birds, humans and other animals. Proc Int Conf Royal Inst Navig, Oxford, Paper 13

Hornung U (1994) Elektrocardiographisch meßbare Reaktionen auf Magnetfeldreize bei Tauben (*Columba livia*). Verh Dtsch Zool Ges 87:246

Hsu CY, Li CW (1993) The ultrastructure and formation of iron granules in the honeybee (*Apis mellifera*). J Exp Biol 180:1–13

Hsu CY, Li CW (1994) Magnetoreception in honeybees (*Apis mellifera*). Science 265:95–97

Ifantidis MD (1978) Wabenorientierung im Nest der Honigbiene (*Apis mellifica* L). Apidologie 9:57–73

Ioalè P (1984) Magnets and pigeon orientation. Monit Zool Ital (NS) 18:347–358

Ioalè P, Guidarini D (1985) Methods for producing disturbances in pigeon homing behavior by oscillating magnetic fields. J Exp Biol 116:109–120

Ioalè P, Papi F (1989) Olfactory bulb size, odor discrimination and magnetic insensitivity in hummingbirds. Physiol Behav 45:995–999

Ioalè P, Teyssèdre A (1989) Pigeon homing: effects of magnetic disturbances before release on initial orientation. Ethol Ecol Evol 1:65–80

Jacklyn PM (1990) Orientation in meridional mounds of *Amitermes meridionalis* (Froggatt) and *Amitermes laurensis* (Mjöberg). PhD Thesis, University of Sydney, Sydney

Jacklyn PM (1991) Evidence for adaptive variation in the orientation of *Amitermes* (Isoptera: Termitinae) mounds from northern Australia. Aust J Zool 39:569–577

Jacklyn PM (1992a) Solar engineering for the blind: „magnetic“ termite mounds of the Top End. North Territory Naturalist 13:9–15

Jacklyn PM (1992b) „Magnetic“ termite mounds' surfaces are oriented to suit wind and shade conditions. Oecologia 91:385–395

Jamon M (1990) A reassessment of the random hypothesis in the ocean migration of Pacific salmon. J Theor Biol 143:197–213

Jander R (1975) Ecological aspects of spatial orientation. Ann Rev Ecol Syst 6:171–188

Johnston JM (1982) Life histories of anadromous cutthroat with emphasis on migratory behavior. In: Brannon EL, Salo EO (eds) Proc Salmon Trout Migratory Behav Symp. University of Washington, Seattle, pp 123–127

Jones DS, MacFadden BJ (1982) Induced magnetization in the monarch butterfly, *Danaus plexippus* (Insecta, Lepidoptera). J Exp Biol 96:1–9

Joslin JK (1977) Rodent long distance orientation („homing“). Adv Ecol Res 10:63–89

Jungerman RL, Rosenblum B (1980) Magnetic induction for the sensoring of magnetic fields by animals – an analysis. J Theor Biol 87:25–32

Kalmijn AJ (1974) The detection of electric fields from inanimate and animate sources other than electric organs. In: Fessard A (ed) Handbook of sensory physiology III, 3: Electroreceptors and other specialized receptors in lower vertebrates. Springer, Berlin Heidelberg New York, pp 147–200

Kalmijn AJ (1977) The electric and magnetic sense of sharks, skates, and rays. Oceanus 20:45–52

Kalmijn AJ (1978a) Experimental evidence of geomagnetic orientation in elasmobranch fishes. In: Schmidt-Koenig K, Keeton WT (eds) Animal migration, navigation, and homing. Springer, Berlin Heidelberg New York, pp 347–353
Kalmijn AJ (1978b) Electric and magnetic sensory world of sharks, skates, and rays. In: Hodgson FS, Mathewson RF (eds) Sensory biology of sharks, skates and rays. Office Naval Res, Arlington, VA, pp 507–528
Kalmijn AJ (1982) Electric and magnetic field detection in elasmobranch fishes. Science 218:916–918
Kalmijn AJ (1984) Theory of electromagnetic orientation: a further analysis. In: Bolis L, Keynes RC, Maddrell SH (eds) Comparative physiology of sensory systems. Cambridge University Press, London, pp 525–560
Kalmijn AJ, Blakemore RP (1978) The magnetic behavior of mud bacteria. In: Schmidt-Koenig K, Keeton WT (eds) Animal migration, navigation, and homing. Springer, Berlin Heidelberg New York, pp 354–355
Karlsson L (1985) Behavioural responses of European silver eels (*Anguilla anguilla*) to the geomagnetic field. Helgol Meeresunters 39:71–81
Karpenko AA (1974) Some features of behavior of planarian *Dugesia tigrina* in geomagnetic and artificial magnetic fields. Vestnik Leningrad Univ 15, Ser Biol 3:5–11 (in Russian)
Katz YB (1978) Direction training of black-headed gulls (*Larus ridibundus* L.). In: Mihelsons H, Blüm P, Baumanis J (eds) Orientazija Ptiz [Orientation of Birds]. Zinatne, Riga, pp 25–30 (in Russian)
Katz YB (1985) Orientation behavior of the European robin (*Erithacus rubecula*). Anim Behav 33:825–828
Katz YB, Liepa V, Viksne J (1988) Orientation research in the Latvian SSR in 1982–1985. In: Ouellet H (ed) Acta XIX Congr Int Ornithol. University of Ottawa Press, Ottawa, pp 1919–1931
Keeton WT (1969) Orientation by pigeons: is the sun necessary? Science 165:922–928
Keeton WT (1971) Magnets interfere with pigeon homing. Proc Natl Acad Sci USA 68:102–106
Keeton WT (1972) Effects of magnets on pigeon homing. In: Animal Orientation and Navigation. NASA SP-262, US Government Printing Office, Washington, DC, pp 579–594
Keeton WT (1973) Release-site bias as a possible guide to the „map" component in pigeon homing. J Comp Physiol 86:1–16
Keeton WT (1974) The orientational and navigational basis of homing in birds. Adv Study Behav 5:47–132
Keeton WT (1980) Avian orientation and navigation: new developments in an old mystery. In: Nöhring R (ed) Acta XVII Congr Int Ornithol, vol I. Verlag Deutsche Ornithologen-Gesellschaft, Berlin, pp 137–158
Keeton WT, Gobert A (1970) Orientation by untrained pigeons requires the sun. Proc Natl Acad Sci USA 65:853–856
Keeton WT, Larkin TS, Windson DM (1974) Normal fluctuation in the earth's magnetic field influence pigeon orientation. J Comp Physiol 95:95–103
Kholodov YA (1966) The effect of electromagnetic and magnetic fields on the central nervous system. Nauka Press, Moscow (in Russian)
Kiepenheuer J (1978) Pigeon navigation and magnetic field. Naturwissenschaften 65:113
Kiepenheuer J (1982) The effect of magnetic anomalies on the homing behaviour of pigeons. In: Papi F, Wallraff HG (eds) Avian navigation. Springer, Berlin Heidelberg New York, pp 120–128
Kiepenheuer J (1984) The magnetic compass mechanism of birds and its possible association with the shifting course directions of migrants. Behav Ecol Sociobiol 14:81–99
Kiepenheuer J (1986) A further analysis of the orientation behavior of homing pigeons released within magnetic anomalies. In: Maret G, Boccara N, Kiepenheuer J (eds) Biophysical effects of steady magnetic fields. Springer, Berlin Heidelberg New York, pp 148–153
Kiepenheuer J, Ranvaud R, Maret G (1986) The effect of ultrahigh magnetic fields on the initial orientation of homing pigeons. In: Maret G, Boccara N, Kiepenheuer J (eds) Biophysical effects of steady magnetic fields. Springer, Berlin Heidelberg New York, pp 189–193
Kilbert K (1979) Geräuschanalyse der Tanzlaute der Honigbiene (*Apis mellifica*) in unterschiedlichen magnetischen Feldsituationen. J Comp Physiol 132:11–25

Kirschvink JL (1980) South-seeking magnetic bacteria. J Exp Biol 86:345–347
Kirschvink JL (1981a) The horizontal magnetic dance of the honeybee is compatible with a single-domain ferromagnetic magnetoreceptor. BioSystems 14:193–203
Kirschvink JL (1981b) Ferromagnetic crystals (magnetite?) in human tissue. J Exp Biol 92:333–335
Kirschvink JL (1982) Birds, bees and magnetism: a new look at the old problem of magnetoreception.Trends Neurosci 5:160–167
Kirschvink JL (1983) Biogenic ferrimagnetism: a new biomagnetism. In: Williamson SJ, Romani GL, Kaufman L, Modena J (eds) Biomagnetism. Plenum Press, New York, pp 501–531
Kirschvink JL (1989) Magnetite biomineralization and geomagnetic sensitivity in higher animals: an update and recommendations for future study. Bioelectromagnetics 10:239–259
Kirschvink JL (1990) Geomagnetic sensitivity in cetaceans: an update with live stranding records in the United States. In: Thomas JA, Kastelein R (eds) Sensory abilities of cetaceans. Plenum Press, New York, pp 639–649
Kirschvink JL, Gould JL (1981) Biogenic magnetite as a basis for magnetic field detection in animals. BioSystems 13:181–201
Kirschvink JL, Kobayashi Kirschvink A (1991) Is geomagnetic sensitivity real? Replication of the Walker-Bitterman magnetic conditioning experiment in honeybees. Am Zool 31:169–185
Kirschvink JL, Walker MM (1985) Particle-size considerations for magnetite-based magnetoreceptors. In: Kirschvink JL, Jones DS, MacFadden BJ (eds) Magnetite biomineralization and magnetoreception in organisms. Plenum Press, New York, pp 243–256
Kirschvink JL, Jones DS, MacFadden BJ (eds) (1985a) Magnetite biomineralization and magnetoreception in organisms. Plenum Press, New York
Kirschvink JL, Walker MM, Chang SB, Dizon AE, Peterson KA (1985b) Chains of single-domain magnetite particles in chinook salmon, *Oncorhynchus tschawytscha*. J Comp Physiol A 157:375–381
Kirschvink JL, Dizon AE, Westphal JA (1986) Evidence from stranding for geomagnetic sensitivity in cetaceans. J Exp Biol 120:1–24
Kirschvink JL, Kobayashi-Kirschvink A, Woodford BJ (1992) Magnetite biomineralization in the human brain. Proc Natl Acad Sci 89:7683–7687
Kisliuk M, Ishay J (1977) Influence of an additional magnetic field on hornet nest architecture. Experientia 33:885
Kitamura S, Iikura T, Ueda K (1985) Magnetic field detection in chum salmon fry – with special reference to the magnetic compass as a navigational cue in open sea migration. Bull Natl Res Inst Aquacult 8:31–42 (in Japanese)
Klimley AP (1993) Highly directional swimming by scalloped hammerhead sharks, *Sphyrna lewini*, and subsurface irradiance, temperature, bathymetry, and geomagnetic field. Mar Biol 117:1–22
Klimley AP, Nelson DR (1984) Diel movement patterns of the scalloped hammerhead shark (*Sphyrna lewini*) in relation to El Bajo Espiritu Santo: a refuging central-position social system. Behav Ecol Sociobiol 15:45–54
Klinkenberg T, Delius JD, Emmerton J (1984) Classical heart-rate conditioning and differentiation of visual CS with an appetitive UCS in pigeons. Behav Proc 9:23–30
Klinowska M (1985a) Cetacean live stranding sites relate to geomagnetic topography. Aquat Mammals 11:27–32
Klinowska M (1985b) Cetacean live stranding dates relate to geomagnetic disturbances. Aquat Mammals 11:109–119
Klinowska M (1986) The cetacean magnetic sense – evidence from strandings. In: Bryden MM, Harrison RJ (eds) Research on dolphins. Oxford University Press, Oxford, pp 401–432
Klinowska M (1987) No through road for the misguided whale. New Sci 113:46–48
Klinowska M (1988) Cetacean navigation and the geomagnetic field. J Navig 41:52–71
Klinowksa M (1989) The cetacean magnetic sense – evidence from drive fisheries. In: Orientation and navigation – birds, humans and other animals. Proc Int Conf Royal Inst Navig, Cardiff, Paper 18

Knorr OA (1954) The effect of radar on birds. Wilson Bull 66:264
Korall H (1987) Der Einfluß statisch verstärkter Magnetfelder auf den Zeitsinn der Honigbiene. Zool Jahrb Physiol 91:377–389
Korall H, Martin H (1987) Responses of bristle field sensilla in *Apis mellifica* to geomagnetic and astrophysical fields. J Comp Physiol A 161:1–22
Korall H, Leucht T, Martin H (1988) Bursts of magnetic fields induce jumps of misdirection in bees by a mechanism of magnetic resonance. J Comp Physiol A 162:279–284
Kowalski U, Wiltschko R, Füller E (1988) Normal fluctuations of the geomagnetic field may affect initial orientation in pigeons. J Comp Physiol A 163:593–600
Kramer G (1949) Über Richtungstendenzen bei der nächtlichen Zugunruhe gekäfigter Vögel. In: Mayr E, Schüz E (eds) Ornithologie als biologische Wissenschaft. Heidelberg, pp 269–283
Kramer G (1950) Weitere Analyse der Faktoren, welche die Zugaktivität des gekäfigten Vogels orientieren. Naturwissenschaften 37:377–378
Kramer G (1951) Versuche zur Wahrnehmung von Ultrakurzwellen durch Vögel. Vogelwarte 16:55–59
Kramer G (1953) Wird die Sonnenhöhe bei der Heimfindeorientierung verwendet? J Ornithol 94:201–219
Kramer G (1957) Experiments on bird orientation and their interpretation. Ibis 99:196–227
Kramer G (1961) Long distance orientation. In: Marshall AJ (ed) Biology and comparative physiology of birds. Academic Press, New York, pp 341–371
Kramer G, von Saint Paul U (1950) Stare (*Sturnus vulgaris* L) lassen sich auf Himmelsrichtungen dressieren. Naturwissenschaften 37:526
Kreithen ML, Keeton WT (1974) Attempts to condition homing pigeons to magnetic stimuli. J Comp Physiol 91:355–362
Kühn H (1919) Die Orientierung der Tiere im Raum. Fischer, Jena
Kullnick U (1991) Influence of weak non-thermic high-frequency electromagnetic fields on the membrane potential of nerve cells. Bioelectrochem Bioenerg 27:293–304
Kuterbach DA, Walcott B (1986a) Iron-containing cells in the honey-bee (*Apis mellifera*). I. Adult morphology and physiology. J Exp Biol 126:375–387
Kuterbach DA, Walcott B (1986b) Iron-containing cells in the honey-bee (*Apis mellifera*). II. Accumulation during development. J Exp Biol 126:389–401
Kuterbach DA, Walcott B, Reeder RJ, Frankel RB (1982) Iron-containing cells in the honey bee (*Apis mellifera*). Science 218:695–697
Lamotte MM (1974) The influence of magnets and habituation to magnets on inexperienced homing pigeons. J Comp Physiol 89:379–389
Larkin RP, Sutherland PJ (1977) Migrating birds respond to Project Seafarer's electromagnetic field. Science 195:777–779
Larkin TS, Keeton WT (1976) Bar magnets mask the effect of normal magnetic disturbance on pigeon orientation. J Comp Physiol 110:227–231
Larkin TS, Keeton WT (1978) An apparent lunar rhythm in the day-to-day variations in initial bearings of homing pigeons. In: Schmidt-Koenig K, Keeton WT (eds) Animal migration, navigation, and homing. Springer, Berlin Heidelberg New York, pp 92–106
Leask MJM (1977) A physico-chemical mechanism for magnetic field detection by migratory birds and homing pigeons. Nature 267:144–145
Leask MJM (1978) Primitive models of magnetoreception. In: Schmidt-Koenig K, Keeton WT (eds) Animal migration, navigation, and homing. Springer, Berlin Heidelberg New York, pp 318–322
Lednor AJ (1982) Magnetic navigation in pigeons: possibilities and problems. In: Papi F, Wallraff HG (eds) Avian navigation. Springer, Berlin Heidelberg New York, pp 109–119
Lednor AJ, Walcott C (1983) Homing pigeon navigation: the effect of inflight exposure to a varying magnetic field. Comp Biochem Physiol 76A:665–672
Lednor AJ, Walcott C (1988) Orientation of homing pigeons at magnetic anomalies: the effects of experience. Behav Ecol Sociobiol 22:3–8
Lemkau PJ (1976) An attempt to condition the green sea turtle *Chelonia mydas* to magnetic fields. Master Thesis, University of Rhode Island, Kingston

Leucht T (1984) Response to light under varying magnetic conditions in the honey bee, *Apis mellifica.* J Comp Physiol A 154:865–870
Leucht T (1989) Magnetic field effects in larvae of *Xenopus laevis.* Fortschr Zool 38:321–331
Leucht T (1990) Interactions of light and gravity reception with magnetic fields in *Xenopus laevis.* J Exp Biol 148:325–334
Leucht T, Martin H (1990) Interactions between e-vector orientation and weak, steady magnetic fields in the honeybee, *Apis mellifica.* Naturwissenschaften 77:130–133
Light P, Salmon M, Lohmann KJ (1993) Geomagnetic orientation of loggerhead sea turtles: evidence for an inclination compass. J Exp Biol 182:1–10
Lincoln JV (1967) Geomagnetic indices. In: Matsushita S, Campbell WH (eds) Physics of geomagnetic phenomena, vol 1. Academic Press, New York, pp 67–100
Lindauer M (1954) Dauertänze im Bienenstock und ihre Beziehung zur Sonnenbahn. Naturwissenschaften 41:506–507
Lindauer M (1975) Verständigung im Bienenstaat. Fischer, Stuttgart
Lindauer M (1976a) Orientierung der Tiere. Verh Dtsch Zool Ges 69:156–183
Lindauer M (1976b) Recent advances in the orientation and learning of honeybees. In: Proc XV Int Congr Entomol, Washington, DC, pp 450–460
Lindauer M (1985) The dance language of honeybees: the history of a discovery. Fortschr Zool 31:129–140
Lindauer M, Martin H (1968) Die Schwereorientierung der Bienen unter dem Einfluß des Erdmagnetfeldes. Z Vergl Physiol 60:219–243
Lindauer M, Martin H (1972) Magnetic effects on dancing bees. In: Galler SR, Schmidt-Koenig K, Jacobs GJ, Belleville RE (eds) Animal orientation and navigation. US Government Printing Office, Washington, DC, pp 559–567
Lindauer M, Martin H (1985) The biological significance of the earth's magnetic field. In: Ottoson D (ed) Progress in sensory physiology 5. Springer, Berlin Heidelberg New York, pp 119–145
Lissmann HW (1958) On the function and evolution of electric organs in fish. J Exp Biol 35:156–191
Lissmann HW, Machin KE (1963) Electric receptors in a non-electric fish (*Clarias*). Nature 199:88–89
Lohmann KJ (1984) Magnetic remanence in the Western Atlantic spiny lobster, *Panulirus argus.* J Exp Biol 113:29–41
Lohmann KJ (1985) Geomagnetic field detection by the Western Atlantic spiny lobster, *Panulirus argus.* Mar Behav Physiol 12:1–17
Lohmann KJ (1991) Magnetic orientation by hatchling loggerhead sea turtles (*Caretta caretta*). J Exp Biol 155:37–49
Lohmann KJ (1993) Magnetic compass orientation. Nature 362:703
Lohmann KJ, Lohmann CMF (1992) Orientation to oceanic waves by green turtle hatchlings. J Exp Biol 171:1–13
Lohmann KJ, Lohmann CMF (1993) A light-independent magnetic compass in the leatherback sea turtles. Biol Bull 185:149–151
Lohmann KJ, Lohmann CMF (1994a) Acquisition of magnetic directional preferences in hatchling loggerhead sea turtles. J Exp Biol 190:1–8
Lohmann KJ, Lohmann CMF (1994b) Detection of magnetic inclination angle by sea turtles: a possible mechanism for determining latitudes. J Exp Biol 194:23–32
Lohmann KJ, Willows AOD (1987) Lunar-modulated geomagnetic orientation by a marine mollusk. Science 235:331–334
Lohmann KJ, Willows AOD (1989) Magnetic field detection and its neurobiological mechanisms. In: Adelmann G (ed) Encyclopedia of neuroscience, Suppl 1. Birkhäuser, Boston, pp 94–97
Lohmann KJ, Willows AOD, Pinter RB (1991) An identifiable molluscan neuron responds to changes in earth-strength magnetic fields. J Exp Biol 161:1–24
Löhrl H (1959) Zur Frage des Zeitpunktes einer Prägung auf die Heimatregion beim Halsbandschnäpper (*Ficedula albicollis*). J Ornithol 100:132–140
Lowenstam HA (1962) Magnetite in denticle capping in recent chitons (Polyplacophora). Geol Soc Am Bull 73:435–438

Lutsyuk BÖ, Nazarchuk GK (1971) Possible orientation of birds by geomagnetic field. Vestn Zool 5:35–39 (in Russian)
MacFadden BJ, Jones DS (1985) Magnetic butterflies: a case study of the monarch (Lepidoptera, Danaidae). In: Kirschvink JL, Jones DS, MacFadden BJ (eds) Magnetite biomineralization and magnetoreception in organisms. Plenum Press, New York, pp 407–415
Madden RC, Phillips JB (1987) An attempt to demonstrate magnetic compass orientation in two species of mammals. Anim Learn Behav 15:130–134
Maier EJ (1992) Spectral sensitivities including the ultraviolet of the passeriform bird *Leiothrix lutea*. J Comp Physiol A 170:709–714
Mann S, Sparks NCH, Walker MM, Kirschvink JL (1988) Ultrastructure, morphology and organization of biogenic magnetite from sockeye salmon, *Oncorhynchus nerka*: Implications for magnetoreception. J Exp Biol 140:35–49
Mann S, Sparks NCH, Board RG (1990) Magnetotactic bacteria: microbiology, biomineralization, paleomagnetism and biotechnology. Adv Microb Physiol 31:125–181
Maret G, Boccara N, Kiepenheuer J (eds) (1986) Biophysical effects of steady magnetic fields. Springer, Berlin Heidelberg New York
Marhold S (1995) Magnetkompaßorientierung bei subterranen Graumullen, *Cryptomys* spec. (Bathyergidae, Rodentia). PhD Thesis, Universität Frankfurt, Frankfurt a.M.
Marhold S, Burda H, Wiltschko W (1991) Magnetkompaßorientierung und Richtungspräferenz bei subterranen Graumullen, *Cryptomys hottentotus* (Rodentia). Verh Dtsch Zool Ges 84:354
Martin H, Lindauer M (1977) Der Einfluß des Erdmagnetfeldes auf die Schwereorientierung der Honigbiene (*Apis mellifica*). J Comp Physiol 122:145–187
Martin H, Lindauer M, Martin U (1983) „Zeitsinn" und Aktivitätsrhythmus der Honigbiene – endogen oder exogen gesteuert? Bayrische Akademie der Wissenschaften, München
Massa B, Benvenuti S, Ioalè P, Lo Valvo M, Papi F (1991) Homing in Cory's shearwater (*Calonectris diomedea*) carrying magnets. Boll Zool 58:245–247
Mather JG (1985) Magnetoreception and the search for magnetic material in rodents. In: Kirschvink JL, Jones DS, MacFadden BJ (eds) Magnetite biomineralization and magnetoreception in organisms. Plenum Press, New York, pp 509–533
Mather JG, Baker RR (1980) A demonstration of navigation by small rodents using an orientation cage. Nature 284:259–262
Mather JG, Baker RR (1981) Magnetic sense of direction in woodmice for route-based navigation. Nature 291:152–155
Mathis A, Moore FR (1988) Geomagnetism and the homeward orientation of the box turtle, *Terrapene carolina*. Ethology 78:265–274
Matis JH, Kleerekoper H, Gensler P (1974) Non-random oscillatory changes in the orientation of the goldfish (*Carassius auratus*), in an open field. Anim Behav 22:110–117
Matsushita S (1967) Solar quiet and lunar daily variation fields. In: Matsushita S, Campbell WH (eds) Physics of the geomagnetic phenomena. Academic Press, New York, pp 301–424
Matsushita S, Maeda H (1965) On the geomagnetic solar quiet daily variation field during the IGY. J Geophys Res 70:2535–2558
Matthews GVT (1951) The experimental investigation of navigation in homing pigeons. J Exp Biol 28:508–536
Matthews GVT (1952) An investigation of homing ability in two species of gulls. Ibis 94:243–264
Matthews GVT (1953) Sun navigation in homing pigeons. J Exp Biol 30:243–267
Maurain C (1926) Les propriétés magnétiques et électriques terrestres et la faculté d'orientation du pigeon voyageur. La Nature (Paris) 54:44–45
McCleave JD, Power JH (1978) Influence of weak electric and magnetic fields on turning behavior in elvers of the American eel (*Anguilla rostrata*). Mar Biol 46:29–34
McCleave JD, Rommel SA, Cathcart CL (1971) Weak electric and magnetic fields in fish orientation. Ann NY Acad Sci 188:270–283
McIsaac HP, Kreithen ML (1987) Attempts to condition homing pigeons to magnetic cues in an outdoor flight cage. Anim Learn Behav 15:118–123

Mead RW, Woodall WL (1968) Comparison of sockeye salmon fry produced by hatcheries, artificial channels and natural areas. Int Pac Salmon Fish Comm Progr Rep 20
Merrill RT, McElhinny MW (1983) The earth's magnetic field: its history, origin, and planetary perspective. Academic Press, New York
Meyer ME, Lambe DR (1966) Sensitivity of the pigeon to changes in the magnetic field. Psychon Sci 5:349–350
Mihelsons H, Vilks I (1975) Methods of investigation of celestial and non-celestial orientation of birds. Zinatne, Riga
Miles SG (1968) Laboratory experiments on the orientation of the adult American eel, *Anguilla rostrata*. J Fish Res Board Can 25:2143–2155
Miller RJ, Brannon EL (1982) The origin and development of life history patterns in Pacific salmonids. In: Brannon EL, Salo EO (eds) Proc Salmon Trout Migratory Behav Symp. University of Washington, Seattle, pp 296–309
Misakian M, Sheppart AR, Krause D, Frazier ME, Miller DL (1993) Biological, physical, and electrical parameters for in vitro studies with ELF magnetic and electric fields – a primer. Bioelectromagnetics Suppl 2:1–73
Moore A, Freake SM, Thomas IM (1990) Magnetic particles in the lateral line of the Atlantic salmon (*Salmo salar* L). Philos Trans R Soc London B 329:11–15
Moore BR (1980) Is the homing pigeon's map geomagnetic? Nature 285:69–70
Moore BR (1988) Magnetic fields and orientation in homing pigeons: experiments of the late W.T. Keeton. Proc Natl Acad Sci USA 85:4907–4909
Moore BR, Stanhope KJ, Wilcox D (1987) Pigeons fail to detect low-frequency magnetic fields. Anim Learn Behav 15:115–117
Moore FR (1975) Influence of solar and geomagnetic stimuli on the migratory orientation of herring gull chickens. Auk 92:655–664
Moore FR (1977) Geomagnetic disturbance and the orientation of nocturnally migrating birds. Science 196:682–684
Moore FR (1982) Sunset and the orientation of a nocturnal bird migrant: a mirror experiment. Behav Ecol Sociobiol 10:153–155
Moore FR (1985) Integration of environmental stimuli in the migratory orientation of the savannah sparrow (*Passerculus sandwichensis*). Anim Behav 33:657–663
Moore FR (1987) Sunset and the orientation behaviour of migrating birds. Biol Rev 62:65–86
Moore FR (1988) Sunset and migratory orientation: observational and experimental evidence. In: Ouellet H (ed) Acta XIX Congr Int Ornithol. University of Ottawa Press, Ottawa, pp 1941–1955
Moore FR, Phillips JB (1988) Sunset, skylight polarization and the migration of yellow-rumped warblers, *Dendroica coronata*. Anim Behav 36:1770–1778
Moreau G, Pouyet JC (1968) Tentative de mise en évidence de l'influence des ondes électromagnétique sur le sens de l'orientation des pigeons voyageurs. Alauda 36:217–225
Munro UH, Wiltschko R (1993a) Clock-shift experiments with migratory yellow-faced honeyeaters, *Lichenostomus chrysops* (Meliphagidae), an Australian day-migrating bird. J Exp Biol 181:233–244
Munro UH, Wiltschko W (1993b) Magnetic compass orientation in the yellow-faced honeyeater, *Lichenostomus chrysops*, a day-migrating bird from Australia. Behav Ecol Sociobiol 32:141–145
Munro UH, Wiltschko R (1995) The role of skylight polarization in the orientation of a day-migrating bird species. J Comp Physiol A 177:357–362
Munro UH, Wiltschko W, Ford HA (1993) Changes in the migratory direction of yellow-faced honeyeaters, *Lichenostomus chrysops* (Meliphagidae) during autumn migration. Emu 93:59–62
Murphy RG (1989) The development of magnetic compass orientation in children. In: Orientation and navigation – birds, humans and other animals. Proc Int Conf Royal Inst Navig, Cardiff, Paper 26
Murray RW (1962) The response of the ampullae of Lorenzini of elasmobranchs to electrical stimulation. J Exp Biol 39:119–128
Neave F (1964) Ocean migration of Pacific salmon. J Fish Res Board Can 21:1227–1244

Nesson MH, Lowenstam HA (1985) Biomineralization processes of the radula teeth of chitons. In: Kirschvink JL, Jones DS, MacFadden BJ (eds) Magnetite biomineralization and magnetoreception in organisms. Plenum Press, New York, pp 333–363
Neumann MF (1988) Is there any influence of magnetic or astrophysical fields on the circadian rhythm of honeybees? Behav Ecol Sociobiol 23:389–393
Nordeng H (1977) A pheromon hypothesis for homeward migration in anadromous salmonids. Oikos 28:155–159
Olcese JM (1990) The neurobiology of magnetic field detection in rodents. Progr Neurobiol 35:325–330
Oldham RS (1965) Spring movements in the American toad, *Bufo americanus*. Can J Zool 44:63–100
Oldham RS (1967) Orienting mechanisms of the green frog, *Rana clamitans*. Ecology 48:477–491
O'Leary DP, Vilches-Troya J, Dunn RF, Campos-Mundoz A (1981) Magnets in the guitarfish vestibular receptors. Experientia 37:86–88
Orgel AR, Smith JC (1954) Tests of the magnetic theory of homing. Science 120:891–892
Orth G, Wiltschko W (1981) Die Orientierung von Wiesenpiepern (*Anthus pratensis* L.). Verh Dtsch Zool Ges 74:252
Ossenkopp K-P, Barbeito R (1978) Bird orientation and the geomagnetic field: a review. Neurosci Biobehav Rev 2:255–270
Ovchinnikov VV, Gleizer SI, Galaktionov GZ (1973) Features of orientation of the European eel (*Anguilla anguilla* L.) at some stages of migration. J Ichthyol 13:455–463
Palinscar EE, Dale PK (1977) Responses of the rotifer *Asplanchna brightwelli* to weak magnetic fields. Trans Missouri Acad Sci 1011: 324
Papi F (1960) Orientation by night: the moon. Cold Spring Harbour Symp Quant Biol 25:475–480
Papi F (1986) Pigeon navigation: solved problems and open questions. Monit Zool Ital (NS) 20:471–517
Papi F, Ioalè P (1986) Pigeon homing: effect of oscillating magnetic fields during flight. Atti Accad Lincei Rend Fis S 8, 80:426–434
Papi F, Ioalè P (1988) Pigeon navigation: new experiments on interaction between olfactory and magnetic cues. Comp Biochem Physiol 91A:87–89
Papi F, Ioalè P, Fiaschi V, Benvenuti S, Baldaccini NE (1978) Pigeon homing: cues detected during the outward journey influence initial orientation. In: Schmidt-Koenig K, Keeton WT (eds) Animal migration, navigation, and homing. Springer, Berlin Heidelberg New York, pp 65–77
Papi F, Meschini E, Baldaccini NE (1983) Homing behaviour of pigeons released after having been placed in an alternating magnetic field. Comp Biochem Physiol 76A:673–682
Papi F, Luschi P, Limonta P (1992) Orientation-disturbing magnetic treatment affects the pigeon opiod system. J Exp Biol 166:169–179
Pardi L, Ercolini A (1986) Zonal recovery mechanisms in talitrid crustaceans. Boll Zool 53:139–160
Pardi L, Grassi M (1955) Experimental modifications of direction-finding in *Talitrus saltator* (Montagu) and *Talorchestia deshayesei* (Aud.) (Crustacea-Amphipoda). Experientia 11:202–205
Pardi L, Scapini F (1983) Inheritance of solar direction finding in sandhoppers: mass-crossing experiments. J Comp Physiol 151:435–440
Pardi L, Scapini F (1987) Die Orientierung der Strandflohkrebse im Grenzbereich Meer/Land. Information Processing in Animals, vol 4. Fischer, Stuttgart
Pardi L, Ercolini A, Ferrara F, Scapini F (1985) Orientamento zonale solare e magnetico in Crostacei Anfipodi litarali di regioni equatoriali. Atti Accad Lincei Rend Sci Fis Mat Nat 76:312–320
Pardi L, Ugolini A, Faqi AS, Scapini F, Ercolini A (1988) Zonal recovering in equatorial sandhoppers: interaction between magnetic and solar orientation. In: Chelazzi G, Vannini M (eds) Behavioral adaption to intertidal life. Plenum Press, New York, pp 79–92
Perdeck AC (1957) Stichting Vogelstation Texel, Jaarsverlag over 1956. Limosa 30:62–75

Perdeck AC (1958) Two types of orientation in migrating *Sturnus vulgaris* and *Fringilla coelebs* as revealed by displacement experiments. Ardea 46:1–37
Perry A, Bauer GB, Dizon AE (1985) Magnetoreception and biomineralization of magnetite in amphibians and reptiles. In: Kirschvink JL, Jones DS, MacFadden BJ (eds) Magnetite biomineralization and magnetoreception in organisms. Plenum Press, New York, pp 439–453
Petterson JR, Sandberg R, Alerstam T (1991) Orientation of robins, *Erithacus rubecula*, in a vertical magnetic field. Anim Behav 41:533–536
Phillips JB (1977) Use of the earth's magnetic field by orienting cave salamanders (*Eurycea lucifuga*). J Comp Physiol 121:273–288
Phillips JB (1986a) Magnetic compass orientation in the eastern red-spotted newt (*Notophthalmus viridescens*). J Comp Physiol A 158:103–109
Phillips JB (1986b) Two magnetoreception pathways in a migratory salamander. Science 233:765–767
Phillips JB (1987a) Laboratory studies of homing orientation in the eastern red-spotted newt, *Notophthalmus viridescens*. J Exp Biol 131:215–229
Phillips JB (1987b) Specialized visual receptors respond to magnetic field alignment in the blowfly (*Calliphora vicina*). Soc Neurosci Abstr 13:397
Phillips JB, Adler K (1978) Directional and discriminatory responses of salamanders to weak magnetic fields. In: Schmidt-Koenig K, Keeton WT (eds) Animal migration, navigation, and homing. Springer, Berlin Heidelberg New York, pp 325–333
Phillips JB, Borland SC (1992a) Magnetic compass orientation is eliminated under near-infrared light in the eastern red-spotted newt *Notophthalmus viridescens*. Anim Behav 44:796–797
Phillips JB, Borland SC (1992b) Wavelength specific effects of light on magnetic compass orientation of the eastern red-spotted newt *Notophthalmus viridescens*. Ethol Ecol Evol 4:33–42
Phillips JB, Borland SC (1992c) Behavioral evidence for use of a light-dependent magnetoreception mechanism by a vertebrate. Nature 359:142–144
Phillips JB, Borland SC (1994) Use of a specialized magnetoreception system for homing by the eastern red-spotted newt, *Notophthalmus viridescens*. J Exp Biol 188:275–291
Phillips JB, Moore FR (1992) Calibration of the sun compass by sunset polarized light pattern in a migratory bird. Behav Ecol Sociobiol 44:796–797
Phillips JB, Sayeed O (1993) Wavelength-dependent effects of light on magnetic compass orientation in *Drosophila melanogaster*. J Comp Physiol A 172:303–308
Picton HD (1966) Some responses of *Drosophila* to weak magnetic and electrostatic fields. Nature 211:303–304
Presti D, Pettigrew JD (1980) Ferromagnetic coupling to muscle receptors as a basis for geomagnetic field sensitivity in animals. Nature 285:99–101
Prinz K, Wiltschko W (1992) Migratory orientation of pied flycatchers: Interaction of stellar and magnetic information during ontogeny. Anim Behav 44:539–545
Quentmeier B (1989) Cardiac and respiratory responses to magnetic fields in pigeons. In: Orientation and navigation – birds, humans and other animals. Proc Int Conf Royal Inst Navig, Cardiff, Paper 15
Quinn TP (1980) Evidence for celestial and magnetic compass orientation in lake migrating sockeye salmon fry. J Comp Physiol 137:243–248
Quinn TP (1982a) Intra-specific differences in sockeye salmon fry compass orientation mechanisms. In: Brannon EL, Salo EO (eds) Proc Salmon and Trout Migratory Behavior Symp. University of Washington, Seattle, pp 79–85
Quinn TP (1982b) A model for salmon navigation on the high seas. In: Brannon EL, Salo EO (eds) Proc Salmon and Trout Migratory Behav Symp. University of Washington, Seattle, pp 229–237
Quinn TP (1984) An experimental approach to fish compass and map orientation. In: McCleave JD, Arnold GP, Dodson JJ, Neill WH (eds) Mechanisms of migration in fishes. Plenum Press, New York, pp 113–123
Quinn TP (1990) Current controversies in the study of salmon homing. Ethol Ecol Evol 2:49–63
Quinn TP, Brannon EL (1982) The use of celestial and magnetic cues by orienting sockeye salmon smolts. J Comp Physiol 147:547–552

Quinn TP, Groot C (1983) Orientation of chum salmon (*Oncorhynchus keta*) after internal and external magnetic field alteration. Can J Fish Aquat Sci 40:1598–1606
Quinn TP, Groot C (1984) Pacific salmon (*Oncorhynchus*) migrations: orientation versus random movement. Can J Fish Aquat Sci 41:1319–1324
Quinn TP, Merrill RT, Brannon EL (1981) Magnetic field detection in sockeye salmon. J Exp Zool 217:137–142
Quinn TO, Wood CC, Margolis L, Riddell BE, Hyatt KD (1987) Homing in wild sockeye salmon (*Oncorhynchus nerka*) populations as inferred from differences in parasite prevelance and allozyme allele frequencies. Can J Fish Aquat Sci 44:1963–1971
Ranvaud R, Schmidt-Koenig K, Kiepenheuer J, Gasparotto OC, Britto LR (1986) Homing experiments with pigeons at the magnetic equator. In: Maret G, Boccara N, Kiepenheuer J (eds) Biophysical effects of steady magnetic fields. Springer, Berlin Heidelberg New York, pp 163–166
Ranvaud R, Schmidt-Koenig K, Ganzhorn JU, Kiepenheuer J, Gasparotto OC, Britto LRG (1991) The initial orientation of homing pigeons at the magnetic equator: compass mechanisms and the effect of applied magnets. J Exp Biol 161:299–314
Ratner SC, Jennings JW (1968) Magnetic fields and orienting movements in mollusks. J Comp Physiol Psychol 65:365–368
Rawson KS (1954) Sun compass orientation and endogenous activity rhythms of the starling (*Sturnus vulgaris* L). Z Tierpsychol 11:446–452
Reille A (1968) Essai de mise en évidence d'une sensibilité du pigeon au champ magnétique à l'aide d'une conditionnement nociceptif. J Physiol Paris 60:85–92
Reiter RJ (1993a) A review of neuroendocrine and neurochemical changes associated with static and extremely low frequency electromagnetic fields. Integr Physiol Behav Sci 28:57–75
Reiter RJ (1993b) Static and extremely low frequency electromagnetic field exposure: reported effects on the circadian production of melatonin. J Cell Biochem 51:394–403
Richardson WJ (1974) Spring migration over Puerto-Rico and the western Atlantic: a radar study. Ibis 116:172–193
Richardson WJ (1976) Autumn migration over Puerto-Rico and the western Atlantic: a radar study. Ibis 118:309–332
Rickli M, Leuthold RH (1988) Homing in harvester termites: evidence of magnetic orientation. Ethology 77:209–216
Rodda GH (1984) The orientation and navigation of juvenile alligators: evidence of magnetic sensitivity. J Comp Physiol A 154:649–658
Rommel SA, McCleave JD (1973) Sensitivity of American eels (*Anguilla rostrata*) and Atlantic salmons (*Salmo salar*) to weak electric and magnetic fields. J Fish Res Board Can 30:657–663
Roonwal ML (1958) Recent work on termite research in India (1947–57). Trans Bose Res Inst (Calcutta) 22:77–100
Rosenblum B, Jungerman RL, Longfellow L (1985) Limits to induction-based magnetoreception. In: Kirschvink JL, Jones DS, MacFadden BJ (eds) Magnetite biomineralization and magnetoreception in organisms. Plenum Press, New York, pp 223–232
Rosengren R, Fortelius W (1986) Ortstreue in foraging ants of the *Formica rufa* group – hierarchy of orienting cues and long-term memory. Insectes Soc 33:306–337
Royce WF, Smith LS, Hartt AC (1968) Models of oceanic migrations of Pacific salmon and comments on guidance mechanisms. Fish Bull 66:441–462
Rubens SM (1945) Cube-surface coil for producing a uniform magnetic field. Rev Sci Instrum 16:243–245
Russo F, Caldwell WE (1971) Biomagnetic phenomena: some implications for the behavioral and neurophysiological sciences. Gent Psychol Monogr 84:177–243
Sadauskas KK, Shuranova ZP (1984) Effect of pulsed magnetic field on electrical activity of crayfish neurons. Biophysica 29:681–683 (in Russian)
Saila SB, Shappy RA (1963) Random movement and orientation in salmon migration. J Conserv Int Explor Mar 28:153–166
Sakaki Y, Motomiya T, Kato M, Ogura M (1990) Possible mechanism of biomagnetic sense organ extracted from sockeye salmon. IEEE Trans Magnet 26:1554–1556

Salmon M, Lohmann KJ (1989) Orientation cues used by hatchling loggerhead sea turtles (*Caretta caretta* L.) during their offshore migration. Ethology 83:215–228
Salmon M, Wyneken E (1994) Orientation by hatchling sea turtles: mechanisms and implications. Herpetol Nat Hist 2:13–24
Salmon M, Wyneken E, Fritz E, Lucas M (1992) Seafinding by hatchling sea turtles: role of brightness, silhouette and beach slope as orientation cues. Behaviour 122:56–77
Sandberg R (1988) Skylight polarization does not affect the migratory orientation of European robins. Condor 90:267–270
Sandberg R (1991) Sunset orientation of robins, *Erithacus rubecula*, with different fields of sky vision. Behav Ecol Sociobiol 28:77–83
Sandberg R (1994) Interaction of body condition and magnetic orientation in autumn migrating robins, *Erithacus rubecula*. Anim Behav 47:679–686
Sandberg R, Pettersson J, Alerstam T (1988a) Why do migrating robins, *Erithacus rubecula*, captured at two nearby stop-over sites orient differently? Anim Behav 36:865–876
Sandberg R, Pettersson J, Alerstam T (1988b) Shifted magnetic fields lead to deflected and axial orientation of migrating robins, *Erithacus rubecula*, at sunset. Anim Behav 36:877–887
Sandberg R, Ottoson U, Pettersson J (1991a) Magnetic orientation of migratory wheatears (*Oenanthe oenanthe*) in Scandinavia and Greenland. J Exp Biol 155:51–64
Sandberg R, Pettersson J, Persson K (1991b) Migratory orientation of free-flying robins *Erithacus rubecula* and pied flycatchers, *Ficedula hypoleuca*: release experiments. Ornis Scand 22:1–11
Sauvé JP (1988) Analyse de l'orientation initiale dans une expérience de retour au gîte chez le mulot (*Apodemus sylvaticus*). Sci Tech Anim Lab 13:89–91
Scapini F (1988) Heredity and learning in animal orientation. Monit Zool Ital (NS) 22:203–234
Scapini F, Ercolini A (1973) Research of the non-visual orientation of littoral amphipods: experiments with young born in capticity and adults from a Somalian population of *Talorchestia martensii* Weber (Crustacea Amphipoda). Monit Zool Ital (NS) 3:23–30
Scapini F, Quochi G (1992) Orientation of sandhoppers from Italian populations: have they magnetic orientation ability? Boll Zool 59:437–442
Scapini F, Ugolini A, Pardi L (1988) Aspects of direction finding inheritance in natural populations of littoral sandhoppers (*Talitrus saltator*). In: Chelazzi G, Vannini M (eds) Behavioral adaption to intertidal life. Plenum Press, New York, pp 93–103
Schietecat G (1988) Meteorologische und geophysikalische Einflüsse bei Brieftaubenwettflügen III and IV. Brieftaube 105:936–938, 972–974
Schiff H (1991) Modulation of spike frequencies by varying the ambient magnetic field and magnetite candidates in bees (*Apis mellifera*). Comp Biochem Physiol 100 A:975–985
Schmidt-Koenig K (1958) Experimentelle Einflußnahme auf die 24-Stunden-Periodik bei Brieftauben und deren Auswirkungen unter besonderer Berücksichtigung des Heimfindevermögens. Z Tierpsychol 15:301–331
Schmidt-Koenig K (1965) Current problems in bird orientation. Adv Study Behav 1:217–276
Schmidt-Koenig K (1970) Ein Versuch, theoretisch mögliche Navigationsverfahren von Vögeln zu klassifizieren und relevante sinnesphysiologische Probleme zu umreißen. Verh Dtsch Zool Ges 64:243–244
Schmidt-Koenig K (1985) Migration strategies of monarch butterflies. In: Rankin MA (ed) Migration: mechanisms and adaptive significance. Contrib Mar Sci (Port Aransas) 27, pp 786–798
Schmidt-Koenig K, Ganzhorn JU (1991) On the problem of bird navigation. In: Bateson PPG, Klopfer PH (eds) Perspectives in ethology. Plenum Press, New York, pp 261–283
Schmitt DE, Esch HE (1993) Magnetic orientation of honeybees in the laboratory. Naturwissenschaften 80:41–43
Schneider F (1957) Die Fernorientierung des Maikäfers während seiner ersten Fraßperiode und beim Rückflug in das alte Brutgebiet. Verh Schweiz Naturforsch Ges Neuenburg 6:95–96
Schneider F (1960) Beeinflussung der Aktivität des Maikäfers durch Veränderung der gegenseitigen Lage magnetischer und elektrischer Felder. Mitt Schweiz Entomol Ges 33:223–237
Schneider F (1963a) Ultraoptische Orientierung des Maikäfers (*Melolontha vulgaris* F) in künstlichen elektrischen und magnetischen Feldern. Ergeb Biol 26:147–157

Schneider F (1963b) Systematische Variationen in der elektrischen, magnetischen und geographisch-ultraoptischen Orientierung des Maikäfers. Vierteljahresschr Naturforsch Ges Zürich 108:373–416
Schneider F (1967) Schwärmbahnen der Maikäfer. In: Hediger H (ed) Die Straßen der Tiere. Friedrich Vieweg, Braunschweig
Schreiber B, Rossi O (1976) Correlation between race arrivals of homing pigeons and solar activities. Boll Zool 43:317–320
Schreiber B, Rossi O (1978) Correlations between magnetic storms due to solar spots and pigeon homing performance. IEEE Trans Magnet Mag-14:961–963
Schulten K (1982) Magnetic field effects in chemistry and biology. Festkörperprobleme 22:60–83
Schulten K, Windemuth A (1986) Model for a physiological magnetic compass. In: Maret G, Boccara N, Kiepenheuer J (eds) Biophysical effects of steady magnetic fields. Springer, Berlin Heidelberg New York, pp 99–106
Scott GR, Frohlich C (1985) Large-volume magnetically shielded rooms: a new design and material. In: Kirschvink JL, Jones DS, MacFadden BJ (eds) Magnetite biomineralization and magnetoreception in organisms. Plenum Press, New York, pp 197–222
Semm P (1983) Neurobiological investigations on the magnetic sensitivity of the pineal gland in rodents and pigeons. Comp Biochem Physiol 76A:683–689
Semm P, Beason RC (1990) Responses to small magnetic variations by the trigeminal system of the bobolink. Brain Res Bull 25:735–740
Semm P, Demaine C (1983) Electrical responses to direct and indirect photic stimulation of the pineal gland of the pigeon. J Neural Transm 58:281–289
Semm P, Demaine C (1986) Neurophysiological properties of magnetic cells in the pigeon's visual system. J Comp Physiol A 159:619–625
Semm P, Schneider T (1991) Magnetic responses in the central nervous system of birds. In: Lieth H (ed) Effects of atmospheric and geophysical variables in biology and medicine. Progr Biometeorol 8:3–13
Semm P, Schneider T, Vollrath L (1980) Effects of an earth-strength magnetic field on electrical activity of pineal cells. Nature 288:607–608
Semm P, Nohr D, Demaine C, Wiltschko W (1984) Neural basis of the magnetic compass: interaction of visual, magnetic and vestibular inputs in the pigeon's brain. J Comp Physiol A 155:283–288
Sherman K (1992) Classical conditioning to magnetic fields of *Apteronotus leptorhynchus*. Master Thesis, State University of New York, Geneseo, NY
Shettleworth SJ (1972) Constraints on learning. Adv Study Behav 4:1–68
Shumakov ME, Vinogradova NV (1992) Interrelations between astronomical and geomagnetic stimuli during migration in scarlet grosbeaks (*Carpodactus erythrinus*). In: Payevsky VA (ed) Problems of birds' population ecology. Russ Acad Sci 247:106–116 (in Russian)
Simpson KS (1979) Orientation differences between populations of juvenile sockeye salmon. Fish Mar Serv Can Tech Rep 717:114
Sinsch U (1987) Orientation behaviour of toads (*Bufo bufo*) displaced from the breeding site. J Comp Physiol A 161:715–727
Sinsch U (1989) Migratory behavior of the common toad *Bufo bufo* and the natterjack toad *Bufo calamita*. In: Langton TES (ed) Amphibians and roads. ACO Polymer Products, Shefﬁeld, pp 113–125
Sinsch U (1990a) The orientation behaviour of three toad species (genus *Bufo*) displaced from the breeding site. Fortschr Zool 38:75–83
Sinsch U (1990b) Migration and orientation in anuran amphibians. Ethol Ecol Evol 2:65–79
Sinsch U (1992) Sex-biassed site fidelity and orientation behaviour in reproductive natterjack toads (*Bufo calamita*). Ethol Ecol Evol 4:15–32
Skiles DD (1985) The geomagnetic field: its nature, history, and biological relevance. In: Kirschvink JL, Jones DS, MacFadden BJ (eds) Magnetite biomineralization and magnetoreception in organisms. Plenum Press, New York, pp 43–102
Smith RJF (1985) The control of fish migration. Springer, Berlin Heidelberg New York

Sokolov LV, Bolshakov KV, Vinogradova NV, Dolnik TV, Lyuleeva DS, Payevsky VA, Shumakov ME, Yablonkevich ML (1984) The testing of the ability for imprinting and finding the site of future nesting in young chaffinches. Zool J (Moscow) 43:1671–1681 (in Russian)
Southern WE (1972a) Magnets disrupt the orientation of juvenile ring-billed gulls. BioScience 22:476–479
Southern WE (1972b) Influence of disturbances on the earth's magnetic field on ring-billed gull orientation. Condor 74:102–105
Southern WE (1975) Orientation of gull chicks exposed to project Sanguine's electromagnetic field. Science 189:143–145
Souza JJ, Poluhowich JJ, Guerra RJ (1988) Orientation responses of American eels, *Anguilla rostrata*, to varying magnetic fields. Comp Biochem Physiol 90A:57–61
Spain AV, Okello-Oloya T, John RD (1983) Orientation of the termitaria of two species of *Amitermes* (Isoptera: Termitinae) from Northern Queensland. Aust J Zool 31:167–177
Stabrowski A, Nollen PM (1985) The response of *Philophthalmus gralli* and *P. megalurus* to light, gravity and magnetic fields. Int J Parasitol 15:551–555
Stasko AB, Rommel SA (1974) Swimming depth of adult American eels (*Anguilla rostrata*) in a saltwater bay as determined by ultrasonic tracking. J Fish Res Board Can 31:1148–1150
Steidinger P (1968) Radarbeobachtungen über die Richtung und deren Streuung beim nächtlichen Vogelzug im Schweizerischen Mittelland. Ornithol Beob 65:197–226
Stutz AM (1971) Effects of weak magnetic fields on gerbil spontaneous activity. Ann NY Acad Sci 188:312–324
Talkington L (1967) Bird navigation and geomagnetism. Am Zool 7:199
Taylor PB (1986) Experimental evidence for geomagnetic orientation in juvenile salmon, *Onchorhynchus tschawytscha*. J Fish Biol 28:607–624
Taylor PB (1987) Experimental evidence for juvenile chinook salmon, *Oncorhynchus tschawytscha* Walbaum: orientation at night and in sunlight after a 7 ° change in latitude. J Fish Biol 31:89–111
Tenforde TS (1979) Magnetic field effect on biological systems. Plenum Press, New York
Tesch FW (1967) Homing of eels (*Anguilla anguilla*) in the southern North Sea. Mar Biol 1:2–9
Tesch FW (1974a) Influence of geomagnetism and salinity on the directional choices of eels. Helgol Wiss Meeresunters 26:382–395
Tesch FW (1974b) Speed and direction of silver and yellow eels, *Anguilla anguilla*, released and tracked in the open North Sea. Ber Dtsch Wiss Komm Meeresforsch 23:181–197
Tesch FW (1978) Horizontal and vertical swimming of eels during the spawning migration at the edge of the continental shelf. In: Schmidt-Koenig K, Keeton WT (eds) Animal migration, navigation, and homing. Springer, Berlin Heidelberg New York, pp 378–391
Tesch FW (1980) Migratory performance and environmental evidence of orientation. In: Ali MA (ed) Environmental physiology of fishes. NATO Adv Study Inst Ser A Life Sci 35: 589–612
Tesch FW (1989) Changes in swimming depth and direction of silver eels (*Anguilla anguilla* L.) from the continental shelf to the deep sea. Aquat Living Resour 2:9–20
Tesch FW, Lelek A (1973a) Directional behaviour of transplanted stationary and migratory forms of the eel, *Anguilla anguilla*, in a circular tank. Neth J Sea Res 7:46–52
Tesch FW, Lelek A (1973b) An evaluation of the directional choice in the eel, in captivity. Arch Fischereiwiss 24:237–251
Tesch FW, Wendt T, Karlsson L (1992) Influence of geomagnetism on the activity and orientation of the eel, *Anguilla anguilla* (L.), as evident from laboratory experiments. Ecol Freshwater Fish 1:52–60
Teyssèdre A (1986) Radio-tracking of pigeons previously exposed to random oscillating magnetic fields. Behaviour 96:265–276
Thompson R, Oldfield F (1985) Environmental magnetism. Allen and Unwin, London
Tomlinson J, Reilly D, Ballering R (1980a) Magnetic radular teeth and geomagnetic responses in chitons. Veliger 23:167–170
Tomlinson J, McGinty S, Kish J (1980b) Magnets curtail honey bee dancing. Anim Behav 29:307–308

Torres de Araujo FF, Germano FA, Gonçalves LL, Pires MA, Frankel RB (1990) Magnetic polarity fractions in magnetotactic bacterial populations near the geomagnetic equator. Biophys J 58:549–555

Towe KM, Lowenstam HA, Nesson MH (1963) Invertebrate ferritin: occurrence in Mollusca. Science 142:63–64

Towne WF, Gould JL (1985) Magnetic field sensitivity in honeybees. In: Kirschvink JL, Jones DS, MacFadden BJ (eds) Magnetite biomineralization and magnetoreception in organisms. Plenum Press, New York, pp 385–406

Tucker DW (1959) A new solution to the Atlantic eel problem. Nature 183:495–501

Ueda K, Kusunoki M, Kato M, Kakizawa R, Nakamura T, Yaskawa K, Koyama M, Maeda Y (1982) Magnetic remanences in migratory birds. J Yamashina Inst Ornithol 14:166–170

Ueda K, Maeda Y, Koyama M, Yaskawa K, Tokui T (1986) Magnetic remanences in salmonid fish. Bull Jpn Soc Sci Fish 52:193–198

Ueno S, Lövsund P, Öberg PA (1986) Effect of time-varying magnetic fields on the action potential in lobster giant axon. Med Biol Eng Comput 24:521–526

Ugolini A (1989) Orientation in the water and antipredatory behavior in sandhoppers. Mar Behav Physiol 14:223–230

Ugolini A (1994) Magnetic orientation in and near the sea: *Idotea baltica*, *Talitrus saltator* and *Talorchestia martensii*. In: Williamson R (ed) Orientation and migration at sea. Marine Biological Association and Society for Experimental Biology, Plymouth, p 10

Ugolini A, Cannicci S (1991) Solar orientation in British sandhoppers. Mar Behav Physiol 19:149–157

Ugolini A, Macchi T (1988) Learned component in the solar orientation of *Talitrus saltator* Montagu (Amphipoda: Talitridae). J Exp Mar Biol Ecol 121:79–87

Ugolini A, Pardi L (1992) Equatorial sandhoppers do not have a good clock. Naturwissenschaften 79:279–281

Ugolini A, Pezzani A (1992) Learned solar orientation in *Idotea baltica*. Mesogee (Bull Mus Hist Nat, Marseille) 52:77

Ugolini A, Scapini F, Beugnon L, Pardi L (1988) Learning in zonal orientation of sandhoppers. In: Chelazzi G, Vannini M (eds) Behavioral adaptation to intertidal life. Plenum Press, New York, pp 105–118

Ugolini A, Felicioni S, Macchi T (1991) Orientation in the water and learning in *Talitrus saltator* Montagu. J Exp Mar Biol Ecol 151:113–119

Vacquier V (1972) Geomagnetism in marine geology. Elsevier, Amsterdam

van den Bercken J, Broekhuizen S, Ringelberg J, Velthuis HHW (1967) Non-visual orientation in *Talitrus saltator*. Experientia 23:44–45

van Riper W, Kalmbach ER (1952) Homing not hindered by wing magnets. Science 115:577–578

Varanelli CC, McCleave JD (1974) Locomotor activity of Atlantic salmon parr (*Salmo salar* L) in various light conditions and in weak magnetic fields. Anim Behav 22:178–186

Vasil'yev AS, Gleizer SV (1973) Changes in activity of *Anguilla anguilla* (L) in a magnetic field. Vopr Ikhtiol 13:381–383 (in Russian)

Viehmann W (1979) The magnetic compass of blackcaps (*Sylvia atricapilla*). Behaviour 68:24–30

Viguier C (1882) Le sens de l'orientation et ses organes chez les animaux et chez l'homme. Rev Phil France Etranger 14:1–36

Viksne J, Blüms P, Baltvilks J (1978) Orientation behavior of the black-headed gull (*Larus ridibundus*) chicks. In: Mihelson H, Blüm P, Baumanis J (eds) Orientazija Ptiz [Orientation of birds]. Zinatne, Riga, pp 7–24 (in Russian)

Vilches-Troya J, Dunn RF, O'Leary DP (1984) Relationship of the vestibular hair cells to magnetic particles in the otolith of the guitarfish sacculus. J Comp Neurol 226:489–494

Visalberghi E, Alleva E (1979) Magnetic influences on pigeon homing. Biol Bull 125:246–256

Vleugel DA (1955) Über die Unzulänglichkeit der Visierorientierung für das Geradeausfliegen, insbesondere beim Zug des Buchfinken (*Fringilla coelebs* L.). Ornis Fenn 32:34–40

von Frisch K (1948) Gelöste und ungelöste Rätsel der Bienensprache. Naturwissenschaften 35:12–23; 38–43

von Frisch K (1950) Die Sonne als Kompaß im Leben der Bienen. Experientia 6:210–221

von Frisch K (1968) Dance, language and orientation of bees. Harvard University Press, Cambridge, MA
von Frisch K (1974) Animal architecture. Harcourt Brace Jovanovich, New York
von Middendorff A (1859) Die Isepiptesen Rußlands. Mém Acad Sci St Petersbourg VI Ser Tome 8:1–143
von Salvini-Plawen L, Mayr E (1977) On the evolution of photoreceptors and eyes. In: Hecht MK, Steere WC, Wallace B (eds) Evolutionary biology, vol 10. Plenum Press, New York, pp 207–263
Wadas RS (1991) Biomagnetism. Ellis Horwood, New York
Wagner G (1972) Untersuchungen über das Orientierungsverhalten von Brieftauben unter Radar-Bestrahlung. Rev Suisse Zool 79:229–244
Wagner G (1976) Das Orientierungsverhalten von Brieftauben im erdmagnetisch gestörten Gebiet des Chasseral. Rev Suisse Zool 83:883–890
Wagner G (1983) Natural geomagnetic anomalies and homing in pigeons. Comp Biochem Physiol 76A:691–701
Walcott B (1985) The cellular localization of particulate iron. In: Kirschvink JL, Jones DS, MacFadden BJ (eds) Magnetite biomineralization and magnetoreception in organisms. Plenum Press, New York, pp 183–195
Walcott B, Walcott C (1982) A search for magnetic field receptors in animals. In: Papi F, Wallraff HG (eds) Avian navigation. Springer, Berlin Heidelberg New York, pp 338–343
Walcott C (1977) Magnetic fields and the orientation of homing pigeons under sun. J Exp Biol 70:105–123
Walcott C (1978) Anomalies in the earth's magnetic field increase the scatter of pigeons' vanishing bearings. In: Schmidt-Koenig K, Keeton WT (eds) Animal migration, navigation, and homing. Springer, Berlin Heidelberg New York, pp 143–151
Walcott C (1980a) Magnetic orientation in homing pigeons. IEEE Trans Magnet Mag-16:1008–1013
Walcott C (1980b) Homing-pigeon vanishing bearings at magnetic anomalies are not altered by bar magnets. J Exp Biol 86:349–352
Walcott C (1982) Is there evidence for a magnetic map in homing pigeons? In: Papi F, Wallraff HG (eds) Avian navigation. Springer, Berlin Heidelberg New York, pp 99–108
Walcott C (1986) A review of magnetic effects on homing pigeon orientation. In: Maret G, Boccara N, Kiepenheuer J (eds) Biophysical effects of steady magnetic fields. Springer, Berlin Heidelberg New York, pp 146–147
Walcott C (1991) Magnetic maps in pigeons. In: Berthold P (ed) Orientation in Birds. Birkhäuser, Basel, pp 38–51
Walcott C (1992) Pigeons at magnetic anomalies: the effects of loft location. J Exp Biol 170:127–141
Walcott C, Green RP (1974) Orientation of homing pigeons altered by a change in the direction of an applied magnetic field. Science 184:180–182
Walcott C, Gould JL, Kirschvink JL (1979) Pigeons have magnets. Science 205:1027–1029
Walcott C, Gould JL, Lednor AJ (1988) Homing of magnetized and demagnetized pigeons. J Exp Biol 134:27–41
Walker MM (1984) Learned magnetic field discrimination in yellowfin tuna, *Thunnus albacares*. J Comp Physiol A 155:673–679
Walker MM, Bitterman ME (1985) Conditioned responding to magnetic fields by honeybees. J Comp Physiol A 157:67–71
Walker MM, Bitterman ME (1986) Attempts to train goldfish to respond to magnetic field stimuli. Naturwissenschaften 73:12–16
Walker MM, Bitterman ME (1989a) Honeybees can be trained to respond to very small changes in geomagnetic field intensity. J Exp Biol 145:489–494
Walker MM, Bitterman ME (1989b) Attached magnets impair magnetic field discrimination by honeybees. J Exp Biol 141:447–451
Walker MM, Bitterman ME (1991) Magnetic sensitivity in honeybees: threshold estimation and tests of a possible sensory mechanism. In: Lieth H (ed) Effects of atmospheric and geophysical variables in biology and medicine. Progr Biometeorol 8:53–66

Walker MM, Kirschvink JL, Chang S-BR, Dizon AE (1984) A candidate magnetic sense organ in the yellowfin tuna, *Thunnus albacares*. Science 224:751–753
Walker MM, Kirschvink JL, Perry A, Dizon AE (1985) Detection, extraction, and characterization of biogenic magnetite. In: Kirschvink JL, Jones DS, MacFadden BJ (eds) Magnetite Biomineralization and Magnetoreception in Organisms. Plenum Press, New York, pp 155–166
Walker MM, Bitterman ME, Kirschvink JL (1986) Experimental and correlational studies of responses to magnetic field stimuli by different species. In: Maret G, Kiepenheuer J, Boccara N (eds) Biophysical effects of steady magnetic fields. Springer, Berlin Heidelberg New York, pp 194–205
Walker MM, Quinn TP, Kirschvink JL, Groot C (1988) Production of single-domain magnetite throughout life by sockeye salmon, *Oncorhynchus nerka*. J Exp Biol 140:51–63
Walker MM, Baird DL, Bitterman ME (1989) Failure of stationary but not of flying honeybees (*Apis mellifera*) to respond to magnetic field stimuli. J Comp Psychol 103:62–69
Walker MM, Lee Y, Bitterman ME (1990) Transfer along a continuum in the discriminative learning of honeybees (*Apis mellifera*). J Comp Psychol 104:66–70
Walker MM, Kirschvink JL, Ahmed G, Dizon AE (1992) Evidence that fin whales respond to the geomagnetic field during migration. J Exp Biol 171:67–78
Walker MM, Montgomery JC, Pankhurst PM (1993) Toward a sensory basis for magnetic navigation by animals. In: Orientation and navigation - birds, humans, and other animals. Proc Int Conf Royal Inst Navig, Oxford, Paper 4
Wallraff HG (1966a) Versuche zur Frage der gerichteten Nachtzug-Aktivität von gekäfigten Singvögeln. Verh Dtsch Zool Ges 59:338–355
Wallraff HG (1966b) Über die Anfangsorientierung von Brieftauben unter geschlossener Wolkendecke. J Ornithol 107:326–336
Wallraff HG (1972) Nichtvisuelle Orientierung zugunruhiger Rotkehlchen (*Erithacus rubecula*). Z Tierpsychol 30:374–382
Wallraff HG (1974) Das Navigationssystem der Vögel. Ein theoretischer Beitrag zur Analyse ungeklärter Orientierungsleistungen. Oldenbourg, München
Wallraff HG (1978) Proposed principles of magnetic field perception in birds. Oikos 30:188–194
Wallraff HG (1980) Does pigeon homing depend on stimuli perceived during displacement? I. Experiments in Germany. J Comp Physiol 139:193–201
Wallraff HG (1983) Relevance of atmospheric odours and geomagnetic field to pigeon navigation: what is the "map" basis? Comp Biochem Physiol 76A:643–663
Wallraff HG, Foà A (1982) The roles of olfaction and magnetism in pigeon homing. Naturwissenschaften 69:504–505
Wallraff HG, Foà A, Ioalè P (1980) Does pigeon homing depend on stimuli perceived during displacement? J Comp Physiol 139:203–208
Wallraff HG, Papi F, Ioalè P, Benvenuti S (1986a) Magnetic fields affect pigeon navigation only while the birds can smell atmospheric odors. Naturwissenschaften 73:215–217
Wallraff HG, Benvenuti S, Foà A (1986b) Attempts to reveal the nature of apparent residual homeward orientation in anosmic pigeons: application of magnetic fields. Monit Zool Ital (NS) 20:401–423
Walmsley DJ, Epps WR (1988) Do humans have an innate sense of direction? Geography 1988:31–40
Walton AS, Herrnkind WF (1977) Hydrodynamic orientation of spiny lobster, *Panulirus argus* (Crustacea: Palinuridae): wave surge and unidirectional currents. Mar Sci Res Lab Tech Rep 20:184–211
Wehner R (1982) Himmelsnavigation bei Insekten. Naturforschende Gesellschaft in Zürich, Zürich
Wehner R, Labhart T (1970) Perception of the geomagnetic field in the fly, *Drosophila melanogaster*. Experientia 26:967–968
Wehner R, Harkness RD, Schmid-Hempel P (1983) Foraging strategies in individually searching ants *Cataglyphis bicolor* (Hymenoptera: Formacidae). Information Processing in Animals, vol 1. Fischer, Stuttgart
Weindler P (1994) Wintergoldhähnchen (*Regulus regulus*) benutzen einen Inklinationskompaß. J Ornithol 135:620–622

Weindler P, Beck W, Liepa V, Wiltschko W (1995) Development of migratory orientation in pied flycatchers in different magnetic inclination. Anim Behav 49:227–234
Wenner AM (1962) Sound production during the waggle dance of the honey bee. Anim Behav 10:79–95
Westerberg H (1982) Ultrasonic tracking of Atlantic salmon (*Salmo salar* L.). I. Movements in coastal regions. Rep Inst Freshwater Res (Drottingholm) 60: 81–101
Williamson DI (1954) Landward and seaward movements of the sandhopper *Talitrus saltator*. Adv Sci (London) 11:71–73
Wiltschko R (1980) Die Sonnenorientierung der Vögel. I. Die Rolle der Sonne im Orientierungssystem und die Funktionsweise des Sonnenkompaß. J Ornithol 121:121–143
Wiltschko R (1981) Die Sonnenorientierung der Vögel. II. Entwicklung des Sonnenkompaß und sein Stellenwert im Orientierungssystem. J Ornithol 122:1–22
Wiltschko R (1983) The ontogeny of orientation in young pigeons. Comp Biochem Physiol 76A:701–708
Wiltschko R (1992) Das Verhalten verfrachteter Vögel. Vogelwarte 36:249–310
Wiltschko R, Wiltschko W (1978a) Relative importance of stars and magnetic field for the accuracy of orientation in night-migrating birds. Oikos 30:195–206
Wiltschko R, Wiltschko W (1978b) Evidence for the use of magnetic outward-journey information in homing pigeons. Naturwissenschaften 65:112
Wiltschko R, Wiltschko W (1981) The development of sun compass orientation in young homing pigeons. Behav Ecol Sociobiol 9:135–141
Wiltschko R, Wiltschko W (1985) Pigeon homing: change in navigational strategy during ontogeny. Anim Behav 33:583–590
Wiltschko R, Wiltschko W (1989) Pigeon homing: olfactory orientation – a paradox. Behav Ecol Sociobiol 24:163–173
Wiltschko R, Wiltschko W (1990) Zur Entwicklung des Sonnenkompaß bei jungen Brieftauben. J Ornithol 131:1–20
Wiltschko R, Wiltschko W (1994a) Avian orientation: multiple sensory cues and the advantage of redundancy. In: Davies MNO, Green PR (eds) Perception and motor control in birds. Springer, Berlin Heidelberg New York, pp 95–119
Wiltschko R, Wiltschko W (1994b) Der Sonnenkompaß bei Brieftauben: Zeitverschiebung induziert nicht immer die erwartete Abweichung. Jahresvers Dtsch Ornithol Ges 127:127
Wiltschko R, Nohr D, Wiltschko W (1981) Pigeons with a deficient sun compass use the magnetic compass. Science 214:343–345
Wiltschko R, Schöps M, Kowalski U (1989) Pigeon homing: wind exposition determines the importance of olfactory input. Naturwissenschaften 76:229–231
Wiltschko R, Kumpfmüller R, Muth R, Wiltschko W (1994) Pigeon homing: the effect of a clock-shift is often smaller than predicted. Behav Ecol Sociobiol 35:63–73
Wiltschko W (1968) Über den Einfluß statischer Magnetfelder auf die Zugorientierung der Rotkehlchen (*Erithacus rubecula*). Z Tierpsychol 25:537–558
Wiltschko W (1974) Der Magnetkompaß der Gartengrasmücke (*Sylvia borin*). J Ornithol 115:1–7
Wiltschko W (1978) Further analysis of the magnetic compass of migratory birds. In: Schmidt-Koenig K, Keeton WT (eds) Animal migration, navigation, and homing. Springer, Berlin Heidelberg New York, pp 302–310
Wiltschko W, Balda RP (1989) Sun compass orientation in seed-caching scrub jays (*Aphelocoma coerulescens*). J Comp Physiol A 164:717–721
Wiltschko W, Beason R (1990) Magneteffekte bei der Heimorientierung von Brieftauben. Verh Dtsch Zool Ges 83:435–436
Wiltschko W, Gwinner E (1974) Evidence for an innate magnetic compass in garden warblers. Naturwissenschaften 61:406
Wiltschko W, Merkel FW (1966) Orientierung zugunruhiger Rotkehlchen im statischen Magnetfeld. Verh Dtsch Zool Ges 59:362–367
Wiltschko W, Merkel FW (1971) Zugorientierung von Dorngrasmücken (*Sylvia communis*). Vogelwarte 26:245–249
Wiltschko W, Wiltschko R (1972) Magnetic compass of European robins. Science 176:62–64

Wiltschko W, Wiltschko R (1975a) The interaction of stars and magnetic field in the orientation system of night migrating birds. I. Autumn experiments with European warblers (Gen. *Sylvia*). Z Tierpsychol 37:337–355
Wiltschko W, Wiltschko R (1975b) The interaction of stars and magnetic field in the orientation system of night migrating birds. II. Spring experiments with European robins (*Erithacus rubecula*). Z Tierpsychol 39:265–282
Wiltschko W, Wiltschko R (1976) Interrelation of magnetic compass and star orientation in night-migrating birds. J Comp Physiol 109:91–99
Wiltschko W, Wiltschko R (1978) A theoretical model for migratory orientation and homing in birds. Oikos 30:177–187
Wiltschko W, Wiltschko R (1981) Disorientation of inexperienced young pigeons after transportation in total darkness. Nature 291:433–434
Wiltschko W, Wiltschko R (1982) The role of outward journey information in the orientation of homing pigeons. In: Papi F, Wallraff HG (eds) Avian navigation. Springer, Berlin Heidelberg New York, pp 239–252
Wiltschko W, Wiltschko R (1987) Cognitive maps and navigation in homing pigeons. In: Ellen P, Thinus-Blanc C (eds) Cognitive processes and spatial orientation in animals and man. Martinus Nijhoff, Dordrecht, pp 201–216
Wiltschko W, Wiltschko R (1988) Magnetic orientation in birds. In: Johnston RF (ed) Current ornithology, vol 5. Plenum Press, New York, pp 67–121
Wiltschko W, Wiltschko R (1991a) Der Magnetkompaß als Komponente eines komplexen Richtungsorientierungssystems. Zool Jahrb Physiol 95:437–446
Wiltschko W, Wiltschko R (1991b) Magnetic orientation and celestial cues in migratory orientation. In: Berthold P (ed) Orientation in birds. Birkhäuser, Basel, pp 16–37
Wiltschko W, Wiltschko R (1992) Migratory orientation: Magnetic compass orientation of garden warblers (*Sylvia borin*) after a simulated crossing of the magnetic equator. Ethology 91:70–79
Wiltschko W, Wiltschko R (1995) Migratory orientation of European robins is affected by the wavelength of light as well as by a magnetic pulse. J Comp Physiol A 177:363–369
Wiltschko W, Wiltschko R, Keeton WT, Madden R (1983) Growing up in an altered magnetic field affects the initial orientation of young homing pigeons. Behav Ecol Sociobiol 12:135–142
Wiltschko W, Wiltschko R, Keeton WT (1984) Effect of a "permanent" clock-shift on the orientation of experienced homing pigeons. Behav Ecol Sociobiol 15:263–272
Wiltschko W, Nohr D, Füller E, Wiltschko R (1986) Pigeon homing: the use of magnetic information in position finding. In: Maret G, Boccara N, Kiepenheuer J (eds) Biophysical effects of steady magnetic fields. Springer, Berlin Heidelberg New York, pp 154–162
Wiltschko W, Wiltschko R, Keeton WT, Brown AI (1987a) Pigeon homing: the orientation of young birds that had been prevented from seeing the sun. Ethology 76:27–32
Wiltschko W, Daum P, Fergenbauer-Kimmel A, Wiltschko R (1987b) The development of the star compass in garden warblers *Sylvia borin*. Ethology 74:285–292
Wiltschko W, Wiltschko R, Walcott C (1987c) Pigeon homing: different effects of olfactory deprivation in different countries. Behav Ecol Sociobiol 21:333–342
Wiltschko W, Daum-Benz P, Munro U, Wiltschko R (1989) Interaction of magnetic and stellar cues in migratory orientation. J Navig 42:355–366
Wiltschko W, Beason RC, Wiltschko R (1991) Sensory basis of orientation: concluding remarks. In: Bell DB, Cossee RO, Flux JEC, Heather BD, Hitchmough RA, Robertson CJR, Williams M (eds) Acta XX Congr Int Ornithol. New Zealand Ornithol Congr Trust Board, Christchurch, pp 1845–1850
Wiltschko W, Munro U, Ford H, Wiltschko R (1993a) Magnetic inclination compass: a basis for the migratory orientation of birds from the northern and southern hemisphere. Experientia 49:167–170
Wiltschko W, Munro U, Ford H, Wiltschko R (1993b) Red light disrupts magnetic orientation of migratory birds. Nature 364:525–527
Wiltschko W, Munro U, Beason RC, Fords H, Wiltschko R (1994) A magnetic pulse leads to a temporary deflection in the orientation of migratory birds. Experientia 50:697–700

Windsor DM (1975) Regional expression of directional preferences by experienced homing pigeons. Anim Behav 23:335–343

Wodzicki K, Puchalski W, Liche H (1939) Untersuchungen über die Orientierung und Geschwindigkeit des Fluges bei Vögeln. V. Weitere Versuche mit Störchen. J Ornithol 87:99–114

Yeagley HL (1947) A preliminary study of a physical basis of bird navigation. J Appl Phys 18:1035–1063

Yeagley HL (1951) A preliminary study of a physical basis of bird navigation. Part II. J Appl Phys 22:746–760

Yorke ED (1979) A possible magnetic transducer in birds. J Theor Biol 77:101–105

Yorke ED (1981) Sensitivity of pigeons to small magnetic field variations. J Theor Biol 89:533–537

Yorke ED (1985) Energetics and sensitive considerations of ferromagnetic magnetoreceptors. In: Kirschvink JL, Jones DS, MacFadden BJ (eds) Magnetite biomineralization and magnetoreception in organisms. Plenum Press, New York, pp 233–242

Zimmermann MA, McCleave JD (1975) Orientation of elvers of American eels, *Anguilla rostrata*, in weak magnetic and electric fields. Helgol Wiss Meeresunters 27:175–189

Zink G (1977) Richtungsänderungen auf dem Zug bei europäischen Singvögeln. Vogelwarte 29 (Special issue):44–54

Zoeger J, Dunn JR, Fuller M (1981) Magnetic material in the head of the common Pacific dolphin. Science 213:892–894

Taxonomic Index of Animal Species

(The family name is given when it is are not derived from the genus name)

Subject Index

MIX
Papier aus verantwortungsvollen Quellen
Paper from responsible sources
FSC® C105338

If you have any concerns about our products,
you can contact us on
ProductSafety@springernature.com

In case Publisher is established outside the EU,
the EU authorized representative is:
Springer Nature Customer Service Center GmbH
Europaplatz 3, 69115 Heidelberg, Germany

Printed by Libri Plureos GmbH
in Hamburg, Germany

MIX
Papier aus verantwortungsvollen Quellen
Paper from responsible sources
FSC® C105338

If you have any concerns about our products,
you can contact us on
ProductSafety@springernature.com

In case Publisher is established outside the EU,
the EU authorized representative is:
Springer Nature Customer Service Center GmbH
Europaplatz 3, 69115 Heidelberg, Germany

Printed by Libri Plureos GmbH
in Hamburg, Germany